TV Fault-Finding Guide

TV Fault-Finding Guide

Selected TV fault reports, tips and know-how from *Television* magazine's popular TV fault-finding column.

Edited by

Peter Marlow

Newnes

OXFORD AUCKLAND BOSTON JOHANNESBURG MELBOURNE NEW DELHI

Newnes
An imprint of Butterworth-Heinemann
Linacre House, Jordan Hill, Oxford OX2 8DP
225 Wildwood Avenue, Woburn, MA 01801-2041
A division of Reed Educational and Professional Publishing Ltd

A member of the Reed Elsevier plc group

First published 2000
Transferred to digital printing 2004
© Reed Business Information Limited 2000

British Library Cataloguing in Publication Data
A catalogue record for this book in available from the British Library

ISBN 0 7506 46330

Library of Congress Cataloguing in Publication Data
A catalogue record for this book is available from the Library of Congress

Composition by Genesis Typesetting, Laser Quay, Rochester, Kent

Contents

About the author

Peter Marlow repaired his first TV sets, procured from jumble sales, at an early age. In his teens he constructed a 3 inch TV set from a design by the great F. J. Camm in *Practical Television Circuits*. At college for his third year project he built a black and white to colour television converter which used a rotating tri-colour disk – it worked but he still has the scars to prove it! He has been a frequent contributor to both *Electronics World* and *Television* magazine since 1986, with articles ranging from a set-top teletext decoder project to an Internet guide for TV and video engineers. In 1992 he set up *SoftCopy* to market the idea of putting trade catalogues and other databases on to floppy disk, CD-ROM and the web, in order to save vast quantities of paper. *SoftCopy* currently sells an index on disk for *Television* magazine and *Electronics World with Wireless World*. Peter lives in Gloucestershire and is married with two children, who understand how to programme the video recorder much better that he does. You can contact him by email at *peter.marlow@softcopy.co.uk*.

List of contributors

Mike Adye
Hugh Allison
Dennis Apple
Ronald Aranha
J. R. Armagh
Tony Ashworth
M. J. Austin
Chris Avis
Richard J. Avis
G. Bakawala
Alan R. Bayly
E. M. Beddow
Steve Beeching
Nick Beer
David Belmont
John G. Bennett
Philip Blundell
Ronnie Boag
Simon Bodget
David Botto
Ian Bowden
Edward D. Branch
Michael Brett
Albert Browne
Roger Burchett
Nigel Burton
Steve Cannon
J. K. Cartet
David A. Chaplin
Paul J. Charlton
Joe Cieszynski
W. H. Clarke
Martin Cleaver
Graham Colebourne
Stephen Cook
John Coombes
L. V. Cooper
David Corcoran

Mark Corner
Michael J. Cousins
V. W. Cox
Ray Crockit
Keith Cummins
J. R. Cutts
Alfred Damp
D. H. Davies
Mervyn Deeley
Glynn Dickinson
Lenny Dinsdale
S. R. Diron
Colin Doman
Michael Dranfield
Dave Dulson
Ray Dunleavy
Mick Dutton
John Edwards
Keith Evans
Geoff Fardon
Adrian P. Farnborough
S. A. Featherstone
K. E. Fellingham
Andrew J. Finn
Russell J. Fletcher
Richard Flowerday
Denis Foley
Gregory C. Foster
Owen Green
J. G. Grieve
Les Grogan
Pete Gurney
Steve Hague
Gordon Haig
Paul Hardy
M. K. Hayter
John Hepworth
Geoff Herbert

Lawrence W. Heslop
G. Hewins
Gerry Hoey
Christopher Holland
John Hopkins
Liz Hopkins
John L. Howard
Shane Humphrey
Phil H. Ireland
Steven Johnstone
C. W. H. Jones
Edward Joyce
Maurice Kerry
Terry Lamoon
Mike Leach
Steven Leatherbarrow
J. LeJeune
George Lithgow
Jim Littler
William G. Lockitt
Robert J. Longhurst
J. R. Lunniss
Dave Mackrill
Hugh MacMullen
Phil Marrison
Robert Marshall
Michael Maurice
Bob McClenning
Blair McEwan
David Minster
R. S. Narwan
Richard Newman
Christopher Nunn
J. Olijnyk
Chris Orr
Mike Orr
Denis R. Parsons
Ramesh C. Patel
S. Pearson
Giles Pilbrow
John Pitt-Francis
Chris Plaice
Martin Pomeroy
Ray Porter

Roy S. Porter
J. K. Potts
John C. Priest
P. Rafferty
Jim Rainey
Mike Rathbone
Graham Rees
Brian Renforth
Graham Richards
John de Rivaz
Aled Roberts
P. J. Roberts
Sergio Roncella
Ed Rowland
J. S. Ruwala
K. W. Saxon
D. W. Sergeant
Alan Shaw
Alan Smith
David Smith
Gerald Smith
Justin Smith
Adrian Spriddell
Steve Stamford
M. Stansfield
Brian Storm
C. R. Taylor
Andrew Tebbutt
Graham Thomson
Alan Travers
J. R. Trimmer
Eugene Trundle
Ray Vesey
Mark Ward
Chris Watton
T. J. Welford
David J. Whilding
Gerald White
Roger F. White
George Whiteside
N. J. Williams
S. Woodbridge-Smith
Andy Worrall
K. Wright

Introduction

Television is a monthly magazine concerned with the technical aspects of domestic TV/video and associated equipment, particularly servicing. It includes monthly fault-finding reports on TV sets, VCRs, camcorders and satellite receivers, and articles on the technology used in servicing particular chassis. Readers are kept up to date with technical developments, reviews on test equipment and new technology. There is also a long distance/satellite TV section.

The notes contained in this Guide are based on material originally published over the last ten years in the TV fault-finding pages that are a regular monthly feature in *Television*. They have been re-edited and collated for ease of reference. We have concentrated on the common models and chassis – over 2000 faults are listed from more than 300 models.

We would like to thank the many engineers who have regularly contributed to the *Television* faults pages over the years – their names appear in the List of Contributors. Grateful thanks also to the editor of *Television* magazine, John Reddihough, for his help and advice.

Indexes for *Television* magazine covering the last 12 years are available in hard or soft copy from SoftCopy Limited, 1 Vineries Close, Cheltenham GL53 0NU, UK. The soft copy version comes on CD-ROM which also contains the text of 12 000 fault reports, covering TVs, VCRs, satellite TV receivers, camcorders, monitors and CD players. See SoftCopy's advert in *Television* magazine or the web site *http://www.softcopy.co.uk* for more information.

Disclaimer

The fault reports listed in this book come from real-life cases as observed in the workshop. There is no suggestion implied that certain models are more fault prone than others or that particular faults are likely. Where we have listed a large number of reports for a set it is because of the popular and widespread use of the set rather than any shortcomings in its design. In some cases we have recommended modifications, some of which come from the manufacturers of the sets. However, there is no implication intended that these modifications are necessarily authorized or condoned. You should not try to modify or repair a set under warranty as this could invalidate the warranty. The equipment should be taken to an authorized dealer.

In some fault reports certain component distributors are named as being able to supply specific spare parts. This is not meant to be an exhaustive list – their names are included for guidance only and are believed to be correct at the time of writing. Of course, circumstances do change and the reader is advised to consult the latest TV & VCR Spares guide as published every spring in *Television* magazine.

There can be high voltages on the chassis of television receivers, even when disconnected from the mains – a repair should only be attempted if you are competent. Many components in TV receivers are designated safety components: they are designed to fail in the event of an overload, e.g. resistors in power supply rectifier circuits and feeds, or to fail safely. Replacements must be parts obtained from or approved by the manufacturer. Failure to use approved components could have legal implications in the event of a subsequent fault occurring.

Neither the editor nor the publisher can accept responsibility for any loss, damage, death, injury or litigation which arises directly or indirectly from the use of information presented in this book. All information has been given in good faith and is believed to be correct at the time of writing.

Note about lamp bulb testing

In the event of an apparently dead chopper power supply it may not be readily apparent whether the cause of the fault is an external overload, usually a short-circuit in the line output stage, or in the power supply itself. A simple check is to disconnect the feed to the line output stage and run the power supply with a dummy load across the h.t. output. A 60 W bulb provides a suitable load. If the power supply produces the correct h.t. when run in this way the cause of the fault is probably in the line output stage. A few chopper circuits, mainly of the series type, will not run with a dummy load.

Akura

Akura CX10

Set cuts out intermittently with no standby light: Check whether R402 has gone high in value.

No picture: Check the efficiency diode FR605.

No picture and no sound, or works intermittently: Change the d.c. converter chip. You can't miss this – it has a massive heatsink. Overheating of this chip is often the problem.

Repeated failure of D410 – going short-circuit: Check the value of R320. It should be 33 Ω – if higher, change it.

Flashing on ITV and BBC-1 which looked like a.f.c. hunting: The fault appears only when the set has been working for about half an hour, and would go away if freezer is applied around the i.f. screening can. When the screen is removed there does not appear to be any faulty components. However, slight adjustment of the a.g.c. potentiometer clears the fault. There seems to be a critical point at which instability occurs.

Dead set symptom: If you get the dead set symptom, sometimes intermittent, check D410 (FR604) in the line output stage. It is advisable to stock this component but two RGP30K diodes in parallel can be used successfully in its place. Makes a change from replacing IC402! Another weakness in these sets is the mains bridge rectifier. If in doubt, replace this item – it can fail intermittently.

No line drive: This set comes out of standby and there is h.t. at the collector of the line output transistor, but apparently no line drive. A scope check at the base of the line driver transistor shows that there is drive here, but at an extremely high frequency. The cause of the trouble is oscillator capacitor C308 (3.3 nF, high stability) which is open-circuit.

Line output transformer overheating: There is normal sound but the line output transistor Q403 overheats and there is no raster. The cause of the trouble is D411 (FR155), which produces the 160 V supply for the RGB output stages. It was very leaky.

Akura CX25

Many symptoms caused by the microprocessor EEPROM: Field faults, brightness faults, no sound, distorted sound, intermittently dead and clicking (very similar to Nicam noise) have all had one cause. In addition, a sort of r.f. patterning, no colour and tuning drift can be added to the list. The micro's EEPROM has always been responsible. It can be obtained direct from Akura, Cumbernauld. The part number for Model CX25 is 13-ONVM3060P25B/00294AA and for Model CX26 13-ONVM30–60PB/00294A.

Reprogramming the EEPROM: For this you need the service manual and the remote control unit. Next time you get a set in and have completed the repair, enter the service mode, scroll down to NVM address and read out the memory. Each memory location has a number in hexadecimal form, from 00 to FF or 0–256, and a setting value, again in hex form. Write the data down and next time you can reprogram the EEPROM yourself. As some locations relate to tuning, picture geometry etc. you will have to make slight adjustments after reprogramming. To get you started, if you have a sound problem with the Nicam Model CX26 scroll down to locations D7, D6 and D5 and enter the following data: location D7, data 3E; location D6, data 09; location D5, data 1F. Location D7 seems to relate to the maximum volume setting. Empty locations are all filled with FF. If you change the data at location D7 to FF the sound is muted – the microcontroller chip sends the sound mute command to the Nicam panel.

Intermittent field collapse: Intermittent field collapse can be cured by resoldering dry joints in the field output stage, including some at the pins of the TDA8172 field output chip. A high wattage, 2.4 kΩ stand-off resistor is mounted near the scan coil connection plug. If its legs are badly dry jointed, and there is board charring around them, it is necessary to remove the resistor and clean up the board. The plug's plastic body can have a deep burn mark caused by heat from the resistor. In case the terminal hidden within its plastic body is damaged, it is advisable to cut the lead closest to the damaged portion of the plug and solder it directly to the board.

Akura CX26

Intermittently switches off after 20 minutes: As the set goes off there is a severe line fold-over. The cause of the problem is the line scan coil plug, which had been very badly soldered. Desoldering all the pins, cleaning and then resoldering them provides a complete cure.

Intermittent faults caused by dry joints: For sets with intermittent faults ranging from text problems to being intermittently dead, the cause is usually dry or poorly soldered joints on the digital signals board, particularly where the top of the board is linked through to the bottom. The most common cause of problems is J746, which is a wire link under IC703. It becomes dry jointed, the result being intermittent line drive. IC703 has to be removed to resolder the through-board links beneath it. Take every precaution against static when you do this – these chips are very sensitive to it.

No f.m. sound with video playback: Although this set would produce stereo sound with a Nicam broadcast there was no f.m. sound with video playback at r.f., nor from a test pattern generator. There was plenty of demodulated f.m. sound at pin 5 of the i.f. chip IC101, but this disappeared into the Nicam board, where all the audio switching takes place, and didn't come out again. Swapping parts with another set proved that the EEPROM chip IC002 was responsible, though we couldn't work out why! When changing this chip much time can be saved by first going into the service mode and noting the various adjustment levels.

Intermittent picture loss: For intermittent loss of the picture to a blank raster, desolder and resolder the mass of tinned copper wire links on the teletext panel.

Alba

ALBA CTV10
ALBA CTV12
ALBA CTV2
ALBA CTV55
ALBA CTV743

Alba CTV10

Wouldn't start: There are two start-up resistors in parallel, R301 and R302, both 330 kΩ, 0.5 W. On this set these had been replaced with a single 0.5 W resistor mounted very casually on the print side of the board – no sleeving on leads almost bridging tracks. Replacing R301/2 with resistors having the correct values and wattage ratings restored the set to health.

Open-circuit mains transformer: Failure of the primary winding of the mains transformer is not uncommon, and often an 'invisible' fuse is looked for without success! Make sure that you also replace the bridge rectifier supplied by the maker if you don't want to see the set back in two weeks' time.

Dead apart from standby relay click: Check the field output chip IC402.

Totally dead set: The 650 mAT fuse F401 blown. Check that the bridge rectifier, D404, is not short-circuit. Also note that the symbols for D403 and D405 are printed the wrong way round on the underside of the panel (at least in the set we had) although they are shown correctly on the component side.

Dead set, although standby LED on: Diode D410 (FR605), which is in series with the h.t. feed to the line output transformer, could be open-circuit, due to transformer failure.

Set goes off after a while: Check the regulator chip with a heat gun (or hairdryer). This chip is prone to break down when hot. When fitting the replacement, remember to put on heat sink cream and tighten it together well.

Alba CTV12

Intermittent dead set: The set would only fail on occasion, yet the power-on light always showed. The circuit is quite involved, but everything seemed to be OK in the switch-mode power supply. TR301 seems to operate in a delayed manner due to R303 charging C307, with discharge via R304 with D306 as a reset between switching off and on. The problem was that R304 had gone high in value. As a result the circuit locked on and starting current didn't reach TR302, via R301–5, to get it all going.

No sound or picture with normal h.t.: The 12 V regulator transistor Q208 was open-circuit base-to-emitter. A TIP41 proved to be a suitable equivalent. These sets are the same as the Lloytron portables.

Dead set although channel indicator alight: The cause was dry joints on the line driver transformer.

Alba CTV2

Trip and shut down after a few minutes: It's quite common for R69 to be the cause of this symptom. Not on this occasion, however. To find out whether the cause of the trouble is in the power supply or the line output stage a good check is to disconnect the 12 V regulator IC6. There is then no line drive. When we did this the power supply ran contentedly, but we couldn't find anything obviously wrong in the line output stage. So a blanket soldering of all the components, including the transformer, was carried out. We then confidently reconnected IC6 and switched on. Yes, you've guessed it, after a few minutes the set tripped. In desperation all the components in the power supply were checked. Nothing wrong here. We switched on again and prodded various components in the hope that the fault could be instigated in this way. But no, after a few minutes the set tripped off without any help from me. We next tried the can of freezer. After switch-on a number of components were squirted. No luck until D16 was squirted. The set then tripped! Now D16 is soldered in series with a resistor that's proud of the PCB. When one end of this combination was disconnected from the board the two components parted company. Closer inspection showed that they had been laid against each other then soldered, the joint being very poor. After twisting the leads together and resoldering the network the set worked happily.

Power supply had self-destructed: The BUV46A chopper transistor TR3 was short-circuit while resistors R97 (4.7 Ω), R92 (1.5 Ω), R93 (1.5 Ω), R86 (4.7 Ω) and R79 (4.7 Ω) were charred and open-circuit. The circuit we were able to obtain from Alba was so poor that the above component values had to be obtained by phone – no charge was made for it, however. We replaced the TEA2018A chopper control chip IC5, checked all the diodes in the power supply circuit and fitted new resistors as necessary. When we switched the set on it worked normally and we were able to adjust the h.t. for the correct reading of 102 V at the cathode of D17. Alba told us that they supply a power supply modification kit for this model. It contains TR3, IC5, numerous resistors and details of how to modify the circuit board. We ordered the kit in case we should need it but don't know how successful it will be as the circuit is again unreadable.

Beware the coffee-coloured resistor!: The only snag is that this set is full of them. You may have come across this type of resistor being supplied to the trade by one of the wholesalers over the past few years. They fail if used in high-voltage situations. In this case the set was dead and R81 (100 kΩ) was open-circuit. It provides a start-up feed for the TEA2018A chopper control chip IC5.

Alba CTV55

Dead set: A check across the line output transistor produced a short-circuit reading. The cause, however, was the FR155 efficiency diode D404 in the series chopper circuit – it's connected between pin 11 and chassis.

No sound or picture: On this model the momentary contacts to bring the set on manually from standby are not in the on/off switch but in a separate button switch just above it. Pressing the mains on/off switch immediately powered the set, however, but with no sound or picture! After removing the fuse wire that had been soldered across the standby relay contacts and pressing the standby button we found that channels could be tuned in with sound and an on-screen status display. But the only 'picture' that could be obtained, when the first anode control was turned up, was a blotchy coloured one – there was no luminance.

(The relay and remote control functions were OK, so the fuse wire remains a mystery.) A common cause of a dark or absent picture with this chassis is an increase in the value of one of the resistors R429 (180 kΩ) or R423 (100 kΩ), usually the latter, in the beam limiter circuit, but both were spot on. The luminance signal passes from the video processing chip IC301 to the matrixing circuit on the tube base via the emitter-follower transistor Q303, which is used for blanking purposes. A scope check showed that the luminance signal was not getting through this stage. Various diodes are connected to the base of Q303. When D309 was disconnected normal luminance was restored. The other (anode) end of D309 is connected to the collector of Q304, where the voltage was high because the transistor's base voltage was low. The value of the base bias resistor R366 had risen from 22 kΩ to 80 kΩ.

Very smeary picture: It looked as if the fault might be in the luminance signal processing chip, but a quick check showed that the voltage at the collectors of the RGB output transistors was only 100 V. Tracing the source of the supply back, we came to the small green choke L501 which had 100 V at one end and 200 V at the other. It was open-circuit, a replacement 330 μH choke from a scrap set curing the fault. Where was the 100 V was coming from?

Alba CTV743

No start-up with channel 7 showing in display: The chopper circuit was OK so we turned our attention to the line output stage where D406 was found to be short-circuit (fit a BY228) and R435 open-circuit (1 Ω, 1 W). Before we could switch on a number of solder splashes and dry joints required attention. All was well when these had been put right.

No sound and field collapse: R435, a 2 Ω safety resistor, was open-circuit because D406 was short-circuit.

Channel numbers lit up but no h.t. supply: D806 in the power supply was short circuit. We fitted a new BYD33M in place of the original 1N4936 and our problem was solved.

Amstrad

AMSTRAD CTV1400
AMSTRAD CTV1410
AMSTRAD CTV2000
AMSTRAD CTV2200
AMSTRAD TVR1
AMSTRAD TVR2
AMSTRAD TVR3

Amstrad CTV1400

Prone to develop dry-joints: These sets are very prone to developing dry joints on one or all of the line driver transformer's pins (T701). If the owner doesn't bother to do anything about the resultant intermittent faults (they very seldom do) the first thing the service engineer knows about it is when he's presented with a set with a dead line output transistor (Q705). The original is a 2SD904 but Amstrad now supply a 2SD822. Incidentally, take circuit details on the panel and in the service manual with a pinch of salt. The LA7800 sync/timebase generator chip IC701 has pin 8 marked correctly on the underside of the panel but pin 16 is marked pin 1. In the manual the voltage at pin 12 of this chip is shown as 1.2 V instead of 12.2 V. Q705's collector connection to the print is via a stud that's soldered in. This is not very reliable. An extra lead from the connecting nut through one of the convenient holes in the board to the print land is a good idea. In our experience these sets benefit from having the board cleaned up and coated with circuit varnish.

Colour flickering and flyback lines at the top of the screen: It was difficult to associate these symptoms, but the fault was in the field timebase. A check on the voltage at the collector of the field output transistor Q702 produced a reading of only 32 V instead of 62 V. C739 (100 μF) was open-circuit, a replacement curing both faults.

Line sync would disappear at channel change: After an unsuccessful prod around with blunt instruments (freezer and hairdryer) we checked with the circuit diagram and noted a couple of likely culprits, the two zener diodes D715 and D704. Replacing them cured the fault. Intermittent lowering of the sound level and/or buzzing is caused by the adjustable capacitor on the sound i.f. subpanel.

Power supply shut down: These very reliable sets are normally strangers to our workshops. When we do see them they generally need the line driver transformer resoldering (intermittent no results) or simply a clean up of the channel pushbuttons. This particular set was more of a problem. The power supply would shut down after approximately three to four minutes. The chopper chip IC502 was touched with a finger but quickly (very) withdrawn: you could now read STR451 on the end! After many inconclusive checks and replacements with no real faults being found the chopper transformer T501 was replaced, curing the problem. It's type RB20826.

Dead set, no line drive: This dead set looked very grotty when we removed the back. The cause of the fault was soon found to be no drive at the base of the line output transistor. The line driver transformer was dry jointed: its legs were so black that it wouldn't solder until they had been scraped clean.

Amstrad CTV1410

Occasional field roll and bright picture: We noticed that occasional field roll coincided with a bright picture. A scope check showed that the TA8701N i.f. chip produced a good output waveform at pin 19. After passing through a 6 MHz crystal filter (CF301) the signal is coupled by C304 (2.2 μF) to an emitter-follower stage. When checked this capacitor was found to be very low in value, causing the loss of field sync. Most TV designs don't use a coupling capacitor here.

Very intermittent field collapse: These sets have proved to be quite reliable. This one had very intermittent field collapse, especially when cold. Replacing IC402 didn't help, but freezing C404 (0.022 μF) produced the fault symptom and a replacement cured it. Another of these sets produced a picture that had all the hallmarks of a low-emission tube. In fact R650 and R651 (both 4.7 kΩ in the on-screen display circuit had gone high in value.

Dead, no h.t.: This set had us fooled for a bit. It was dead with no h.t., i.e. no voltage could be measured at the cathode of the 112 V supply rectifier D904. The h.t. supply is switched on/off by relay RLY901, but the supply's 100 μF, 160 V reservoir capacitor is on the output side of the relay. Hence the no-voltage situation at the cathode of D904 – things would have been clearer if we had made a scope check at the cathode of D904 first. The cause of the problem was RLY901's contacts, which were open-circuit.

Power supply faults: If the 2SD1545 chopper transistor has blown, always check/replace Q902 (2SB774) and the 9.1 V zener diode ZD902 as these are usually damaged as well. If you have a dead set with the −30 V and 16 V supplies OK but the 112 V h.t. supply appears to be missing, check relay RLY901 first for open-circuit contacts. Since the reservoir capacitor comes after the relay, no voltage can be measured at the cathode of the h.t. rectifier D904 when the relay's contacts are open-circuit.

Vague raster: There was no picture although a vague raster could be seen when the first anode control's setting was advanced. A video waveform was present at the input to IC301 but nothing came out. The cause was loss of the sandcastle pulses at pin 17, because R316 (12 kΩ) was open-circuit. The same chassis is used in the Alba CTV840 and Hinari HIT14RC.

Amstrad CTV2000

Dead, fuse blown: This set came from the shop with the complaint 'dead' and it sure was! The 2.5 A mains fuse was open-circuit well blackened – so attention was centred on the chopper transistor TR6. A check revealed that it was short-circuit all ways round. We also found that D13 in its base circuit had shorted. After replacing these two items everything seemed to be OK so we switched on. Bang! There was a flash that made the three fluorescent lights look dim and FS1 and TR6 had again bitten the dust. This time D13 had survived but the h.t. rectifier D15 was short-circuit, which it hadn't been two minutes previously. A new BY299 h.t. rectifier was fitted and the other items were replaced, along with D13 for good measure the set would now run – for about two and a half minutes. The whole lot had gone again! Now the 119 V h.t. supply is used by the tuning circuit as well as the line output stage, and although the aerial had been plugged in there had been no tuning. Interesting! The ZTK33ADPD tuning voltage regulator IC2006 was suspect but the culprit turned out to be the nearby transistor TR2016 (BF460) which was short-circuit collector-to-emitter. It's associated with the memory in IC2007. When this was put right the set continued to run.

Reduced height and fold-over: This set suffered from reduced height with a degree of fold-over. The field output stage is conventional, using a pair of transistors that are fed from a high-voltage supply. The supply decoupler C720 (4.7 μF, 160 V), which decouples Q702's collector, was found to be open-circuit. As the set is now about nine years old we replaced C718 and C739 as well for good measure.

No sound, field collapse and channel LEDs didn't light: This set came in because of field collapse, but on futher investigation we found that there was no sound and the channel LEDs didn't light up. A start was made at the TDA3652 field output chip TC801. Pin 9 had the correct 28 V but there was no field drive at pins 1 and 3. This took us back to the LA7800 timebase generator chip which had two supplies, one at pin 15 for the line generator and one at pin 12 for the field generator. There was no voltage at pin 12 because R840 (10 Ω, 2 W) was open-circuit. Replacing this resistor restored the set to life. While it was on the soak table next day, however, R840 again failed. This time a short-circuit could be measured to chassis. Several electrolytics were checked to no avail, then the short disappeared. Until next day that is, when R840 once again went open-circuit. This time the cause of the mystery short was found amongst the spaghetti of wires that makes this set so difficult to work on. A 2 A, 20 mm fuse was found wedged between some components in the corner of the chassis. We can only assume that a previous repairer had dropped it and been unable to find it. The fault showed up after the set had been disturbed whilst moving house.

Amstrad CTV2200

Self-destruction caused by dry joints: The main reason why the set self-destructs at switch-on or soon after, blowing Q501 (2SC3156), Q802 (2SD139B) and sometimes Q503 (2SA916), is dry joints on the line driver transformer T801. They can also be responsible for reduced width at one side, sometimes intermittently. Also check the h.t. smoothing capacitor C520 (100 μF) which works very close to its tolerance of 160 V. If you don't have a 2SC3156 to hand a BU208A will work quite happily as a replacement.

Dead set, no standby light: These sets often come in dead. No shorts this time, however. The standby light was inoperative because the 12 V supply to the front panel was missing. This is provided by IC103, which is a standard three-leg regulator. Its input, of around 16 V, comes from a bridge rectifier/capacitor combination on the main chassis and was correct. IC103 was open-circuit, a replacement bringing the set to life.

This wasn't to be the last time we saw the set, however. On its return IC103 had again failed and a check showed that the reservoir capacitor for the 16 V supply, C524 (470 μF, 35 V), was virtually open-circuit. IC103 obviously didn't like the 100 Hz ripple at its input. We subsequently had this fault on a couple of other occasions.

Field collapse: The cause of field collapse was traced to C828 (100 μF, 35 V). IC801 is normally responsible for this fault, but not on this occasion.

After a few hours the top inch of the picture would become compressed: The cause of the fault was traced to C812 (0.018 μF, 160 V) which is in the feedback circuit between the TDA3652 field output chip and the LA7800 timebase generator chip.

Field jitter/picture rolling, possibly intermittently: Replace C814 (470 μF, 35 V). It dries out because of the heat from a nearby wirewound resistor.

Amstrad TVR1

Took a long time to come on: It was the TV section of this TV/VCR combination that wasn't working correctly, taking anything from a quarter to half an hour to come on. Voltage checks around the STK7348 power supply chip IC1501 showed that the start-up feed voltage at pin 7 was low – only 0.7 V. A resistance check on the start-up feed resistors R1503/4, both 270 kΩ, showed that they were of the correct value, leaving only the 0.22 μF electrolytic capacitor C1507. This didn't read leaky but fitting a replacement restored normal operation.

Dead machine, power supply fuse blown: This machine came in dead. The mains supply was present up to the relay in the TV section but the relay didn't operate when the monitor button was pressed. In this model the standby power supply comes from the VCR section, so we started our investigation here. Fuse F604 (630 mA) in the power supply had blown and a replacement blew at switch-on. We disconnected plugs and found that the short-circuit was across the all 5 V line. The output from the 5 V regulator transistor Q603 goes to the syscon PCB, but extensive checks here failed to reveal the cause of the short-circuit. We then discovered that the fault disappeared when one of the interconnecting plugs between the VCR and the TV sections was disconnected. This enabled us to establish that the short was on the tuning preset PCB. Desoldering various suspect components brought us to C1109 (100 μF, 10 V) which decouples pin 1 of IC1102. When this had been replaced the fuse held and the TV and VCR were back in working order. If the TV section won't switch on when the monitor button is pressed although the monitor-on LED lights up, check whether the timer record button is pressed in. It will save you having to strip down the VCR.

Would not warm-start properly: This set always came on from cold. If it was switched on when warm, however, it would start all right but after only a very short time the picture would go off and the power transformer would whine. Replacing D1507 and C1511 (100 μF, 160 V) cured the fault.

TV section dead: Replacing the STR7348 regulator, C1509 (3300 pF, 1 kV) and the 8.2 Ω 5 W surge limiter and 33 Ω 10 W resistors restored normal operation.

Dead or intermittently dead: Check C1507 first for low value. This can save a lot of time.

Amstrad TVR2

No results symptom: Check whether the surge limiter R1501, 8.2 Ω 5 W, is open-circuit. This may be fairly obvious, but if you rush out and buy a service manual it will cost you a cool £24.

Line output transistor shorted: The set came in with the complaint that it was dead. There was a measurable short across the collector and emitter of the line output transistor. D1606, type FR304, was short-circuit.

Wouldn't switch-on: The TV section of this unit wouldn't switch on. The bridge rectifier produced a d.c. output which was present up to the power supply chip. Replacing this item didn't provide a cure and as we had no manual we resorted to component checking on spec. When C1507 (1 μF) was replaced we were rewarded with a healthy line whistle and the rustle of e.h.t. We also replaced the other two electrolytics in the power supply as they appeared to be in a distressed state.

When the aerial was connected the set went dead: Investigation revealed that when the fault occurred, the output from the power supply was normal but the line drive disappeared. As most of the signal processing is carried out in the μPC1420 chip we checked the supplies then replaced this i.c. This made no difference. Chassis isolation problems were next suspected, but this couldn't be the cause as the set went dead or oscillated on and off if different lengths of wire were used as the aerial. We eventually traced the cause to a faulty 500 kHz oscillator associated with the μPC1420 chip. It was unstable as the 0.0015 μF series capacitor was defective. While on the subject of this i.c., if the resistor that supplies pin 37 is faulty the result is no sync. You may find it to be 120 kΩ or 150 kΩ. We've also found the chip to be responsible for **no colour or loss of one colour.**

Sound and vision took 1 hour to appear from switch-on: We didn't bother to check this but went straight for C1507 (0.47 μF). After fitting a 1 μF capacitor in this position the equipment fired up straight away. Removing the TV chassis is fairly easy provided you first remove the VCR unit in order to gain access to the small screw located beneath the front of the main PCB. Before returning one of these sets to the customer a point worth checking is that the 'repeat' switch at the rear of the receiver is in the 'off' position. Otherwise a call-back is a certainty – with the switch in the 'on' position the machine will carry out only the play function. Another thing to catch out the unwary is the separate mains on/off switch at the back.

Took half an hour to come on: C1517 was low in value.

Dead set, C1507 leaky: Several of these sets have come in either dead or intermittently dead. In just about every case the cause has been that C1507 (1 μF, 50 V) was either leaky or open-circuit. One set, however, had a job ticket which said that the set was dead although it wasn't the 1 μF capacitor. The cause of the problem was that the mains relay wasn't being energized because there was no 5 V output from the power supply. In fact the fault had nothing to do with the TV side of the combination: a fuse in the VCR section had blown.

Wouldn't come out of standby: The TV section wouldn't come out of standby. We found that C1507 (1 μF, 50 V) on the main transformer was open-circuit.

STK7348 chopper chip blows: If the STK7348 chopper chip blows at switch-on or has a very short life, replace C1509 (3300 pF, 1 kV) in the snubber network. If you have any old Rank T20/T22 chassis power supply panels you'll find an excellent, more robust replacement for C1509!

Supposed to be intermittently dead: Several days of soak testing failed to reveal any fault, but a week later the set came back permanently dead. Checks showed that the primary side of the line driver transformer T401 was open-circuit. A replacement (part number 152591) put matters right.

Amstrad TVR3

Dead set: We've repaired quite a few of these TV/VCR combinations in the past. Frequently the dead set condition means that R301 (10 Ω, 7 W) has gone open-circuit. This seemed to be the case with the latest TVR3, but after fitting a replacement there was a cloud of smoke from C310 (3300 pF, 1 kV) at switch-on. R301 then promptly burnt out again. We consulted the circuit diagram then removed and checked IC301 (STK7348). There was a short-circuit between pins 8 and 10. Replacing IC301, R301 and C310 restored normal operation.

TV section dead – no output from IC301: The TV section was dead. There was h.t. across C306, the mains rectifier's reservoir capacitor, but no output from the STK7348 power supply chip IC301. Before condemning the chip I checked the rectifiers on the secondary side of the chopper circuit and found that D306 (FR304) was short-circuit. It provides the 120 V supply for the line output stage. A new diode restored normal operation.

Whine from power supply, no picture or sound: A loud whine came from the power supply and there was no picture or sound. No shorts were present on the supply lines but a visual inspection showed that C310 (3300 pF, 1 kV) had cracked, going short-circuit.

TV section dead: The TV section of this unit was dead. We found that the 3.3 nF snubber capacitor C310 and the STK7348 chip in the power supply were both short-circuit. As the replacement capacitor supplied by Amstrad didn't look capable of withstanding 100 V, let alone 1 kV, we made up a replacement consisting of two 1.5 nF capacitors connected in parallel. The set bounced within two weeks. This time we obtained the chip from another source. After a long soak test the set was pronounced fit. It could be that there's a bad batch of STK7348s around.

Line disturbance on dark scenes: The cause was found by connecting the scope to the cathode of the h.t. rectifier D306. The display showed a smooth d.c. supply when the picture was bright. When the picture was dark, however, there were negative-going spikes (gaps) in the supply. Replacing the diode and the associated capacitors made no difference. The cause of the trouble was the STK7348 chip IC301.

Wouldn't tune: We found that there was no 12 V supply at pin BU of the tuner. The culprit turned out to be the LA7913 band-switching chip IC2.

Odd reaction after switch-off: There was an odd fault with this set. Everything worked all right until it was switched off. When the operate button was pressed, the set switched to the VCR channel and stayed on. After some careful study we found that the contacts of the standby relay remained closed. They remained closed with the voltage across its coil down to 1 V. We assumed that the relay was magnetized, as a 24 V relay shouldn't hold its contacts closed with only 1 V across the coil. The collector of the transistor that controls the relay switching is also connected to the sound mute pin of the AN5265 chip, via a 22 kΩ resistor (R155). We traced the source of the 1 V to this chip. When we increased the value of R155 to 47 kΩ the voltage across the coil fell to 0.7 V and the relay dropped out when asked. Another dealer we know has had this problem. A new relay would probably cure the fault, but this simple modification is cheaper.

Severe line corrugation from cold: Go straight to C91 (22 μF, 16 V). We've recently had a run of these models with this fault.

Very poor off-air reception and recording quality: With pre-recorded tapes the picture quality was OK. The aerial input is on the TV section. It consists of a coaxial socket/phono socket adaptor. The phono plug on the lead that feeds the aerial signal to the tuner is plugged into the back of this adaptor, then soldered in place. The solder had cracked and no longer held the phono plug. When the aerial had been plugged in, the phono plug had been pushed back, taking with it the centre contact tube of the coaxial socket. Reception was back to normal when a new aerial socket assembly had been fitted and securely soldered in position.

B & O

B & O 33XX series

Thin black lines would flash: Thin black lines would intermittently flash across the screen, and occasionally there would be loss of line and field sync. The cause was traced to 1TR11 on the tuner/i.f. panel. It was going leaky intermittently – when leaky its bias altered and the video signal sync pulses were almost cut off.

Set wouldn't start: The usual dry joints had been attended to so further investigation was required. Q3 on the power supply panel turned out to be open-circuit.

Grey-scale tracking variations: Caused by 4R18 (1.8M Ω) which had gone high in value. It's mounted on the c.r.t. base panel.

Sound mute didn't work: The remote control handset volume button worked correctly but sound mute didn't. Replacing diode 1D4 on the tuner/i.f. panel cured the trouble.

Pulsing on and off: Traced to loose screws on the chopper transistor 6TR1.

No luminance: Can be caused by high-resistance contacts on the tuner/i.f. contact strip or the same item in the decoder. It can also, often intermittently, be caused by dry joints on the inductors that form the luminance delay line.

Insensitive remote control handset: Due to one of the LEDs being open-circuit.

Teletext, channel display and selection problem: There were three problems with this set. First, when text was selected the required page number couldn't be obtained – you simply got 100. Secondly, the on-screen channel number couldn't be brought up with the remote control unit's TV button. And finally when trying to select a channel above nine the on-screen display would just flash on then off instead of staying there for three seconds so that you can select the second digit of the channel number. All three problems were caused by a faulty 4015 shift register chip (8IC1) on the remote control decoder panel.

Set stuck in standby: The cause was an excessive line output stage load. A check showed that the output transistor was OK and subsequent tests revealed a primary-to-secondary short in transformer 5T11.

B & O 4402

No results, C10 and C12 open-circuit: The complaint with this set was no results. C10 and C12 (both 47 μF, 40 V) were, as so often, open-circuit but replacing them failed to restore normal operation. We found that there was a break in the print at the cathode of SCR1.

Width varied intermittently: This was accompanied by a nasty crack of e.h.t. followed by line collapse. A dirty connection on the scan coil plug, at the coil end, was responsible. It had started to arc and burn the plastic plug casing.

No results, 17 V on h.t. rail: We don't see many B and O sets and it took us a while to sort this one out. The problem was no results, with only 17 V on the h.t. rail and the voltages in the control section of the power supply approximately correct. The cause of the fault was 5C10 (47 μF) which had gone open-circuit.

B & O MX1500 (78XX series)

Raster but no sound or vision: There was a raster but no sound or vision. The 19 V and 95 V rails were present and correct, as you would expect, but the 8.5 V supply was missing because ICP1640 (630 mA) was open-circuit. This supply is derived from the line timebase. No reason for the ICP failure could be found – this is very often the case with these devices, as in Technics disc players where removing the lid seems to blow one or two of them! The set was also afflicted by a broken level-up button in its Beolink 1000 remote control handset. This is quite common and a new mat put matters right – interesting that this seems to affect only the Link 1000 and not the earlier A, V and A/V terminals of the same design.

Set couldn't be tuned: When the sweep tuning was activated you could see where there were transmissions as the snow would momentarily change to a virtually blank screen but the set wouldn't stop at any of them. We connected an external d.c. supply to the tuner's varicap bias input and set it to a point where very weak vision could be seen, then used a scope to check the output (pin 16) from the i.f./detector can. This showed that a very noisy signal with no sync pulses was being fed to the BC548 video emitter-follower transistor 7070. A second video output (pin 18) from this can is fed to the scart socket. We moved the scope to this pin and found that the signal here was perfect. The next clue to the cause of the trouble was the fact that the 5 V d.c. level at pin 16 seemed to be excessive. When the pin was disconnected from the print the 5 V was still present at the print. This suggested that the emitter-follower transistor 7070 was faulty – in fact it was leaky from collector to base. After fitting a replacement transistor the set was tuned up and left on soak test. Next morning we found that all the preset stations had gone – the rechargeable memory supply battery was short-circuit.

Would intermittently go into search: At intervals that varied from a few minutes to many days or weeks, this set would suddenly start off up the band, searching and seeking through the v.h.f. and u.h.f. channels, then just as suddenly stop and behave itself. The cause of the fault was eventually traced to leakage in the momentary-contact start-up switch associated with, and mounted alongside, the mains on/off switch.

B & O MX2000 (31XX series)

No sound: After much checking around – access is not good with this chassis – we discovered that the 6 MHz coil LD40 was open-circuit. Due to the cramped layout alignment was very difficult. This chassis is manufactured by Thomson (NordMende) and many spares are not available from Bang and Olufsen.

Severely reduced height: A colleague was dealing with this set, the initial problem being severely reduced height. The cause was soon traced to IG01 (TDA4950) which was short-circuit internally and RL28 (1.5 Ω) which had gone high in value. This situation is not unknown. Destruction of these devices is nearly always caused by dry joints on the

EW modulator diodes DG05/6. He attended to this and was rather distressed to switch on and find that there were severe EW distortion problems – the picture was bowed in significantly at either side and the correction controls had no effect. He rechecked the diodes and their joints, changed a resistor that had gone slightly high in value, then checked other likely items. Getting even more upset and being already harassed by a number of Pioneer compact disc players that wouldn't set up very well he asked us to take a look. Bearing in mind our colleague's thorough and logical checks we took the view that the fault was probably due to something unusual. A scope check on the three relevant waveforms showed that they were all of excessive amplitude. So it seemed to me that the series coil LG02 had to be short-circuit or low resistance, which proved to be the case.

Low, distorted sound: This NordMende-made set had very low, distorted sound. The audio circuit is pretty complex, with pseudo-stereo and audio enhancement included. I checked the supplies to the various stages and found that these were all in order. Next we injected a signal into the output chips. This produced an ear-splitting sound as the volume was still right up to hear the distorted off-air sound . . . Assuming that the fault was earlier on in the sound channel we decided to freeze the f.m. detector chip ID35. This proved to be a fruitful move but the chip itself wasn't the cause of the trouble – the culprit was the 6 MHz filter directly behind it.

No go, chopper not working: Only the channel number indicator was alight. The chopper circuit wasn't working and the $0.39\,\Omega$ feed resistor RP14 to the line output stage was open-circuit. The obvious conclusion to draw was that the line output transistor was short-circuit, and indeed an in-circuit check showed a very low resistance across all its pins. The transistor itself was OK, however, the cause of the fault being an internal short from pin 8 of the line output transformer to chassis. A new transformer restored the set to working order.

No picture: As usual the cause was dry joints on the c.r.t. base socket. Rather pleased to have an easy one for a change, we wrapped it up and took it over to the soak test bench. When plugged in and switched on the standby light remained off. The 12.5 V secondary supply was missing – the yellow lead from the transformer PCB makes a good test point – as the $0.47\,\Omega$ series resistor RP43 had burnt out. The bridge was OK, but there was a very heavy load across its output. Lifting the yellow lead didn't remove this load. The only other connection is to the microcomputer panel PCB8, where we found that CR67 (1000 μF) was extremely leaky.

Severe field contraction: This set would intermittently show severe field contraction, with no sound and the raster at peak white. If almost any part of the chassis was touched the fault could be made to come and go. Some careful prodding was required to find the most sensitive area. We then found that there was a dry joint at RL28, the 1.5 Ω 13 V supply feed resistor.

Beko

Beko 12220

No colour at switch-on: At switch-on there was no colour in the display. But if the set was switched to standby and back, or the channel was changed, the colour would appear and remain until the set was switched off with the mains switch. The cause of the fault was traced to dry-joints on the main board, where the colour decoder panel plugs into it.

Tripped: When this set was powered it tripped off. With a 60 W bulb as the load, we found that the h.t. was 125 V. Further checks brought us to the BY299 diode D801 which was short-circuit. A replacement restored normal operation. If D801 fails at switch-on the LOPT is suspect.

Raster in top quarter of screen: This set's symptoms were strange. The top quarter of the screen was lit up but contained no vision information. The rest of the screen was blank. According to the customer the height had started to decrease before this happened. After much headscratching we found that R717 (1 MΩ) was open-circuit. It's connected to pin 5 of the TEA2029 timebase generator/power control chip. For good measure we also replaced R722 (1 MΩ), which is in series with R717. Pin 5 of the chip is connected to the field generator circuit: there should be a 4 V peak-to-peak field sawtooth waveform at this pin.

Binatone

Binatone 01

Failure of line output transformer: A very common fault with these 14 inch colour portables is failure of the line output transformer – it usually burns out in a big way! Be sure to get the right one when ordering a replacement. Two types were fitted. They can be identified by the numbers CF82 or CF65A. The two are interchangeable but give different focus voltages depending on the type of tube fitted. If you fit the wrong one the result will be a blurred picture.

Start-up problem: These nice little sets are very common to us. A recurrent problem we have is with the start-up circuit. There are two 180 kΩ, 0.5 W resistors here, R622 and R633, and one invariably succumbs. We fit more manly replacements.

Bush

BUSH 1452T
BUSH 2020
BUSH 2057NTX
BUSH 2114T
BUSH 2321T
BUSH 2520

Bush 1452T

Dead set, blown mains fuse: This receiver was dead, with a blown mains fuse (F901), as the 2SD1554 chopper transistor Q404 was short-circuit. When these items had been replaced we powered the set via a variac. The power supply now produced an h.t. output, but there was no regulation and the line output stage-derived 12 V supply was missing. The 12 V supply was missing because the 12 V, 1.3 W zener diode ZD402 was short-circuit. This had seen off R425 (5.6 Ω, 3 W). After much checking we traced the cause of the failure to regulate to C911 (47 μF, 50 V) in the chopper transistor's base circuit. It had gone low in value. We decided to replace C909 (47 μF, 50 V) as well.

Surge limiter resistor open-circuit: If you find that the 5.6 Ω, 3 W surge limiter resistor R425 is open-circuit, check the 12 V zener diode ZD402 as well. The cause of the failure may be high or intermittently high h.t., so replace C911 (47 μF, 25 V) and possibly the 9.1 V zener diode ZD902. See also note about C911 (polarity and rating) below.

Blown power supply: If you get one of these sets with a blown power supply, you will probably find one or two capacitors with lifted lids. After replacing them, make sure that C911 has its negative end connected to the base of the chopper transistor. It's sometimes fitted the wrong way round during manufacture, or the panel may be incorrectly marked. It is also wise to uprate it to 63 V.

Picture rolling and jumping: The picture was rolling and jumping and the top half was distorted. We replaced the field output and jungle chips, and several electrolytic and tantalum capacitors in the field timebase, before we eventually traced the cause of the fault to R413 in the field feedback/correction circuit. It had risen in value from 6.8 kΩ to 125 kΩ.

Bush 2020

No colour: Checks showed that normal chroma was present at the input to the TDA3562A colour decoder chip (pin 4) but there was no output at pin 28, which feeds the delay line circuit. A new chip put matters right. Another common cause of no colour with these sets is R251 (18 kΩ) going open-circuit. This removes the bias from the chip's colour control pin 5.

Vertical black line on screen 2 inches from right: The TDA3562A colour decoder chip used in this set is the Telefunken version. You can fit other makes such as Philips if R513 is changed from 180 kΩ to 68 kΩ and R515 from 120 kΩ to 150 kΩ. If you fit a TFK chip later the values of these resistors must be restored to the original ones. An interesting fault occurs when R419 (470 kΩ, 0.5 W) goes open-circuit: you get a vertical black line on the screen, 2 inches from the right-hand side. This resistor is in the line flyback pulse feedback path to the TDA2579 timebase generator chip. If the set jumps into and out of standby when a channel change or adjustment is being made, change the MDA2061 EEPROM chip IC2.

Intermittent fault; either a blank screen or thin red line on screen: This set suffered from an intermittent fault. There would sometimes be a blank screen, but on occasions this would have a thin red line across it, as if there was field collapse. The fault was so intermittent that it could take anything from minutes to weeks for it to recur. We found that touching the board almost anywhere when the set was warm would produce the fault, which was thus very difficult to localize. Eventually we found a bad joint on C307, which is partially hidden by a plastic support strut.

Front panel buttons had no effect: A glance at the manual showed that column d on the key scan wasn't working. This line is also connected to the HEF4066 chip IC3, which was found to be faulty. The customer also asked if we could enable the three-band tuning for him. After some experimentation we found that bit 5 of option byte 3 toggles this function. The set used an SAA1293A-03 tuning/remote control decoder chip.

Bush 2057NTX

Dead set, standby light on: Checks showed that the 12 V supply was missing because zener diode ZD402 was short-circuit and R423 (0.68 Ω safety) open-circuit. When these items had been replaced the set worked but couldn't be switched into standby. The fairly hefty 2SC2335 transistor Q907 in the standby switching circuit was found to be leaky. A replacement restored correct operation.

Dead set subjected to mains surge: Various problems needed to be sorted out. The 2SC2482 line driver transistor Q401 and its feed resistor R416 (2 Ω, 5 W) were faulty. The cause of field collapse was simply failure of the TDA3653B field output chip IC401. While the cause of no sound or vision was the 12 V, 1 W zener diode ZD402 being short-circuit with its supply resistor R422 (3.3 Ω, 2 W) open-circuit (these components are situated just to the right of the scart socket). Make sure that the HT voltage is 112 V at TPB+ or you may lose all the work you've done.

Crackle of e.h.t., no picture: When this set was switched on at the mains there was a rather large crackle of e.h.t. but no picture. The h.t. was found to be over 170 V instead of 115 V because C910 (10 μF) and C909 (47 μF) in the power supply had dried up. On advancing the tube's first anode control a blank, unmodulated line appeared and the set wouldn't go to standby. The standby fault was caused by Q907 (2SC2335) which was short-circuit. Replacing ZD402 (12 V, 1 W) and R442 (3.9 Ω, 1 W fusible) brought back the signals, while the field scan was restored by replacing the TDA3563B field output chip IC401.

Bush 2114T

Wouldn't come on when asked: There was no output from the power supply. Replacing C801 (47 μF) and C802 (100 μF) cured the trouble. We uprated them both to 63 V.

Poor at starting: These little teletext portables, with a TDA4602 type power supply, are quite often poor at starting. The cause is usually faulty electrolytic capacitors in the power supply. In this particular set, however, the small choke L801 in the chopper transistor's base drive circuit was intermittent.

Colour in text but only monochrome in TV mode: Replace the TDA3562 colour decoder chip. But be sure to use only the Telefunken version. Others are not compatible.

Dead set, power suppy overloaded: This set was dead because the power supply had supplied too much power to the rest of the set – the h.t. voltage was at around 150 V. The basic cause of the trouble was a defective capacitor, C818 (1 μF, 63 V), which is part of the power supply feedback system. This is in fact quite a common fault, but further troubles were being caused by IC301 (TDA3563), IC601 (TDA2006), R607 (10 Ω), C806 (1000 μF, 25 V) and C808 (1000 μF). When the set eventually came on the display showed 6.6. We found that the SAA1293A-03 tuning/remote control decoder chip and the three-pin supply regulator IC5 were faulty.

No picture or sound, but a high pitched whistle: We found that the h.t. was at 172.5 V instead of 112 V. Replacing C801, C802, C817 and C818 in the power supply cured the problem. Unusually, there was no other damage.

No channel change with teletext: When we pressed the teletext button the picture remained but the TV channel couldn't be changed. Although the set was in the teletext mode, text couldn't be displayed. The first thing to do in this situation is to check at pin 22 of the DPU2540 chip IC903. If the 5 V field pulse is missing, resolder all 12 through-the-board links on the double-sided print text PCB.

Bush 2321T

No colour: The cause turned out to be an open-circuit connection at plug 10. However, the sound became uncontrollable when after we put the first fault right. The culprit this time was the SAA1293 chip.

No picture or intermittent loss of picture due to poor soldering on the teletext board: This fault is becoming quite common with these sets now, due to poor soldering on the teletext module. These are a number of through-the-board links and need to be resoldered. On a couple of sets, however, the same symptom has been caused by the crystal being dry-jointed.

No picture or sound: There was no picture or sound and the standby colons faded a few seconds after switching on. The chopper transistor's base drive coupling capacitor C802 (100 μF, 16 V) was faulty.

Intermittent behaviour: The symptoms produced by this set were most peculiar. Sometimes it would come on normally and stay on; sometimes it would come on, trip off and go totally dead; usually it would start up then, after just a couple of seconds and often before the picture put in an appearance, it would trip to standby with faint thumps from the loudspeaker. In fact there were two faults competing for our attention! The mains input plug to the mother board was badly crimped. And C802 (100 μF, 16 V), which couples the drive to the base of the chopper transistor, had gone low in value.

Bush 2520

Sound but no picture: On investigation we found that the line timebase was inoperative – there was no e.h.t. and the tube's heaters were out. As we had no circuit diagram we used an Avo to check the resistors in the line timebase. This revealed that R605 (5.6 kΩ) had gone open-circuit. A replacement restored normal operation.

Needs several attempts to switch on: We found that the positive temperature coefficient thermistor R802 in the power supply was dry-jointed, while R800 (820 kΩ) read just under 2 MΩ. When these components had been replaced the set started up with no further problems. Another of these sets suffered from low-gain, noisy reception, as if the tuner had failed. The cause of the fault was the TDA4505 chip IC100, however.

Would not switch on warm: This set would switch on OK from cold, but if it was switched back on when warm there were no results. The set would come on if the posistor in the start-up circuit, R802, was frozen or tapped. A replacement posistor cured the fault. These sets are fitted with the Indiana 100 chassis.

Cathay

Cathay CTV3000

Set dead, not starting up: R504 (150 kΩ) was open-circuit.

No results, no bias to Q501 base: No results with no bias at the base of Q501 was traced to R504 (150 kΩ) which was open-circuit. We weren't out of the woods yet, however, since the set would intermittently shut down and refuse to operate. After a thorough search we found a crack in the track from D301 to the line output transformer. A piece of tinned wire put things right.

Line drift when warm: The problem with this portable was line drift when warm. We found that the area around the TA7698A chip was very sensitive to heat. The cause of the trouble was C232, which is connected to pin 33.

Two hum bars on screen: Replacing the main smoothing block C507 (100 μF, 400 V) cured the fault.

Contec

Contec KT8135

Top of picture stretched out: Although we've had a number of these sets in I've never come across this fault before – the top of the picture was stretched out and the bottom was cramped, while a 4 inch hum bar punctuated and severely distorted the display. We found that C336 (220 µF, 50 V) had gone low in value, a replacement putting matters right. It's quite common to come across one of these sets with an intermittently grainy picture. The cause is a defective or dry jointed SAW filter. We've never had to change a tuner.

Field cramping when hot: This was found to be the result of C336 (220 µF, 50 V) developing a 10 kΩ leak.

Very bright picture with flyback lines: The first anode control R406 had no effect as R474 (390 kΩ) on the tube base board was open-circuit.

No luminance: The cause was a poor plug and socket connection for the luminance drive on the c.r.t. base panel.

Very bright raster with flyback lines: R474 (390 kΩ) on the tube base panel was found to be open-circuit. As a result the tube's first anode voltage was high.

Dead set: As there was 320 V across the reservoir capacitor C501, the mains input and rectifier circuits were obviously OK. As we didn't have a circuit diagram we carried out cold checks in the power supply and line output stages. The silicon all seemed to be OK, so we made a start on the resistors in the power supply. R501, R502 (both 330 kΩ) and R503 (1 MΩ) were all faulty. Replacing them restored normal operation.

Decca

Decca 100 chassis

Bottom half of scan missing: It was as if one of the field output transistors wasn't working. The cause of the trouble was reverse leakage in D302 (1N4148) which is in the sync pulse feed to the field oscillator circuit. One thing we noticed about this fault was that adjusting the field hold control VR302 moved the section of scan still present up and down – the hold control had the effect of a shift control!

Severe ringing and poor luminance: It looked very like a mistuned i.f. stage but we eventually found that the 47 μF luminance coupling capacitor C208 on the colour decoder panel had gone very low in value.

Line sync jitter: The horizontal phase was not working. We eliminated the field/line timebase panel, and traced the fault to C403 (10 nF) in the line output stage. It's part of the flyback tuning network, along with C402 and C404 (both 100 nF). We changed all three capacitors as a precaution. We had a number of these sets that all required a replacement line output transformer. We also had a number of surplus transformers for the ITT CVC25/30 chassis. These appeared to be suitable replacements. On investigation we found that the pin configuration is the same, but unless a slight modification is carried out the c.r.t. will have a short life. The original Decca transformer has a coil in the heater supply, mounted beneath the transformer. Reposition this beneath the line output stage PCB by cutting the print between pin 3 of plug PLA and pin 2 of the transformer. Also remove the line shift winding at the top of the transformer, otherwise it gets in the way of the screening can.

Decca 120 chassis

Loud low pitched whistle from LOPT and uncontrollable sound not tuned to a station: Rather misleading this. It's a line output stage fault: change the tripler.

Muted sound: The audio output was all right with no signal present, but as soon as an aerial was connected there was only muted sound. The sound detector coil L601 was open-circuit. Fortunately it could be repaired, by resoldering the legs inside.

The power supply would pump at switch-on from cold: After about 10 minutes the supplies would become steady and no amount of cooling would bring the fault back. The set had to be switched off and left until next day. The culprit turned out to be the mains bridge rectifier's reservoir capacitor C804 (220 μF, 385 V). It was open-circuit.

Dull picture: The contrast control had virtually no effect. When the beam limiter line was disconnected (at the CRT base) the contrast was restored, albeit with no black-level clamping. We found that the two beam limiter biasing resistors R425 and R426 (both 68 kΩ) were open-circuit. When they had been replaced the contrast could be set up correctly.

Decca/Tatung 130 chassis

Black level of one colour varies: Check the value of R226 (red), R244 (green) or R251 (blue) – this assumes that the c.r.t. is OK. These resistors provide feedback in the cascade RGB output stages and often go high in value (the correct value is 100 kΩ).

Very dim picture: The cause was no c.r.t. first anode supply. The supply is derived from the focus module, however – a self-contained unit with four leads. The prospect of a spares order and a wait didn't appeal, so a further check was made. Sure enough there was no voltage at the c.r.t. pins and it did appear that the focus unit was responsible. On disconnecting the lead to the c.r.t. base, however, ample voltage was found to be present. A resistance check was then made from the first anode supply to chassis. As the reading was 1 kΩ and there's a 1 kΩ resistor in series with the three c.r.t. first anode pins one of them was clearly shorting to chassis. Pin 5 was found to be responsible for the trouble. It was shorting to the ring in the c.r.t. base – this acts as a spark gap. Stripping the base down and gently bending the ring slightly away provided a cure.

Set stuck on channel 1: The cause of the fault was traced to R003 (10 kΩ, 1 W) being open-circuit – it feeds the 33 V supply to the tuning PCB. The cause was given away by the fact that channel 1 couldn't be tuned in.

Field collapse: The TDA1670A field timebase chip IC301 was short-circuit while R438 (1.2 Ω) which provides its 23 V feed was open-circuit. Replacing these items restored normal operation.

Intermittently going into standby: This set had been in several times for intermittently going into the standby mode. Several things had been tried – the tripler, the BU426A chopper transistor, the TDA4600 chopper control chip etc. As the fault was of such an intermittent nature we decided to change the line output transformer, suspecting an internal arc. This put matters right.

Intermittently green picture: The set ran for some time without playing up, though we noted that there was a faulty seven-segment channel indicator LED which was soon replaced. Then the picture started to go green. The 100 kΩ resistors on the c.r.t. base panel had already been changed so we dived for the RGB teletext interface panel to look for dry-joints, particularly as it was below where we'd been working on the display. The joints on the plug and socket were poor but the fault persisted after these had been attended to. The c.r.t. also looked poor, as most A56–540Xs do now, so we put the tester on to have a look. This showed that there was an intermittent heater-cathode short.

Looked as if tube emission was low: The cause was traced to R425 and R426 (both 68 kΩ). They are in the beam limiter circuit and were both open-circuit.

Decca/Tatung 140 chassis

Wouldn't switch on from standby: Because there were none of the usual problems on the main PCB we powered the set directly from the mains supply, bypassing the remote control circuitry. The set then worked, so we had a remote control fault. The best bet was the SAA1351 chip, and when this was replaced normal operation was restored. On previous occasions we've had the tripler upset this chip, so we disconnected it before testing. When we reconnected it the SAA1351 was affected, so a new tripler had to be fitted as well.

Intermittent no go: This turned out to be due to R805 which had risen in value from 390 kΩ to about 600 kΩ. It supplies pin 5 of the TDA4600 chopper control chip. The voltage at this pin was very low and there was very low drive at the base of the chopper transistor Q801. Watch out for R802 (10 kΩ) being dry jointed.

No results: H.t. was present at the line output transistor (ouch!) but there was no other activity. It didn't take us long to find that there was no 18 V supply. The connection to the rectifier (D811) from the chopper transformer appeared to be perfectly soldered but the diode's cathode lead had never been pushed through the board fully.

No tuning: Checks showed that there was no tuning voltage at pin 4 of the tuner. This set has voltage-synthesized tuning. An i.c. produces a variable mark–space ratio square wave which is fed to a transistor to produce the appropriate proportion of the regulated 33 V supply. There was 33 V at the collector of this transistor but nothing at its base and emitter because the bias resistor RR70 (33 kΩ) was open-circuit.

Produces loud hum at switch-on: We found that C808 and C810 (both 100 μF, 25 V) in the chopper circuit had dried out. Replacements cured the hum but the verticals were ragged. This was cured by replacing the h.t. reservoir and smoothing capacitors C822 and C826 (both 47 μF, 250 V).

Picture went red intermittently: We traced the cause to the 1 kΩ preset R224 in the red output stage. Replacement and setting up produced a good picture.

Decca/Tatung 160 chassis

Dead set: Our first check showed that there was h.t. at the collector of the chopper transistor. So we checked the 12 kΩ start resistor R802, which is a common failure. It was OK, and the voltages around the TDA4600/2D chopper control chip were also correct. Scope checks showed that the drive output was present at pin 7 of the chip but not at the base of the chopper transistor. L802 in the coupling circuit had gone open-circuit, a replacement putting matters right.

Dead set: The h.t. was present and correct but there was no 11.5 V supply to the signal circuits. This supply is provided by a very simple regulator consisting of transistor Q501 and a zener diode. Q501 was without base bias because R508 (10 kΩ) was open-circuit.

Sound and picture disappears intermittently: This set had been in two weeks previously with field collapse. We'd replaced the TDA3651 field output chip to put that right. Now it was back in the workshop again, this time with the complaint that the sound and picture disappeared intermittently. The cause was a dry joint at the emitter of the chopper transistor Q801. It would seem that replacing the field output chip had contributed to the problem as the chopper and line output transistors are mounted on the same heatsink as the chip. Thus flexing the panel during the first repair could have brought on the second fault. For good measure we resoldered the legs of both transistors.

Picture rolls when hot: Sure enough only a few seconds with the hairdryer were needed before line lock was lost. Beside the TDA4503 chip we found a tower of ceramic capacitors connected in parallel, giving a total value of 2400 pF. Fitting the correct 2700 pF Suflex type capacitor in position C113 restored temperature stability.

Ferguson

FERGUSON 16A2 (TX90 CHASSIS)
FERGUSON 20E2 (TX90 CHASSIS)
FERGUSON 51P7 (TX98 CHASSIS)
FERGUSON A10R (TX80 CHASSIS)
FERGUSON A51F (IKC2 CHASSIS)
FERGUSON A59F (ICC7 CHASSIS)
FERGUSON B14R (TX90E CHASSIS)
FERGUSON ICC5 CHASSIS
FERGUSON ICC7 CHASSIS
FERGUSON ICC8 CHASSIS
FERGUSON ICC9 CHASSIS
FERGUSON IKC2 CHASSIS
FERGUSON TX100 CHASSIS
FERGUSON TX85 CHASSIS
FERGUSON TX86 CHASSIS
FERGUSON TX89 CHASSIS
FERGUSON TX90 CHASSIS
FERGUSON TX90 CHASSIS (20″)
FERGUSON TX98 CHASSIS
FERGUSON TX99 CHASSIS

Ferguson 16A2 (TX90 chassis)

Only snow when the set first comes on: When we tried the set we found
that this description was correct, but the customer had neglected to tell
us that the fault lasted for only about thirty seconds. We tried a new tuner,
then the M923 tuning chip, but neither of these was responsible. The set
started to take longer and longer to come on, and at last we were able to
get some sensible readings from the test equipment. The tuner unit's
tuning pin was permanently high at 31 V. A check at pin 19 of IC902
(M923) with the oscilloscope then revealed that there were no tuning
output pulses. At this point the set returned to normal working and had
to be left for another day. The next time the fault occurred we
immediately checked the 5 V supply to the chip. It was missing from the
5V regulator, which was without its 12 V input. This comes from the main

board and was present at the output from the 12 V regulator. A check on the main board print showed that there was a hairline crack around the solder pad where the 12 V supply is taken off to the tuning board.

Terrible field linearity: The cause was the 68 V zener diode D137 between the collector of TR105 and chassis – it was leaky. This zener diode is not present in the smaller-screen versions of the chassis.

Dead set, but the power supply was working: The cause of the fault was absence of the 150 V h.t. supply because of an open-circuit between pins 6 and 9 of the line output transformer. A new transformer brought the set to life, but though stations could be searched for and found the set wouldn't memorize them. The M293B1 chip IC902 had failed. All was well when this chip had been replaced.

Ferguson 20E2 (TX90 chassis)

Tuning problems: We've had tuning problems with two of these sets. The first wouldn't search. We found that the 1 kΩ safety resistor R986 in the tuning circuit on the remote control panel was open-circuit. This was in turn due to the 150 V tuning regulator supply being high because R56 on the line linearity/width panel had fallen in value from 2.4 kΩ to less than 1 kΩ. When this had been attended to we discovered that channels could be found but not memorized. A new M293B1 memory chip (IC902) put that right. The second 20E2 found channels up to around number 30, but anything higher produced motorboating and again there was no memory. That's right, the voltage across the ZTK33B regulator D909 was low at 25 V. But a different diode was to blame: D908, the 33 V zener diode that stabilizes the voltage at the base of transistor TR916, was leaky.

No remote control functions: This set worked perfectly except that it wouldn't respond to remote control commands. Our tester showed that the handset was transmitting all right so attention was turned to the set. As pin 7 of the T9005N remote control decoder chip IC901 was receiving an input at around 8 V peak-to-peak the IR receiver and buffer stages were assumed to be OK. This seemed to suggest that IC901 was responsible for the problem, but chips of this sort rarely fail. So the handset was investigated further. The manual gives setting-up instructions for a preset labelled 'factory set freq adj' inside the handset. When this adjustment had been carried out everything worked fine. The handset had probably been dropped, causing the potentiometer to shift slightly.

No memory on channels 1–4: The M293 chip was faulty.

Ferguson 51P7 (TX98 chassis)

Failure of line output transformer: A problem that's becoming quite common is failure of the line output transformer – usually the grey Philips type. The set pumps and doesn't work. You can usually confirm that the power supply is working all right by disconnecting the feed to the line output transformer's primary winding. When ordering a replacement the correct type must be specified: a number of different types have been used, and a list of modifications/checks comes with the replacement and must be carried out.

Channels would change on their own: This set had a number of problems the channels would change on their own, the sound would mute and the text would do odd things. We had it on test for several days without any fault putting in an appearance, but the customer insisted that it was faulty. So we phoned a Ferguson expert who suggested an answer. All you have to do is to cut the print at pin 11 of the infrared preamplifier chip and add an 0.1 μF capacitor across the cut. This was done and the set was sent back, never to return. So the modification worked. If it doesn't you replace the tube base.

Power supply hiccup after 2 hours: After about 2 hours the power supply would hiccup: when the set came on again it was on a different channel. In fact it always came back on channel 1, the customer using only channel 3. She said that it always went off in the middle of her favourite programme and came back in the middle of her most hated one! We became overnight heros after curing the fault by replacing the TBA8138 multi-voltage regulator chip IC11.

Intermittently cuts out: We nearly gave up with this bouncer and refunded the customer his money! The original complaint had been 'intermittently cuts out'. We resoldered a number of joints, then soak tested the set for five days. No problem. But after returning it the customer was on the phone within 4 hours! The set was intermittently resetting itself to programme 1. Unfortunately no amount of flexing, tapping, cooling or heating would instigate the fault. But we discovered that running the set for a few hours covered with a blanket would produce a reset action. During the following week we replaced the IR receiver diode, the IR amplifier chip, undertook the capacitor modification in the remote control input line and resoldered every single joint that looked even halfway suspicious. Still no joy. As a last attempt, we replaced the TDA8138 5V/12V regulator chip IC11. We then covered the set with a blanket and put it on test, taking a tentative look each hour. It stayed resolutely on programme 3. Hindsight is wonderful. When we re-examined the circuit diagram it became obvious that IC11 also has a reset function.

Ferguson A10R (TX80 chassis)

Intermittently red picture: This can be caused by an open-circuit track between RV110 and the emitter of TV101. We've had three sets recently with this fault: you can't see the track break.

Bright red cast with flyback lines: A picture was present but it had a bright red cast with flyback lines on top of it. We looked at the R – Y drive signal at the base of the red output transistor TV101 on the c.r.t. base panel and found that it was very similar to the other colour-difference drive signals. A check at its collector showed that there was just a steady d.c. level, however. RGB matrixing is carried out in the output stages in this chassis, the luminance signal (–Y) being fed to the emitters of the output transistors via preset potentiometers. A check across the red preset PV104 showed that there was no voltage drop across it, as there was across the equivalent presets in the G and B stages. Close examination of the print revealed that there was a tiny fracture between the end of RV110, which is in series with PV104, and TV101's emitter. With this linked across normal operation was restored.

Bad intercarrier buzz: This receiver was intermittent to start with but later the fault became constant. We eventually traced the cause to the 6 MHz ceramic filter QI02 which is adjacent to the M52038SP chip IL01 beneath a screening can.

Intermittent results: The cause turned out to be the 7 nF, 1.6 kV flyback tuning capacitor CP18. It was badly swollen and one leg was charred. Note that the value is different with Model A14R.

Ferguson A51F (IKC2 chassis)

Intermittent revert to standby: The customer said that 'it went off every five minutes', but it took over a week of soak testing for the problem to show up in the workshop. When it did, we found that the line oscillator's supply at pin 40 of IV01 was missing, though the voltage at the power control pin of the microcontroller chip IR01 was low – the correct state for a running set. Attention was turned to the standby switching transistors TR16/17. These turned out to be open-circuit, although freezing it brought the set back to life. A replacement transistor cured what was an extremely vexing problem. Incidentally TR18 is not to be found in the set although it's shown in the circuit diagram.

Loud plop from the speaker when the set was switched back to standby: The owner hadn't complained about this and we were not sure whether it was normal or not. So we looked at the circuit and found the power muting system. Checks here brought us to CA15 (10 μF) which you could say was open-circuit – it had never been fitted! According to the circuit the power muting system is not fitted in the 17 inch Model 41P3.

Set was tripping: We found that TV01 (BC558C) in the electronic trip circuit was leaky. When we'd replaced this item the set still tripped, but at a slower rate. The cause was excessive current being drawn by the thyristor field output stage. To cure the trouble we had to replace the TL082 field oscillator/driver chip IF01. For a quick check, just remove it. In this particular set it was providing excessive drive.

Arcing in line output transformer: A new transformer put this right but the set tripped, with field collapse. On advice from Ferguson we disconnected the field sensing circuit (TV01) and connected pin 20 of the microcontroller chip IR01 to chassis to force the set out of standby. Having got the set to stay on, it didn't take me long to discover that the BC548B field drive shaping transistor TF25 was the cause of the problem.

Ferguson A59F (ICC7 chassis)

Poor focus: A number of early production sets suffered from poor focus because the focus lead connector came adrift from its socket on the c.r.t. base panel. Some customers have complained about lack of straightness of the verticals and horizontals with the very flat tubes fitted. There doesn't seem to be anything one can do about this.

Line driver transistor TL17 dies as soon as the line output stage gets going: Replace DF16 (BA157) in the field output stage. You will find that it's short-circuit.

Power supply shuts down at switch-on: Following instructions from Ferguson, the TDA8178F field output chip had been replaced. But on this occasion the solution hadn't worked. The cause of the trouble turned out to be field related, however. There was a short on the UL2 (65 V) feed to the field output stage because the BA157 rectifier diode DL09 was short-circuit. It's connected to pin 12 of the line output transformer.

Ferguson B14R (TX90E chassis)

Stuck in standby: On discovering that the S2000A3 line output transistor TP10 was short-circuit we assumed that the repair would be a very simple one, but the set remained in the same state when a replacement had been fitted. Further checks showed that the BA157 diode DP08 was short-circuit.

Blank screen with no on-screen graphics: The cause is leakage in the PCB material around the data and clock lines between the micro-controller chip and other chips under its command. Cleaning with a good aerosol safety solvent followed by a thorough drying out will usually restore normal operation.

Dead, standby light on, rumbling: The red standby LED was alight but there was a rumbling noise from the set which would do nothing else. A check showed that the start-up voltage at RP71 (9.1 kΩ) was very low, at about 1.1 V. Rather interestingly, RP71 had fallen in value to about 8 kΩ. But the cause of the fault was DP36 (BA157) which was short-circuit.

Dead set, stuck in standby: There was only 2 V at the bottom end of the 9.1 kΩ, 9 W start-up resistor RP17. The cause of the problem turned out to be DP36 (BA157) which was short-circuit.

Safety message: A word of warning on these sets. The mains bridge rectifier's reservoir capacitor remains charged for a long time after the set has been switched off. Recently we found that there was still 100 V across the capacitor despite the set having been off for 17 hours.

Ferguson ICC5 chassis

Set failed at switch-on: You could hear the power supply pulse three times before shutting down. The power supply's h.t. output would pulse up to around 130 V followed by a second pulse up to about 80 V and a third up to only 40 V. A voltmeter connected to pin 28 (safety sensing input) of the power processor chip IL14 revealed nothing, i.e. no high input was apparent. We checked for shorts across all the outputs from the chopper power supply, then all the rectifier diodes, without finding anything amiss. So we disconnected the supply to the diode-split line output transformer LL53 and switched on again. This time the h.t. came up – in fact it rose to some 170 V, but the power supply didn't shut down, pointing to a fault somewhere in the line output stage. All the secondary supplies from LL53 were checked for shorts, then the rectifiers were checked. This revealed the culprit. DL55 (BA157) which is used to provide the 200 V supply was short-circuit.

No sound or raster: When the e.h.t. attempted to rise the receiver tripped and went into the shutdown state. The 8 V supply reservoir capacitor CP37 (4700 μF) was leaky. For no sound check whether resistor RS13 (4.7 Ω) which provides the 30 V feed to the TDA2030A audio chip IS11 is open-circuit.

Field collapse: The trouble with this set was field collapse – there was a single straight line about 2 inches from the top of the screen. We decided to check the components connected to pin 5 of the TEA2029C timebase generator chip as these are associated with the field oscillator. RF01 (3 MΩ) had gone open-circuit.

Intermittent loss of colour: The problem got worse the longer the set was left on. After a couple of hours the fault was almost permanent so, armed with the trusty freezer, we delved into the colour decoder subpanel. Unfortunately no amount of freezing revealed anything, so out came the meter and scope. As the supply and the chrominance input to IV01 were OK, a check was made on pins 12 and 13 of this chip – this is the reference oscillator section. There was no problem here. We next checked for line pulses at pin 15. They were missing. Tracing the circuit then brought us to DV21, which separates the line pulse from the sandcastle waveform. There was a perfect waveform at its anode but nothing at its cathode. A replacement 1N4148 diode provided a complete cure.

Set would trip three times and shut down: This set would trip three items then shut down completely, as it's supposed to do under overload conditions. We disconnected the various supplies obtained from the line output transformer in an attempt to isolate the cause of the overload but had no luck. Replacing the transformer itself cured the fault.

Random flashing lines on the screen: We found that a signal fed in via the scart socket produced a perfect picture. As we've had similar faults caused by the tuner this was our first suspect. We hooked up a rotary tuner from an old monochrome set as a check. This cured the fault, so a new tuner was obtained and fitted – only to find that the fault remained as before. A check on the 11.5 V supply to the tuner showed that it was stable and ripple-free. The voltage at the tuning pin was also stable but the scope showed that there was a low-frequency noise component at some 10–20 mV peak-to-peak. Our first thought was that this was audio noise, but decoupling various points in the audio output circuit made no difference. The surface-mounted capacitors CT01, C103, C105 and C109 in the tuning voltage supply were changed but this had no effect on the fault. What next? In desperation I cooled the U6316 PLL chip IT20. This provided a cure. When IT20 was then heated the fault was back again. A new PLL chip restored correct operation.

In shutdown state: When switched on the set would trip three times then become lifeless. That's the usual trip sequence with this chassis. It can sometimes be a nightmarish situation as the fault can be almost anywhere – the protection circuits are very sensitive. As a start we disconnected pins 8 and 10 of the line output transformer, but the set still tripped. Next, various secondary supplies from the chopper transformer were disconnected including, inadvertently, pins 17 and 18. A note was then made in the service manual warning never to do this or else the power supply will suffer impending doom. As a result of our miscalculation a new BU508 chopper transistor was next fitted. Back to the original fault. A glimmer of light appeared at the end of the tunnel when pin 22 of the chopper transformer was disconnected the set powered up with a raster. There was no sound though as pin 22 is the 36 V supply to the stereo audio board. When the audio output chips IS40 and IS41 were checked the left-hand channel one was found to be low resistance from its supply input to chassis. A new TDA2030A42 chip put matters right.

Severe EW bowing and line pairing: We found that the TDA4950 EW modulator driver chip was too hot to touch. A replacement cured the fault. Comparison with a new chip showed that the defective one had a short between pins 5 (EW output) and 4 (chassis).

Set was tripping: Raising the PCB to the vertical position to view the underside proved to be worthwhile – a semi-carbonized dry joint could be seen on one leg of the flyback tuning capacitor CL48. We cleaned the joint and resoldered it, then dealt with the pins of the line output transformer. The line output transistor was then checked and we weren't surprised to find that it was short-circuit all ways. A replacement brought the set back to life.

Right-hand speaker cut out intermittently: When we removed the back we found that there were two huge dry joints on the right-hand speaker socket. After resoldering these the set was returned to the customer, only to come back next day with the same complaint. The cause of the fault took some time to locate, because the symptom occurred only momentarily. There turned out to be a dry joint on the front-mounted headphone socket.

Sound fault: This Nicam set (Model 51K5) had a sound fault. After 2/3 hours the sound would crackle very loudly, so loudly in fact that the set had to be switched off. This also made fault-finding more difficult. On top of this the fault would show up only every few days. To cut a very long story short, the cause of the fault was traced to the smaller of the two screening cans on the Nicam PCB. Pin 24 of the main edge connector is linked to one of the screening can's legs. This then distributes the D earth line to various points around the panel. In order to avoid future problems we resoldered all the screening can's connections.

Would trip three times then die: This set made a loud tripping noise that came from the line output stage. It would trip three times then die. Apart from lots of dry joints, we soon found that the EW correction transformer LG11 was short-circuit all round. So we ordered a new one plus a TDA4950 driver chip (IG01) as this had also suffered. When the parts arrived we fitted them and also replaced a couple of charred resistors. At switch-on we were rewarded with . . . nothing, or rather a very soft tripping noise three times before the set once again died. We cold checked all the diodes in the line output stage, also the output transistor, but could find nothing wrong. With the line output stage disconnected and a dummy load connected between the cathode of DP41 and chassis the power supply ran quite happily. So what was wrong? For want of something better to do we replaced the line output transistor, which was type S2000A3. Not having one of these I tried a BU508AF. The set then worked normally and continued to do so. We later fitted the correct type, but couldn't find anything wrong with the original one when carrying out meter checks on it. Maybe it broke down under load. Something to bear in mind in future.

Line output transistor's short life: Check CL48 in the flyback tuning circuit by replacement. The value varies with different models. It was 11 nF in this case (Model 51K7).

Dead, '88' showing in LED display: Apart from '88' showing in the LED display this set appeared to be dead. We traced the cause of the fault to RP42 (1 Ω, safety) in the power supply being open-circuit. This resistor provides a kick-start supply for the line driver stage: once the set gets going the supply for the driver stage comes from the line output stage.

Failure of EW loading coil: Failure of the EW loading coil LG11 in models fitted with 110° tubes can produce the dead set symptom. In addition to replacing LG11, the following items must be replaced, using Ferguson approved components: CL44 330 nF; RL44 56 Ω + 120 Ω fusible; J134 22 Ω fusible; IG01 TDA4950 EW modulator driver chip. Note that RL44 consists of two resistors in series. Also check the 1 nF ceramic capacitor CG11 which may well have gone short-circuit.

Tuner cutting out occasionally: Fortunately the cause was easy to rectify: there was a dry joint in the tuner.

Picture would go in at the sides very intermittently: This was caused by cracks in the print around the line output transformer.

Tuning problems: The tuning mode could be entered and the correct channel number could be set. But there was no variable tuning voltage at the input to the tuner and thus no signals. The voltage just stuck at 20 V. There was plenty of activity at pin 9 of the phase-locked loop chip IT20, but because TT12 (BC547) was open-circuit it couldn't affect the tuner unit. A new transistor restored all stations.

Ferguson ICC7 chassis

Two faults: power supply and line driver: Two of these sets have given us problems recently. The power supply in the first one didn't work although the start-up supply was present at pin 16 of the TEA2261 chopper control chip IP01. The cause of the trouble was an internal breakdown within IP01, as a result of which there was no voltage at pin 15. This pin provides the supply for the chopper drive stage within the chip. The BSR51 line driver transistor TL17 was short-circuit in the second set. When this had been replaced we found that there was field collapse. The cause was failure of the BA157 diode in the field output stage: it had gone short-circuit, putting 65 V on the 24 V line.

Dead set, mains fuse open-circuit: This set was dead with its 1.6 AT mains fuse FP01 open-circuit and the BUH515 chopper transistor TP29 short-circuit. Checks failed to reveal any obvious causes for the failure of these devices. On replacing them the power supply still didn't work but at least there were no fireworks. As there were no shorts across the secondary lines we looked at the drive to TP29. This is produced by 1P01 (TEA2261), which proved to be faulty. There's a d.c. connection between the base of TP29 and IP01, which was presumably killed when

TP29 went short-circuit – all ways. After replacing this device the h.t. appeared but the set still refused to produce results. RP62 (0.1 Ω) in the 24 V supply was open-circuit. A replacement finally brought the set back to the land of the living.

Repeated failure of chopper transistor: To prevent repeated failure of the BUH515 chopper transistor TP29, change the value of CP29 from 2.2 nF to 3.3 nF. An alternative chopper transistor, type S2000AF3, can be used if necessary. If the S2000AF line output transistor TL19 has failed, change the value of RV01 from 4.7 kΩ to 5.6 kΩ and add a 560 Ω resistor (RV03) in series with the anode of DV01.

Line output transistor short-circuit: As we couldn't find an obvious cause for this we fitted a replacement transistor and switched on. Ten minutes later the transistor failed again. The cause turned out to be a poor connection at the line driver transformer.

Set is dead but not tripping: You'll probably find that RP62 (0.1 Ω) is open-circuit and the BA157 rectifier diode DL09 short-circuit. DL09 takes its feed from the line output transformer, generating the 65 V supply across CL09. RP62 is in the chopper circuit.

I.f. instability at switch-on: The symptoms were either a negative picture with poor sync or a display that consisted of just short, white streaks. Tapping the i.f. unit made the fault come and go. The cause of the trouble was C525 (1 μF) which was dry jointed.

Keeps blowing its mains fuse: Four had blown at about six monthly intervals. The set was brought to us to see if we could cure the problem, as its fuse had blown again. When we checked the fuse's end caps, spark erosion could be seen. The fuse holder wasn't gripping the fuse tightly. To ensure a sound repair we fitted a new fuse holder.

Would not come on: Red LED, green LED then the sound of EHT followed by silence and no lights were the symptoms with one of these sets. The cause was traced to transistor TP53 (BC557) which was leaky collector-to-emitter. It's the voltage error-sensing transistor in the power supply.

Intermittent loss of signals: A common fault with this chassis and similar ones. The usual effect is a blanked-out screen – black, but with very regular white patterning. The cause is dry joints within the i.f. unit, particularly at CS32 which is associated with the LA7550 chip IS10.

Wouldn't start every time: This set had two faults. Another dealer had replaced the mains switch, but said that the set still wouldn't start every time. The replacement switch was faulty: another one put this right. The second fault was also intermittent. The picture and sound would both go, leaving odd black and white dots and lines on the screen. When the fault was present, we found that flexing the PCB in the area of the tuner made the fault come and go. As there were no dry joints on the main board the tuner and i.f. modules were removed and dismantled. The tuner seemed to be OK, but the i.f. module had a single coil with two of its pins dry jointed.

Ferguson ICC8 chassis

Intermittent fuse blowing: The cause was leakage in the blue protection capacitors across the bridge rectifier diodes.

No picture: No picture was the complaint with this set, though it worked fine when put on test. The next day it failed: there was no vision although the tube's heater and e.h.t. supplies were present. Vision was restored when the video drive board CD17000 was flexed. On removing this plug-in subpanel we found that every through-the-board joint had been resoldered, as had the socket connections to the mother board. The cause of the trouble, however, was that the MELF devices CD41, RD42 and CD12 had not been correctly soldered from new: solder was present, but it had not taken to the pads. Correcting this cleared the fault.

Dead set, red LED was alight: The set attempted to start up, but immediately went to standby. The cause of the trouble was loading on the line output transformer. We found that the focus pin at the base connector on the c.r.t. panel was corroded – it had turned green inside. As the set lives in very damp conditions we checked inside the focus/A1 control and found signs of arcing there. Replacing these two items cured the fault, but the set will probably be back – the customer doesn't believe that his house is damp!

Dead set with the line output transistor short-circuited: When a replacement transistor had been fitted the set was still dead, because there was no line drive – the BSR51 line driver transistor TL17 was short-circuit. The set worked when a new BSR51 had been fitted, but after only a few minutes' use the line output transistor was overheating and in danger of failing again. Checks around TL17 showed that its collector was at 50 V instead of 23 V. The collector supply comes from the U124V rail, which was also at 50 V, though the other three supplies from the chopper circuit were all normal. How could this be? The 24 V supply and the line output stage-derived 65 V supply meet at the field output stage, where DF16 (BA157) was found to be short-circuit. Thus the two supplies were shorted together. As the 65 V supply was providing more current than usual, the voltage had fallen to 50 V.

Stuck on channel 2 with snowy raster: A new microcontroller chip and 8 MHz crystal were tried but this made no difference. As checks in the microcontroller circuit suggested that everything was OK, attention was turned to the text board. After unloading IV02's SCL and SDA lines a picture came up. A new SDA5243 chip restored normal operation.

Picture darkens and the grey scale changes: We discovered that RT24 on the tube base panel had increased in value from 39 kΩ to 300 kΩ. It's the red channel feedback resistor in the version of the panel that has a TEA5101A RGB output chip (IT01).

Ferguson ICC9 chassis

Picture flashed and had red tint: When this set was switched on the picture flashed and pulsated and had a very red tint. As it warmed up the fault cleared. We suspected the TEA5101 RGB output chip on the tube base panel – cooling it down brought the fault symptom back. As we didn't have a TEA5101 in stock we decided to make a few more checks before placing an order. This brought us to RB24 (39 kΩ, 1 W) which provides d.c. feedback in the red amplifier channel. It was open-circuit. After fitting a replacement the picture remained the same no matter how cold the chip was.

Intermittent field faults: This set suffered from various intermittent field faults, such as top cramping or expansion of the centre of the scan as the set warmed up. Freezing almost any component in the field output stage would clear the fault. The cause of the trouble turned out to be the BA157 diode DF31, which is in the flyback boost voltage generator circuit. It was going high resistance.

Would come on for a second with the rustle of e.h.t. then die: There were no shorts, so we decided to disconnect the TDA8172 field output chip IF01. The result was field collapse with no tripping. A replacement TDA8172 chip restored the picture.

No or intermittent sound: You will probably have to look no further than socket BA05 at the back of the main panel. It's quite a tall socket which tends to 'wobble' when the PCB or chassis is moved. As a result there is a tendency for dry joints or cracked print where the socket is connected to the PCB.

Blank raster and no sound: A blank raster and no sound were the symptoms with one of these sets (Model B59F). As we've had this before, we headed straight for TX07, a surface mounted BC846B transistor which is fitted on the print side of the PCB. It measured short- circuit as fitted, but OK when it was removed. A replacement cured the fault.

Ferguson IKC2 chassis

Wouldn't come out of standby: When the field scan coil plug was temporarily removed the set came back on, indicating that the protection circuit was sensing a field fault. We carried out some d.c. checks in the field output stage and found that DF16 (BY398) was short-circuit.

Would come out of standby for only about a second: The cause of the trouble was that the TL082 field timebase generator chip IF01 was short-circuit, dragging down the line output stage derived U5 (13 V) supply. While tracing this short-circuit we found that DF16 (BY398), which is in parallel with the field output thyristor, measured a bit odd. So we changed that as well.

Intermittently going to standby: The fault might occur once an hour or a few minutes from switch-on, and then the set would be OK for the rest of the day. Heating and freezing didn't seem to have any effect, neither did blanket resoldering on the main PCB. We disconnected the field scan plug, turned down the first anode voltage to protect the tube, and left the set running. It operated for a couple of days without going to standby. The most likely suspects in the unusual thyristor field output stage seemed to be the TL082C driver chip IF01 and the BY398 diode DF16. We replaced these two items and left the set on test for a further two days. Everything now seemed to be OK.

Trips off for a second: This sometimes occurred with a channel change, or with a rapid scene or brightness level change. There's a modification to deal with this: change the value of RP26 from 2.2 kΩ to 1.5 kΩ.

Ferguson TX100 chassis

Dead set: A Ferguson 22D1 was pushed on to the bench, reported 'dead'. No trouble was expected: these sets are normally reliable and easy to service. So we folded the circuit diagram along the line above the line output transformer and laid it on top of the cabinet. With the line output stage feed disconnected and a 60 W bulb to check the chopper circuit the power supply was found to be in order. The line output transistor and the diodes fed from the line output transformer were OK, so the transformer was replaced. Still the power supply tripped out. After tinkering about a bit the whole circuit from the line driver through to the line output stage was checked. We found the BC372 driver transistor short-circuit and its feed resistor R143 open-circuit. Naturally replacing these items didn't help. The circuit slipped off the top of the set and I saw the separate line output stage bit above the fold – the bit applicable to 110° models. There's a fat diode, D28 (BY299), in series with the h.t. feed. This turned out to be short-circuit. Replacing it brought up the h.t. and e.h.t., but there was still no raster. The three outputs from the TDA3562A colour decoder chip were very low at about 1 V. Video in was OK and there were voltages from the controls. We tried a new i.c., but again no luck. Then we did what we all know to do on Grundigs with auto grey scale, check the first anode voltage at the c.r.t. The reading was around 200 V instead of 350 V, so I flicked the transformer mounted potentiometer up full. This produced a picture and flyback lines. Turning it down again flooded the raster with white, with no picture. Panic, what had we done?! The TDA3562A outputs were now at 9 V. Greater panic. The set was switched off. When I switched on again there was a black raster. This is where we came in! All was well when the first anode voltage was set correctly, for 350 V. So the morals are (a) don't fold the circuit diagram when working, and (b) if the set has auto grey scale don't let the voltages fool you if the first anode voltage is not set correctly.

No tuning: None of the channels could be tuned in, and when the set was switched on from cold the mute LED was permanently on. A known good set was used to make comparative voltage checks. This revealed that in the faulty set the voltage at pin 13 of IC2007 on panel PC1546, i.e. one of the crystal oscillator pins, was 0 V instead of 0.2 V. Replacing C2022 (330 pF) connected to this pin cured the fault.

Intermittent dead symptom: This was soon traced to an intermittent mains supply to the PCB. When the on/off switch was broken open it was seen to have blackened and badly burnt contacts. The set came back a few days later, however, with a note saying 'same fault'. This time the power supply outputs were all present when the fault eventually occurred. The cause of the trouble was an intermittently open-circuit line driver transistor, TR8 type BC372.

Dead, 2.5 A mains fuse open-circuit: This set came in with the complaint 'dead' and it sure was! The 2.5 A mains fuse was open-circuit well blackened – so attention was centred on the chopper transistor TR6. A check revealed that it was short-circuit all ways round. We also found that D13 in its base circuit had shorted. After replacing these two items everything seemed to be OK so we switched on. Bang! There was a flash that made the three fluorescent lights look dim and FS1 and TR6 had again bitten the dust. This time D13 had survived but the h.t. rectifier D15 was short-circuit, which it hadn't been 2 minutes previously. A new BY299 h.t. rectifier was fitted and the other items were replaced, along with D13 for good measure, and surprisingly the set ran – for about two and a half minutes. The whole lot had gone again! Now the 119 V h.t. supply is used by the tuning circuit as well as the line output stage, and although the aerial had been plugged in there had been no tuning. Interesting! The ZTK33ADPD tuning voltage regulator IC2006 was suspect but the culprit turned out to be the nearby transistor TR2016 (BF460) which was short-circuit collector-to-emitter. Its associated with the memory in IC2007. When this was put right the set continued to run.

No red: This was because the 100 kΩ feedback resistor R609 in the red output stage had gone open-circuit.

Sound but no raster (black screen): The e.h.t. and first anode supplies were OK but the c.r.t.'s cathodes were at 200 V. The outputs from the TDA3562A colour decoder chip were low at only about 1 V, so the chip was as usual changed. No good. Perhaps the field timebase/c.r.t. protection circuit had come into operation? A check revealed that there was no drive from the field output chip though its supply was present. So this chip was replaced again as usual. Again no good! Then we found that there was no input from pin 1 of IC4. We followed the same routine: supply OK so fit new chip, but still no raster. As I was pressed for time I phoned our ever helpful Ferguson distributor. He said he hadn't had this one, but shouldn't I try the field feedback circuit? Well there are quite a few components here. I decided to follow the old principle of checking high-value resistors and low-value electrolytics first. My initial check on C101 hit the nail on the head (its exact value depends on the type of tube). Nowadays a lost raster is often caused by field collapse. It's a good idea to read and note the first anode voltage, then turn the first anode control hard up in case a tell-tale white line appears. In this particular case switching off and on again inside a second or so produced a clean, bright line which stayed on.

Failure of line output transformer: Here's a point worth noting when servicing this chassis. Failure of the line output transistor/transformer is quite a common problem, replacement of one or both of these items restoring normal operation. Something that's often overlooked, however, is the BC372 line driver transistor TR8. If you have to replace the line output transistor/transformer in one of these sets the drill is to connect your meter to the case of TR8. The reading should be 0.8 V. Any steady or significant increase with the meter left on for 10 minutes or so means that TR8, which is a Darlington device, should also be replaced. Incidentally. Ferguson now supply an improved device.

Would go off after 2 hours: The 119 V h.t. supply switching relay chattered rapidly on and off. If the set was left switched off for a couple of minutes the fault would correct itself and the set would run for some time. With the fault present a check was made on the 119 V and 20 V supplies, which are derived from secondary windings on the chopper transformer. They were both at approximately half the correct value. which pointed to a fault in the primary side of the circuit. A d.c. check showed that the TDA4600–2 chopper control chip's 12 V supply was low at about 7 V. With the aid of a scope we found that it was actually pulsing on and off at 50 Hz. As a result the chopper transistor's output was also pulsing at this rate. This pulsing was due to the start-up circuit (thyristor SCR1 and the associated components) operating. When the set is running normally the TDA4600–2's 12 V supply is produced by rectifying the output from winding 7–10 on the chopper transformer. The rectifier was OK, but the d.c. resistance of the winding measured 50 Ω. After it had been left for a few minutes to cool the resistance had fallen to the correct figure of 0.5 Ω.

Flicker in brightness level: After successfully repairing a power supply fault (short-circuit chopper transistor) in this set we found that there was another fault. It appeared only when the set was tuned to BBC-1, and was most noticeable on darker scenes. The symptom was a flicker in the brightness level in the darker areas. After spending some time checking components in the i.f. and video areas we arrived at the colour decoder chip IC3 (TDA3562A). A replacement was tried just in case, and much to our surprise it cleared the fault.

No go, LOPT primary short: This set wouldn't go – it pumped as though the line output transistor had failed, as tends to happen with this chassis. Indeed it read short-circuit all ways round, until removed from the circuit. With the line output transistor end of the line output transformer's primary winding (pin 10) disconnected there was a short across the 119V line. This short disappeared when the top end of the primary winding was lifted. The short was within the transformer, a replacement restoring normal operation.

JVC set with low h.t.: This was a JVC set fitted with the TX100 chassis. It was 'dead' with only 56V on the h.t. line, though the 12V and 20V supplies were all right. This proved that the power supply was working. There was no short-circuit across the h.t. line and the line output transistor and h.t. rectifier were both OK. We next checked the h.t. reservoir capacitor C121 which turned out to be open-circuit, a replacement restoring normal operation. The line driver transistor in this chassis is a BC372 Darlington device. We've found that a TIP110 works very well in this position.

Lack of width: The fault with this 110° set was lack of width. Both the width and the EW presets worked and the h.t. was correct at 119V, but the picture could not be made to fill the screen. The 110° version of this chassis has a separate EW/width correction panel fitted: it plugs into the main board. We eventually found that the cause of the trouble was that the width coil had been fitted – it should be present only in the 90° version. Linking it out brought full width. Watch out for that one!

Dead set symptom: Dealing with this fault is usually fairly straightforward with this chassis. In this case a new BU508A line output transistor and BC372 line driver transistor brought the screen back to life, but with lack of width – about an inch down each side. Unfortunately the line output transformer proved to be at fault, making it an expensive repair.

Intermittent channel changing and going into standby: On the bench the set wouldn't come out of standby. We found that the voltage at pin 20 of the remote control processor chip IC901 was very low at 2V instead of 9V. When we checked back to the regulator transistor TR901 we discovered that all three legs were dry jointed. Resoldering them provided a cure to all the intermittent problems.

Sound sometimes disappeared: The customer's complaint was that the sound sometimes disappeared completely, a sharp tap on the cabinet bringing it back. Checks showed that R95 (2.2 Ω) was dry jointed at one end – in fact it had never been soldered. It's surprising that the set had worked for several years without giving trouble.

Field collapse: A common fault with these sets is field collapse due to failure of the TDA3652 field output chip. As this chip is no longer available a TDA3654 has to be fitted. We've started to experience problems when this is done: the results can be no scan at all, severe cramping or inability to set up the field hold correctly – despite changing the value of the resistor (R96) in the drive line from 6.8 kΩ to 3.3 kΩ as suggested in the sheet that comes with the chip. This chip is also used in some Sony sets, but in this case with a 1.5 kΩ resistor in series with the drive. So we've tried 1.5 kΩ in the Ferguson chassis and have found that the chip then works perfectly. When contacted Ferguson commented that R96 can be reduced to 2.2 kΩ if problems are experienced but that the TDA3654 must be obtained from them, also that they've had a lot of complaints about TDA3654 chips obtained from other sources. Sony, however, said it didn't matter where the chip came from as long as the value of the resistor was altered.

Intermittent chopper transistor failure: This set had been a long-term problem in the workshop. It blew the chopper transistor intermittently and would sometimes go dead with no chopper transistor drive. During a rare occasion when it stayed dead for some minutes we were able to establish that the TDA4600–2 chopper control chip had a supply and a reference voltage but no output was present while the voltage at pin 4 was higher than normal. This is the pin to which the famous 300 kΩ (270 kΩ or two separate resistors in some designs) resistor is connected. We've all in the past had trouble with this charging resistor. On this occasion, however, the associated charging capacitor C118 (8.2 nF) had gone open-circuit. The value of the replacement capacitor should be 6.8 nF as in later production.

Set goes dead on channel select: If the standby light works but the set goes dead when a channel is selected and all the lights go out, check for a 12 V drop across RL1. If this is absent replace TR9 (BC107).

Dead set, power supply problem: A recent case of a dead set was caused by C115 in the power supply being short-circuit. In this condition the outputs from the chopper circuit are at only 10 per cent of the normal level. No sound with a remote control plus teletext set led us to pin 10 of the SAA5012 chip on the remote control panel. The response here to remote commands was normal but was not reaching the audio circuit because D123 was open-circuit.

Distorted sound: The complaint was of distorted sound. We traced the cause to C82 (5.6 pF) which is connected between pin 4 of the audio output chip and chassis to prevent h.f. oscillation within the chip. It had gone open-circuit.

Occasional green picture: The picture would occasionally turn a distinct shade of green. This usually showed up after a quarter of an hour or so. It was noticed that from switch-on the green output took a shade longer to appear than did either red or blue. The culprit was C51 (0.47 μF).

No signals: This set was fitted with the PC1564–311 remote control/ sweep tune panel. The fault was no signals, with the 30 V rail measuring just that. Tuning is carried out by the M293B1 chip IC2001 which has a 455 kHz crystal connected across pins 13 and 14. The crystal oscillator was working normally and the two 5 V regulators on the panel were both OK. The next check should be at pin 19, which was ramping correctly under the action of the tuning system. A check on transistor TR2018, which produces the tuning voltage from this ramp, proved that it was open-circuit. A BC639 turned out to be a suitable replacement. While the channels could now be tuned in, pressing the store button made them disappear again! We eventually found that there was a splash of solder across pins 5 and 6 of PL175. This effectively shorted pins 22 and 24 of IC2001 when store was pressed, initiating the search down command. Some stupid engineer had probably done this when he changed TR2018.

Set dead, relay click at switch-on: When we switched this set on the relay clicked but the set remained dead, with no standby or channel indicators alight. A check was made on the voltage at the collector of the BU508A line output transistor TR10. It was low at only 80 V. The cause of the trouble was the 119 V supply reservoir capacitor C121 (47 μF, 160 V).

Intermittent low contrast: Check whether the tube's heaters are going out. The solid-cored wires can break where they go into the plug connector by the line output transformer.

Field collapse after half an hour: This wasn't a field fault, but a rather curious power supply fault. When we'd had the set on for about half an hour a slight whining noise from the power supply developed. This increased over a period of 15 minutes, then the field collapsed. Checks showed that something was wrong with the supply to the field output stage – in fact all the secondary outputs from the power supply were quite low. By this time the power supply sounded like a jumbo jet taking off about 30 yards away. The set was duly turned off and, after a bit of thought, we switched on again and went in search of the cause of the offending commotion. Although the sound came from the chopper transformer it seemed unlikely that this was responsible: we could count on one hand the number of faulty chopper transformers we've had in sets of any make. But all fingers pointed at it, so out it came. A larger, black type from a scrap chassis was half-heartedly fitted – a bit of a crush as some redundant pins for a non-existent socket got in the way. We soldered the transformer in at 30° to the PCB just to prove the point, and prove the point it did. Not a squeal was heard and the voltages remained steady for hours. It was time to order a new transformer from Ferguson. After all, it's a safety component that has to be the same as the original. When the package arrived and was opened my face fell in despair. The transformer was of the larger, black type that just didn't want to go in. I was about to discard the jiffy bag that had contained it when something fluttered out – it was the modification kit. The packet contained a few components that had to be changed, and one sentence in the fitting instructions met with great elation. 'Remove the redundant pins, if fitted, near the transformer mounting holes.' They were out in a flash, the new transformer was fitted and the modification kit worked perfectly. Another 'top banana' repair had been successfully completed.

Relay chattering: The voltage at the 119 V output from the chopper power supply was slightly negative because the 47 μF, 160 V reservoir capacitor C121 was faulty. If the relay chatters all the time the line output transistor is usually short-circuit. You can risk fitting a replacement without a repeat.

Bright raster with flyback lines: Checks showed that the 200 V and first anode supplies were correct so we decided to investigate the beam-limiter circuit where transistor TR60 (BC308) was found to be short-circuit.

Tripped when the rotary brightness control was advanced: As the on/ off switch had just been replaced we checked to see whether there was something amiss in this area. There was! The four rotary customer controls have exceptionally long, bendable legs. Two of the brightness control's legs were touching. Separating them cured the fault, but it's easy to cause this sort of trouble by accidental mishandling in this area – it could happen to the volume, contrast or colour.

Relay chattering, TR10 is OK: If the relay is chattering and the line output transistor TR10 is OK, go straight to the BY299 diode D28 in the h.t. feed to the line output stage. Check whether it has gone short-circuit. If it's all right, suspect the line output transformer. Note that D28 is present in the 110° version of the chassis only.

Faulty line output transformer: A faulty line output transformer has been the cause of no sound or vision in a number of these sets. You find that the voltage on the 119 V line, which feeds the line output stage, is low and that the line output transistor gets very hot.

No sound or raster: Checks showed that the power supply was working all right, with the correct 119 V and 20 V outputs. But the line driver stage didn't receive its regulated 15 V supply because IC9 (MC78M15CT) was dry jointed. A clean-up and resolder brought the set back to life.

Dead set: A dead TX100 with a difference this time. C118 (6.8 nF) was short-circuit, taking pin 4 of the TDA4600–2 chopper control chip to chassis. C118 is part of the chopper transistor collector current simulation circuit.

Goes off after a few minutes: The h.t. dropped, and the set appeared to trip. This was not because of an overload: the cause was a high-resistance connection on the chopper transformer, which had to be replaced.

Picture would pulsate: On high-brightness scenes the picture would pulsate, especially when the set was cold. The cause was C117 (100 μF, 16 V) which couples the drive to the chopper transistor TR6.

Lack of width: About 2 inches lack of width at either side of the screen, was the problem with this set. As the h.t. voltage was correct at 119 V, we decided to try a replacement plug-in EW panel. This made no difference, neither did replacement EW modulator diodes. We then found that the 148 V supply obtained from the line output transformer was low at 129 V. In view of this we decided to replace the transformer. This cured the fault.

Reverts to standby after half an hour: If this set was left to cool it would run for another half an hour before giving a repeat performance. As we've had this fault before we fitted a new chopper transformer. The clue is that the power supply will run all right until it's loaded.

Intermittent low gain: The tuner and various components in the i.f. strip had been replaced, but the cause of the fault turned out to be the SL1432 i.f. preamplifier chip.

Low sound: There was no sound until the volume was turned right up. We wasted a lot of time on the remote control PCB before tracing the cause of the fault to the TDA2578 sync/line generator chip IC4, where the 'valid signal' voltage at pin 13 was slightly low. As a result, the sound was partially muted. A new chip put matters right.

Dead, relay click from power supply: When the set was switched on a single click could be heard from the relay in the power supply. As a new LOPT had recently been fitted, this possibility was ruled out. We checked the h.t. and found that it was low at 74 V. After a few minutes it had risen to 79 V. A capacitor close to the relay looked somewhat distressed. It was hot to touch and obviously the culprit. The offending item was the 47 µF, 160 V h.t. reservoir capacitor C121. A replacement restored the h.t. and normal operation of the set.

Dead set, no output from choppers: The mains rectifier was producing a 330 V supply but there were no outputs from the chopper circuit. H.t. was present at the collector of the chopper transistor TR6, and there were voltages at the pins of IC7 but they were not as shown on the circuit diagram. A check at the anode of SCR1, the thyristor in the start-up circuit, produced a reading of only 49 V instead of 115 V. Replacing IC7 and SCR1 made no difference. The cause of the trouble was eventually traced to C117 (100 µF) which couples the drive to the base of TR6.

Set took a very long time to produce the correct grey scale: When switched on it would produce a green picture for 10 minutes, followed by 5 minutes of red, then a normal display would appear. C52 (1 µF), which decouples pin 19 (grey-scale reference) of the TDA3562A colour decoder chip, was open-circuit.

Dead set, start-up resistor R116 hot: A scope check at pin 8 of the TDA4600 chopper control chip showed that it was producing a drive output, but this didn't arrive at the base of the BU508A chopper transistor because choke L15 was open-circuit. A replacement from a scrap set restored normal operation.

'Deadish', faulty LOPT: This monstrous old 26 inch set came in deadish, with a faulty line output transformer. When a replacement had been fitted the set ran up but the sound was permanently muted. Lifting the various lines that lead to muting brought me to the SAA5012 remote control decoder chip, which turned out to be faulty.

No EW or width correction: D701 (SK4G4/04) on the little EW modulator PCB was the cause. It was open-circuit – in fact it was split wide open. A replacement restored full width.

'Death rattle:' We are so used to the 'death rattle', meaning a faulty line output transformer, that it's easy to get caught out. In this case the transformer was innocent. The cause of the trouble was C119 (22 μF) in the start-up circuit. It had gone low in value.

Line tearing from cold: Many possible causes are listed, and we checked out every one – with no success. Matters were made worse by the fact that the fault would reappear only when the set had been off for quite some time. Finally, more out of desperation than anything else, we decided to replace the line driver transformer. We then waited for two hours before switching on again. This time everything was OK.

Set blowing chopper transistor: When we say that this set was blowing its BUS08A chopper transistor TR6 we mean physically blowing it apart. The TDA4600–2 chopper control chip IC7 failed as well. The cause of the trouble was the chopper transistor's base drive coupling capacitor C117 (100 μF, 16 V).

Dead set, apart from relay clatter: No problem here: check the line output transformer, which had failed. The BU508A line output transistor had survived, which is often the case. The next step was to look up the transformer required. All of our suppliers' catalogues listed the green spot type, so one of these was obtained and fitted. But when we switched on there was a small raster with severe EW distortion. After a few seconds the picture faded to a blank screen. The BU508A transistor and D26 (BY228) in the EW diode modulator circuit had burnt out. Considering that it had been in operation for only a few seconds, the LOPT was very warm. Maybe it was faulty in some way. As I removed it, I was embarrassed to see a little blue sticker on the line output transistor's heatsink. Oh well, we all make mistakes. After fitting a blue spot transformer, along with a new BU508A transistor and BY228 diode, I switched on and was amazed to find that the same fault symptoms were present, along with very bad line pairing, field roll and a definite smell of something cooking. The set was quickly switched off – but not quickly enough to save the transistor and diode. I double checked several components, but couldn't find anything wrong. While looking at the original LOPT, and praying for divine intervention, I decided to order a transformer with identical markings, i.e. forget blue and green spots, just order as originally fitted. It was marked FAT3758. All my suppliers offered a green spot type, but a feeling I had made me insist on a Konig 3758 type. Chas Hyde and Son came up with the required LOPT and when it was fitted, along with another transistor and diode, the set sprang to life with a normal picture, sound etc.

Screen just full of snow: 'Loss of the picture, with just a screenful of snow', said the report that came with this remote control set. And the fault didn't occur very often! Various thrashings were handed out to the tuner and the remote control panel, but no disturbance testing would instigate the fault. Heat and freezing didn't help either. After a couple of hours, however, the signals did disappear and we found that the 9 V supply to the remote control panel was missing. The unusual cause was that winding 3/4 on the chopper transformer was going open-circuit intermittently.

'Bang, and a smell of burning': A visual inspection revealed that the h.t. smoothing capacitor C129 (100 μF) had exploded. A replacement restored normal operation.

Ferguson TX85 chassis

Dead set: The mains input fuse had blown, the TIPL791A chopper transistor had gone short-circuit collector-to-emitter and the TEA2018A chopper control/driver chip had an internal short from its 11 V supply input (pin 6) to its non-isolated chassis connection (pin 2). The fuse, transistor and chip were replaced and the set was switched on. It appeared to be working normally, apart from a scream from the power supply. So the set was quickly switched off. A meter was connected to the h.t. output (97.5 V) from the chopper power supply to check that the regulation was correct and the set was switched on. The h.t. was correct, but after about 15 seconds the fuse blew. We found that the transistor and the chip had once more failed. A thorough check was then made on the components in the power supply. This revealed that R101 (1.2 kΩ, 5 W) in the snubber network across the chopper transistor had gone open-circuit. When all these items had been replaced the set was OK.

Line driver stage problem: These sets employ a rather unusual line driver stage with three transistors. Two dead sets recently had problems here. The first simply had a dry joint on one of the 100 Ω, 12 V supply resistors. With the second set we had to replace all three transistors: two were short- and one was open-circuit.

Fault caused by a thunder storm: This set had apparently gone off with a bang during a thunder storm. The 800 mA mains fuse had blown, the chopper transistor TR6 was short-circuit and the control chip IC4 (TEA2018) had an internal short between pin 6 and chassis. After replacing these components the set failed to start – a quiet pulsing could be heard coming from the power supply. The cause of this was the 15 V supply rectifier D13 (RGP30) which was short-circuit.

Dead set, L10 open-circuit: Initial checks showed that there was normal h.t. right up to the line output transistor. But there was no 15 V output from the chopper circuit. We resoldered the various dry joints you get around the chopper transformer and were rewarded with a good picture and sound. These disappeared as quickly as they had come. The cause of the problem is that L10 in the rectifier circuit that feeds the 12 V regulator has a habit of going open-circuit in an irregular manner before it fails permanently. What can sometimes be a little disconcerting is that in L10's high-resistance condition the 12 V regulator still receives an input of 12–15 V, yet the set remains dead. This can lead you to suspect the regulator etc.

Field linearity fault: It is quite common to get one of these sets in with a field linearity fault, usually top fold-over and cramping. The thing to check is transistor TR7, which is type TIP112H or T0167 V. It's the lower transistor in the field output stage and is a Darlington device, so it should be checked by substitution.

Mains fuse blown: If you find that the 800 mAT mains fuse FS1 has blown, replace the TIPL791A chopper transistor TR6, its TEA2018A driver chip IC4 and the three 1N4001 diodes D8, D10 and D23. If R101 (1.2 kΩ) in the snubber circuit has ruptured replace the associated 1 nF capacitor C75 and uprate FS1 to 1.25 A.

Dead set with faulty snubber network: The TIP791A chopper transistor TR6 was short-circuit and both the 2.2 Ω surge limiter R88 and the 800 mAT mains fuse FS1 were open-circuit. These items and the TEA2018A chopper control chip IC4 were replaced. At switch-on everything seemed to be fine, with a good picture etc., although there was a faint whistle from around the power supply. But 5 minutes later the power supply had self-destructed again. TR6 felt very hot and because of the previously mentioned faint whistle, the penny dropped. Check the snubber network, we thought. Sure enough R101 (12 kΩ, 5 W) was open-circuit. When this and the other items had been replaced the set finally worked correctly.

Repeated failure of line output transformer: This can be caused by a break in the track of the h.t. preset control. As a result the h.t. suddenly goes sky high. This seems to be a common fault, so it would be wise to replace the preset as a precaution whenever one of these sets comes in for repair.

Dead or intermittently dead set: Check for dry joints on the vertical add-on filter panel next to the line output transformer, especially the lead-out wires. They tend to pull away from the print because of stress caused by the cable tie around the transformer being too tight.

Intermittent goes into standby or appears dead: We found that the BD385 9 V regulator transistor TR901 on the remote control and sweep-tune board was dry jointed.

No sound or vision: We've had no sound or vision with several of these sets recently. One with remote control had no unregulated 16 V input to the 12 V regulator because TR902 (ZTX753) on the remote control/sweep-tune board was open-circuit. In other sets the 12 V regulator IC6 has been dry jointed or open-circuit.

No line drive and no display: The power supply was running but there was no line drive and no display. A check on the two voltage regulators on the small board to the left of the chassis revealed some fine dry joints.

Power supply blow-up: There's a tendency for power supply blow-ups with this chassis. The usual cause is a faulty degaussing posistor (Z2): it's good practice to replace this item whenever one of these sets comes in for repair. With this particular set, however, the cause of the problem was the fact that R97 (10 Ω) had risen in value to 11 kΩ. As well as the obviously faulty components (the mains fuse, the surge limiter R88, the chopper transistor etc.) we also replaced the mains rectifier diodes, the TEA2018 chopper control chip and the three series biasing diodes D8/D10/D23 (type 1N4002).

Intermittently comes out of standby: The complaint with this set was that it would come out of standby of its own accord. Noise from the SL486 IR preamplifier chip was found to be the cause of the trouble.

Dead set: The mains fuse had blown and the chopper transistor TR6 was short-circuit. R101 in the snubber network was dry jointed, so this was resoldered. Resistance checks around the TEA2018A chopper control chip then showed that the earth print from the emitter of TR6 to pin 2 of the chip via pin 2 of the chopper transformer was open-circuit: the break was near the transformer. After linking across the break we continued our tests. R98, which is in series with TR6's base, was open-circuit; D20 (1N4148) in the error feedback network was short-circuit; and the TEA2018A chip had blown its top. After replacing these items we connected the set to the mains supply via the variac and carefully increased the input. Fortunately the set was now OK.

'Went bang': Sets fitted with this chassis have a habit of going bang. Replacing the obviously damaged components will often get the set going again, but with an unhappy power supply that goes bang soon after. For a reliable repair, replace TR6 (TIPL791A), IC4 (TEA2018A), R102 (1.2 Ω, 3 W) and FS1. This fuse must be a 1.25 A anti-surge, not a slow-blow, type – you may find that an 800 mA fuse is fitted. Solder the chopper transformer and mains switch tags – replace the latter if in doubt. Then check the following: R101 (1.2 kΩ, 5 W) – this is vital!; D8, D10 and D23 (all type 1N4002); D3–6 (all type BY133); R95 (100 Ω) and R97 (100 Ω). Power the set via a variac, with the degaussing coils disconnected. It should now work. If the picture is blotchy, replace the degaussing thermistor.

Chopper transistor short-circuit: If the TIPL791A chopper transistor TR6 is short-circuit, check R101 (1.2 kΩ, 5 W) in the snubber network. You will usually find that it has gone open-circuit. Replace these items and also R102 (1.2 Ω WW), R88 (1.8 Ω WW) and IC4 (TEA2018A). Check D8, D10 and D23 (all type 1N4001) and R95 (100 Ω). One nasty point with this chassis is the use of live heatsinks for the chopper and line output transistors.

Bad contrast: As the setting of the contrast control was turned up towards the normal level it produced the effect of progressive picture overloading. At the maximum setting the picture was overlaid with black horizontal bars and there was also an audio buzz. After some time we found that there was a break in an earthing print track, where it leaves the customer controls adjacent to the colour control.

Ferguson TX86 chassis

Dead set: The TIPL791A chopper transistor TR6 and its TEA2018A control chip IC4 had both failed, as did the replacements. The cause of the trouble was eventually traced to the 1N4001 diode D8 being short-circuit. It's in the current sensing network connected to pin 3 of the chip. Presumably failure of D10 or D23, which are in series with D8, would have the same effect.

Set tripping: This was the result of the TIPL791A line output transistor TR13 being short-circuit.

Line collapse: This was caused by a dry-joint on the 330 nF line scan coupling capacitor C94.

Lack of height: A check at the collector of the upper transistor TR8 in the field output stage showed that only 39 V was present here. Its 27 Ω feed resistor R62 was open-circuit.

Power supply wouldn't start: The cause was traced to the TEA2018A chopper control chip IC4, which was faulty. It's always worth checking the soldering of R61 in the power supply in these sets, as a dry-joint here can destroy the line output transistor.

Power supply failure: The TIPL791A chopper transistor TR6 and the TEA2018A chopper control chip IC4 were both short-circuit, while the 1.8 Ω surge limiter resistor R88 was open-circuit. The cause of the trouble was a dry joint at pin 4 of the line deflection coils socket 13.

Ferguson TX89 chassis

Dead set: This set was dead with R88 and R102 open-circuit and the TIPL791A chopper transistor TR6 short-circuit. We replaced these items, also the TEA2018A chopper control chip IC4, but didn't get sound or a picture when we switched on. At least the set didn't go with a bang. This time we found that the TIPL791A line output transistor TR13 was short-circuit. A replacement restored the sound and picture but the height and width were varying. A check on the h.t. line showed that the voltage was varying over the range 90–105 V. D7 (BYD33G) which provides the supply for IC4 was found to have a slight leak. Replacing this diode cured all the problems.

Set was tripping: This set was tripping. It didn't take us long to find that the main reservoir capacitor C69 (100 μF, 400 V) was open-circuit.

Pulsating white raster, sound hum: When switched on from cold this set produced a pulsating white raster with a hum from the speaker. Within seconds the picture would begin to appear, the pulsations started to abate, the hum died away and normal sound could be heard from the speaker. After 30 seconds or so the set worked normally. The fault would not show up again until the set had been left switched off for at least 6 hours. After much testing and component swapping we found that an area on the component side of the PCB, in the vicinity of R100 and C90, was conductive. This allowed current leakage between the isolated and non-isolated sections of the chassis. When warm the resistance reading was 5 MΩ, falling to 150 kΩ when the set was cold. No spillage could be seen, but the leakage had gone once the board had been cleaned.

No sound or vision: There was a distinct lack of snow. We eventually found that C13 (220 nF) was short-circuit. It's connected to pin 9, one of the i.f. input pins, of the TDA4501 multifunction chip.

Inability to memorize programmes and other functions: This is usually a symptom of an exhausted back-up battery. That was not the case with this rarely seen chassis. The memory chip IC14 thus became the prime suspect. A replacement proved the point.

Ferguson TX90 chassis

Mains transformer replacement: This set had been got at before it came to me. First we had to replace the mains transformer and the line output transistor. After this everything went well for about 30 seconds, then TR107 (BD839) in the boost regulator went short-circuit and the overvoltage trip operated. TR107 was very hot. All the transistors, diodes and resistors in the regulator section were checked and found to be blameless, so TR107 was replaced and the set was powered via a variac to limit the dissipation in TR107 while further checks were made. Scope checks revealed that the waveforms around TR107 and TR108 were horrible. The amplitudes were correct but the switching peaks were ill-defined and there appeared to be a lot of what looked like ringing between the peaks. This was why TR107 was getting so hot: it was not being turned on and off properly. The ringing led to about the only component that hadn't been changed, the series choke L120 which had shorted turns. At first glance this choke appears to be just a bit of smoothing, but it's an essential part of the regulator arrangement. This is the first time we've had to get involved with this rather unusual regulator system, so the repair took a considerable time. One thing is certain: fault-finding would have been impossible without the use of a variac. The choke had probably failed as a result of the action of the previous repairer who had increased the ratings of the fuses connected to the mains transformer to 'cure' fuse blowing. The fuse failure had been caused by shorted turns in the transformer. When an uprated fuse eventually blew as more and more turns joined the short there must have been a brief but substantial overload. Regulation problems can also be caused by the four parallel-connected 47 kΩ resistors R208, R209, R210, R221. Apparently their end caps sometimes make intermittent contact.

Faulty c.r.t. produces three different symptoms: Three separate sets had three different faults, but one component was the common cause. The sets were of both the basic and remote control type and the symptoms were intermittent 'clicking', intermittent field roll and loss of tuning memory. The set with the loss of memory was initially repaired by replacing the memory chip but came back a week later with the report 'same as before' – the chip had again failed. In each of these sets the c.r.t. was faulty, with a flashover in the neck between pin connections. The tubes were all Mullard ones.

Intermittent colour: No amount of heat or freezer would induce the fault. We noticed that the colour was always correct when the set was first switched on. It would then go into bars and finally off. From this it seemed likely that the problem was around the colour decoder reference oscillator. We started to change capacitors and when C155 (47 pF) was replaced the colour stayed on. We used a 100 pF component as in later production.

No picture: We advanced the first anode control and got a blank white raster. Checks were then made around the decoder chip. The –0.8 V that should have been present at pins 19 and 23 was found to be missing. A check with the scope showed that pulses from the line output transformer were arriving at R171 (270 kΩ) but there was nothing at the chip side. Replacing R171 cured the problem.

Fuseholder FS102 tends to overheat: The fuseholder for FS102, the secondary fuse on top of the mains transformer, tends to overheat and cause intermittent operation. It's of the type with a Paxolin base, and you'll find this scorched. As this is the fuse that blows when the mains rectifier diodes short – often intermittently – the problem could be due to regular fuse changing before the fault is tackled properly, i.e. change all four diodes to type BYD33G. The fuse has been uprated from T1A to T1.6A.

Patterning but no picture: The TDA4500 chip was changed but the trouble was due to C121 (0.01 µF) being leaky.

Expanded teletext lines with flyback lines above going down the centre of the picture: R194 (243 kΩ, 1 per cent) in the field feedback circuit had gone high in value – to over twice its correct value in this case.

Weak sound: The sound was also distorted at the centre of the range: when the volume was turned up the sound disappeared. Adjustment of the sound discriminator coil L114 had no effect. The sound discriminator circuit is coupled to pin 13 of the TDA4500 chip IC102 by C114 (0.01 µF) which was the cause of the trouble.

No sound or picture: The line output transistor was short-circuit. When this was replaced we found that the boost line regulator transistor TR107 got extremely hot – so hot in fact that the solder was melting. The series choke L120 (5 mH) had shorted turns.

Tripping: We'd not previously had this problem with the TX90, but when we checked the power supply we found that the h.t. was high. A resistor that had gone high in value, causing poor regulation, was suspected. We were surprised to find that R225 (33 kΩ) and R222 (10 kΩ) were both open-circuit.

Intermittent lack of height: This was of course due to dry joints on the field output transistors, but when setting up the contrast etc. we noticed that there was a dark band down the left-hand side of the picture and severe ringing on low-contrast scenes. We decided to bridge the electrolytics in the video circuits and while looking for C132 saw a capacitor with electrolyte leaking from its top – C132 (10 µF, 50 V) in fact. It had a small hole in the underside, visible only after it was lifted upright. C132 decouples pin 4 (brightness control) of the µPC1365C colour decoder chip IC103.

Set would go off changing channel: We found that there were dry joints on both the field output transistors. It looked as though the heatsink on which they were mounted had received a knock, forcing the legs of the transistors further through the board than they ought to have gone.

No results, supplies present: The h.t. and 12 V supplies were both present and correct. A scope check on the line drive waveform came next and we were rewarded. It was missing at the secondary of the driver transformer but was present at the primary. A base-emitter short in the line output transistor perhaps? No, simply an open-circuit wire on the driver transformer where it joins the print connection. We were able to remove and repair the transformer.

Field scan shrink: The problem with this 20 inch model was that the field scan shrank after a couple of hours. We eventually found that R187 (6.2 kΩ) in the chain of resistors that form the load for the field driver was going high in value when warm. It read correctly with no voltage applied.

Band that moved up screen: The picture produced by this set had a band that moved up the screen, with picture tearing within this band. Replacing C189 (22 μF, 50 V) in the boost regulator circuit provided a cure.

Slight ripple on picture: There was slight ripple on the picture. We found that the h.t. voltage was slightly low at 85 V and that the h.t. preset R224 was inoperative. TR107 was open-circuit base-to-emitter and short-circuit base-to-collector. When this was replaced the h.t. could be adjusted but the transistor soon failed. The power supply storage choke L120 was short-circuit.

Occasional partial field collapse: After some very inconclusive cold checks we took some voltage readings. The upper transistor TR104 in the field output stage is biased by a chain of four resistors. The third resistor in this chain, R187 (6.2 kΩ), was found to be going open-circuit. When removed its body had a tell-tale ring around it.

Low gain: We found that the cause was the SL431 i.f. preamplifier chip – a squirt of freezer would restore the signals for a couple of minutes.

No field output: Check the joints on TR105 (TIP112H) at the rear of the central heatsink.

Field collapse: We thought it would be a quick resoldering job but the joints were perfect and there was h.t. at the field output stage. The mid-point voltage was low at 22 V, however. Drive from the TDA4501 chip was fine. The cause of the problem was R198 (6.8 kΩ) in the chain that supplies the base of TR104. It had gone high in value, upsetting the field output stage biasing.

Line output transistor shorted: A replacement was fitted and when the set was switched on there was a smell of burning. Before we could locate its source the transistor blew again. Suspecting the line output transformer we removed it and provided the replacement transistor with a resistive load, so that the drive waveform could be checked. All was well. Then the transformer's primary winding was reconnected, via flying leads, but the transistor again failed. After fitting a new transformer and transistor we ran up the set using a variac. The correct h.t. voltage was obtained with only 220 V input. With 240 V input the verticals were bent. Also we found that the h.t. potentiometer didn't work. Checks in the regulating circuit showed that Tr107 and Tr108 were short-circuit. They'd probably been damaged by the faulty transformer. Replacing them restored normal operation.

The set had no picture: When the first anode control was turned up we got a faint, watery image on the screen. Voltage checks around the BC307B beam limiter transistor Tr114 showed that its base was at 0.6 V instead of 5.8 V. The cause was eventually traced to the 143 kΩ resistor R231 which had gone open-circuit.

No colour or didn't store channels: We've had two of these sets in recently. The problem with the first one was no colour. D103 (BAV20), which is in series with the sandcastle pulse feed to the colour decoder chip, was open-circuit. The second set wouldn't store channels. IC902 (M293B1) had failed.

Ripple on picture, no h.t. adjust: Transistor TR111 (BC307B) was short-circuit emitter-to-collector.

Rolling picture, horizontal ripples across the screen: This is a common fault with these sets. The usual cause is that R236 or R241 in the pulse feed to the TDA4500/S1 chip IC102 is high in value or open-circuit. This time, however, the identical symptoms were accompanied by a clatter when the set was turned over, caused by the ferrite core that had fallen out of the line driver transformer T103. Memories of the 3000 chassis!

No colour: This is usually a very easy fault to cure these days. This one tripped us up, however. The job was given to an apprentice who checked the supply to the μPC1365C decoder chip then, finding it to be correct, replaced the chip and its reference oscillator crystal. There was still no colour, however. Scope checks were then carried out on the waveforms around the decoder chip. We found that the line pulse at pin 23 was low in amplitude and a little distorted. Replacing the BAV20 diode D103 in the feed with a 1N4148 gave us unlocked colour but thankfully restored the correct pulse amplitude. Another new crystal put matters right, the first one being faulty.

Dead set, fuse FS51 blown: The cause of a dead set that blows fuse FS51 on the secondary side of the mains transformer is usually a leaky line output transistor (TR112): check by removing it. If the set 'hums' at switch-on when a new line output transistor has been fitted it means that the line output transformer also needs to be replaced. If the set trips when a new transformer has been fitted it's because the power supply is at maximum and can't be set. Replace TR107 – we usually use a BD809 here. Note that the secondary fuse is FS102 in the small-screen models.

Grainy picture: This set produced a grainy picture, as though the tuner's gain was low. As a new tuner made no difference attention was turned to the SL1432 chip and the SAW filter. These two items weren't responsible either! We eventually found that C106 (10 nF) which couples the output from the tuner to the SL1432 chip was faulty.

Low h.t.: This fault is beginning to occur with these sets. The symptoms are that the h.t. preset won't increase the h.t. voltage above 93 V while verticals cog and bow dependent on beam current. In the regulator section of the circuit there's a chain of 270 Ω resistors numbered R206, R207, R211 and R212. In all the cases we've had so far one of them has read anything from 6 kΩ to open-circuit.

Would trip on scene changes: We found that the h.t. voltage was high at 126 V instead of 115 V. The cause of the trouble was the fact that the two resistors on the chassis side of the set-h.t. control had gone high in value: R225 measured slightly high at 49 kΩ instead of 47 kΩ while R229 measured 49 kΩ instead of 39 kΩ.

Blackened mains fuse: As the fuse on the secondary side of the mains transformer was blackened it was no surprise to find that two of the rectifier diodes and the T9064 V line output transistor were short-circuit. When these had been replaced and the set was switched on the fuse blew again. The culprit turned out to be the posistor in the degaussing circuit – it was arcing internally.

Ragged verticals at high contrast: We all know by now that when C189 (22 μF) in the 20 inch version of this chassis dries up the result is ragged verticals at high contrast levels. In 14 inch sets, however, it causes a vertical herring-bone patterning that's most noticeable at very low contrast settings.

Severe hum, now totally dead: The 1 A fuses associated with the mains transformer were both open-circuit and one of the BYD33G mains rectifier diodes was short-circuit – so the final fault could have occurred when the set was switched off.

No sound or vision: The bases, emitters and collectors of the transistors in the boost voltage regulator circuit were all at about 95 V. The T9064V line output transistor TR112 was open-circuit.

Sound OK but no sync: The picture looked like that produced by a poorly tuned set. Sound was OK but there was no sync. A check on the video waveform showed that it was highly distorted. The video demodulator coil L105 required adjustment. According to the customer the fault had got gradually worse.

Couldn't adjust h.t. voltage: The h.t. voltage was a little high and couldn't be adjusted, giving the trip symptom with the appearance that the line speed was incorrect. R229 (47 kΩ) in the power supply had gone high in value. To be on the safe side we also replaced R225 (33 kΩ) which is in parallel with it.

Dead set with FS102 fuse blown: Checks showed that the line output transformer was short-circuit between the primary winding (pins 3, 5, 10) and chassis – it's the first time we've had a dud line output transformer in one of these sets. When a new transformer had been fitted a slightly undersized picture with a ripple was produced. The electrolytics were all OK, the cause of this fault being the BD839 regulator transistor TR107 which was open-circuit base-to-emitter.

Secondary-side mains fuse blown: The fuse on the secondary side of the mains transformer was open-circuit. As the rectifier diodes were all OK I fitted a new fuse and switched on. This produced an uncontrollable hum from the speaker. The cause was quickly traced to the line output transistor, which was short-circuit base-to-collector, and the line output transformer's chassis pin (6) which was badly dry-jointed. The set worked normally after fitting a new line output transistor and resoldering all the pins of the transformer, but the customer was disappointed about the continuous buzz that could be heard even with the set switched off. We explained that this was because of the design, with the mains transformer on the live side of the on/off switch, but the customer was not impressed. To tell the truth, neither were we.

Speaker produced a loud buzz: Operation of the volume control made no difference, and although the e.h.t. was present there was no raster. The cause of all this was eventually traced to a hairline crack in the print that goes to one of the legs of C181 (2200 μF), which is the reservoir capacitor for the 18 V supply.

Field collapse or linearity faults: Replace D106 if leaky. Its type varies between different models.

Bad ripple on picture and hum on sound: Check the 12 V regulator IC105's output. This must be spot on at 12 V. If not, replace it.

Ragged verticals: This condition improves if the brightness/contrast is reduced: Replace C189 (22 μF, 50 V) in the power supply. It dries up.

Intermittent failure of R4050 (Model 20E1): Check/replace TR107, TR108, D112, D114, D110 and R267 in the power supply, D113 and D125 associated with the line output transistor, and check for dry-joints at the line scan coupling capacitor C193. This last item may have changed value because of heating and thus be in need of replacement.

Field collapse or partial collapse with no video information displayed: Check at the collectors of the RGB output transistors to see whether the supply contains line-frequency hash. If line ripple is present here replace C190 (22 μF), the reservoir capacitor for the 175 V supply. In addition to powering the RGB output transistors this supply is used to bias the field output stage.

No colour: Disable the colour-killer by fitting a shorting link across R156. If this produces unlocked colour bars, check whether R171 (270 kΩ) is open-circuit.

Dim picture: If the contrast and brightness controls are having no effect the beam limiter is operating. Check whether R233 (470 kΩ) is open-circuit.

No signals: Try changing the SL1432 SAWF driver before condemning the tuner.

No field sync, line scan jitter: One of these sets had no field sync while the line scanning suffered from jitter at the top of the picture only. The problem could be familiar to those who deal with Ferguson sets regularly, but was one we've never come across before. We carried out various checks around the TDA4500 i.f./timebase generator chip, and eventually replaced it and various other components that could have been relevant. But things still weren't right. When we up-ended the set something fell out of the cabinet on to the bench. It was the ferrite core from the line driver transformer. When the core was replaced in the hole it obviously came from normal operation was restored. An unlikely fault, but one that we'll remember.

Wouldn't power up properly: This set incorporated the PC1139 remote control panel. It wouldn't power up unless the on/off switch was held on. If the switch was released, the set lapsed back to standby. Checks revealed that the 9 V supply to the TMS1000N2LL microcontroller chip

IC901 was missing. Of the components involved in providing this supply, TR901 was short-circuit and TR906 open-circuit.

Blank raster and no sound: This set produced a blank raster with no sound apart from a slight vision buzz when the local relay was tuned in. After eliminating the tuner our suspicions fell on the multipurpose TDA4500 chip, which amongst other things incorporates the i.f. strip. A replacement put matters right.

Intermittently dead: The fault responded to tapping, and I eventually found that the line driver transformer was very sensitive to pressure. Its primary winding was going open-circuit at the base, where the fine wire is wrapped around one of the mounting pins. After fixing this I was left with poor audio quality, which varied with the volume setting – the higher the volume, the worse the distortion. This was caused by a faulty TDA4500 chip. As the volume increased, the negative-going half of the audio waveform gradually disappeared.

Very loud humming noise and hum bar: The picture was almost obscured by a massive hum bar. This fault is often caused by the BD839 transistor in the power supply or the mains rectifier diodes, but a scope check showed that the ripple on the h.t. line was acceptable. The cause of the fault was the 12 V regulator chip IC105, which was almost short-circuit. Fortunately no other damage had occurred.

Set dead, h.t. OK: The cause of the problem was no line drive. In this chassis overvoltage protection is carried out by a couple of transistors that short out the line drive. One of them, TR110, was short-circuit collector-to-emitter.

No picture or sound, e.h.t. present: We've had a couple of these sets with no picture or sound but e.h.t. present and the standby light on. The cause has been in the line output stage, where the BA157 25 V supply rectifier DL38 has been short-circuit and the associated 0.22 Ω, 0.5 W safety resistor RL41 has been open-circuit.

Ferguson TX90 chassis (20″)

Lack of height: Check whether D106 is leaky.

Excessive height, poor linearity: The height control had minimal effect. The cause of the trouble was that D137, a 68 V zener diode, was faulty. It didn't measure short- or open-circuit, however, nor did it have any reverse leakage. But a replacement cured all the symptoms.

Occasional partial field collapse: After some very inconclusive cold checks we took some voltage readings. The upper transistor TR104 in the field output stage is biased by a chain of four resistors. The third resistor in this chain, R187 (6.2 kΩ), was found to be going open-circuit. When removed its body had a tell-tale ring around it.

Line output transformer and transistor short-circuit: The line output transistor TR112 and line output transformer T102 were both short-circuit. After fitting replacements, along with a new 1.6 AT mains input fuse, we switched on. The result was a pulsating picture, with low h.t. and incorrect line frequency. TR107 (BD839) and D109 (BYD33G) in the regulator circuit were both leaky. Failure of the line output transformer always seems to damage these two components. After replacing them we had good pictures and sound. All that was left to do was to check and adjust the h.t. (115 V with a 20 inch tube).

Non-linear field scan: The field scan was stretched across the top and bottom but cramped across the middle. After a few minutes the fault would clear, but a squirt of freezer on the 68 V zener diode D137 would bring it back. The diode had a 200 kΩ leak when cold.

Dead set, 1.6 AT fuse had blown: The R4050 line output transistor was also leaky. These items were replaced but the set then tripped rapidly. In this chassis the trip measures the 175 V RGB output h.t. supply, killing the line drive if the voltage is excessive. The first thing we did was to measure the boost voltage, which was correct at 115 V. As the e.h.t. was heard to rustle up between the on/off tripping we next checked this. The reading was around 40 kV. No wonder the set was protesting! Only one component could be responsible for this, the 5.6 nF line flyback tuning capacitor C194. When we removed it we found that it was open-circuit. A replacement restored normal operation.

Intermittently rippling narrow picture: When the fault was present the h.t. was low and the set-h.t. control RV224 had no effect. Inspection showed that heat had darkened the PCB in the vicinity of the boost regulator circuit. More to the point, the track between diodes D112 and D114 had cracked. When the track had been repaired the h.t. was reset to 115 V. All was then well.

Power supply trip on scene change/contents: We found that the set-h.t. control wouldn't reduce the h.t. below 125 V (it should be 115 V). The culprit was R229, which had risen in value from 39 kΩ to 50 kΩ.

High e.h.t.: After replacing the line output transformer, the line output transistor and the power regulator transistor we found that the e.h.t.

was spectacularly high. The cause of this was the 5.6 nF flyback tuning capacitor C194, which was open-circuit.

No front controls or remote control: This was a 20 inch set with remote control. After routine repairs to cracked joints the set refused to respond to either its front controls or the remote control unit, though this was working. In addition the sound was muted, but came up briefly when the set was turned off. We were led to the cause of the trouble when we found incorrect voltages in the TR910 area of the auxiliary PCB: the momentary contacts in the mains switch were stuck closed. A new switch restored normality.

Ferguson TX98 chassis

Dead set, mains fuse FS1 blown: This set was dead with the mains fuse FSI blown. It was no surprise to find that the chopper transistor TR3 was short-circuit. We replaced it along with the TDA4600–2 control chip and all the usual things that can cause this type of chopper circuit to fail. As there didn't seem to be anything else amiss we switched on. The 115 V line was now present but there were no 12 V and 5 V supplies. Circuit protector ICP1 had gone open-circuit. We applied 17 V to the TBA8138 l.t. regulator chip IC11 from the bench power supply, with its outputs disconnected. This proved that the regulator was faulty as the 2 A bench supply overloaded. Fitting a new TBA8138 restored the supplies but the set was still dead. Scope checks around the TDA4505E-N1 chip IC2 showed that the line oscillator was working, but there was no line drive output at pin 26 even with the line driver transistor TR7 disconnected. At this point we discovered that TR7 was short-circuit. Replacing IC2 and TR7 restored a stable picture, but the sound didn't appear until the TDA2611A audio output chip IC4 had been replaced. So much for all the overload protection that manufacturers boast about in modem power supplies!

Dead set. Faulty regulator chip: This set was dead with 17 V at the input to the TDA8138 regulator chip IC11 but no 5 V output at pin 9 and no 12 V output at pin 8. The chip itself was faulty – this is becoming quite a common fault.

Intermittently cuts out (Model 51P7): Not dry-joints in this case. Replace the TDA8138 5 V/12 V regulator chip IC11 – it has a reset function.

Goes to standby or intermittent channel 1: Deal with the numerous dry joints first then replace the TDA8138 multi-regulator chip IC11. This should provide a cure.

Ferguson TX99 chassis

Picture disappeared after switch-on: The customer complained that the picture usually disappeared shortly after switch-on, though the set would sometimes run all evening without trouble. We put the set on soak test but it ran all day without the fault showing up. Next day we switched on and within minutes the picture had disappeared, the sound continuing. Removal of the back cover while the fault was present showed that the c.r.t.'s heaters were out. This was due to dry joints on the line output transformer: resoldering restored normal operation.

E.h.t. came on but no picture: When this set was switched on the e.h.t. came up but there was no picture, just the occasional flicker of flyback lines. There was no sound and it was impossible to change from channel 1. R239 (100 Ω) was open-circuit due to an internal short in the M494B1 tuning/standby chip IC241.

Loss of colour or intermittent colour when warm: A new TDA3301B colour decoder chip appeared to cure the fault but the set came back next morning. The cause of the trouble turned out to be 4.433 MHz oscillator drift. We replaced the crystal XL1, R60 (100 Ω) and C63 (22 pF).

No remote control operation: The on-board controls worked normally and a test showed that the handset was transmitting. So attention was turned to the infrared receiver in the set. When this was removed the cause of the fault was immediately obvious. One of the three leads connected to the module was broken where it's soldered to the PCB.

No sound or vision: A quick check showed that the h.t. supply to the line output transformer was missing. The smoothing choke L21 in the power supply was open-circuit.

H.t. voltage low: After repairing the power supply we found that the h.t. voltage was low. The reservoir capacitor C103 (47 μF, 160 V) was to blame, though it measured OK when checked with a capacitance meter.

Loss of sound after an hour: A soak bench test confirmed that this was so. After checking the loudspeaker we turned our attention to the audio amplifier, but touching R72 at its input with a finger produced a healthy buzz from the speaker. Our next check was at the volume control pin (11) of the TDA4505 jungle chip IC2. As the voltage here didn't change when the volume control button was pressed we moved to the M494F chip IC241 on the remote control board. But surprise, surprise, pin 34 and the output from the emitter-follower TR246 ramped up and down. The only components between the emitter of TR246 and pin 11 of the TDA4504 chip are C66 and R12, which were both OK, and the red LED D5. This was the culprit, going open-circuit when hot. It's easy to find, being mounted just behind the tuner. When working, its brightness intensifies as the volume is increased. In fact you can see it through the cabinet back. Remember the green LED in the TX10 chassis? Ferguson seems to like using LEDs!

Field collapse: This is quite a common problem with these sets. You will usually find that one or other of the series-connected 11 Ω resistors R134-R137 are burnt up. My advice is to replace both the output transistors TR5 (TIP29B) and TR6 (TIP111 – a Darlington device) as well. Other components we replace as a matter of course are R109 (33 Ω safety), R108 (91 Ω safety) and the 56 V zener diode D30.

Shorted line output transistor: The cause of this set's failure was traced to a short-circuit line output transistor. When a new R4050 had been fitted it worked perfectly. Ever mindful of intermittent drive faults, however, I left the set on soak test. After about an hour the width would jump in and out intermittently. The eventual cure was to resolder all the legs of the TDA4505 chip.

Fidelity

FIDELITY AVS1600 (ZX3000 CHASSIS)
FIDELITY AVS2000
FIDELITY CTV140 (ZX4010 CHASSIS)
FIDELITY CTV14R (ZX2000 CHASSIS)
FIDELITY F14 (ZX4000 CHASSIS)
FIDELITY ZX3000 CHASSIS

Fidelity AVS1600 (ZX3000 chassis)

Shorted line output transistor: Two reports below.

The original problem had been a shorted BU508 line output transistor. No cause could be found, and a replacement restored the picture. A couple of days later, however, we received a report that the TV was flashing over. Removing the back (18 screws) and switching on produced a single crack from the focus unit in the line output transformer, followed by normal operation. Again nothing else could be found, so a new line output transformer was fitted. This time switching on produced a few seconds of fireworks before we could reach the mains switch. A check on the h.t. line revealed that we now had over 130 V instead of 119 V, so we gave up and reluctantly transported the monster to the workshop. Those familiar with the AVS cabinet will sympathize deeply. The culprit was found to be C91 (1 μF, 63 V) in the power supply, but the damage had been done. To restore full operation the TDA3562A decoder chip, the TDA8190G audio chip, the TDA2070 field timebase chip and even one of the TDA1908 audio output chips had to be replaced.

This set came in with a short-circuit line output transistor (Tr5, BU508A). A replacement was fitted and the set was switched on. There was a loud crack accompanied by a blue flash from the c.r.t.'s final anode and the new transistor went the way of its predecessor. We then did what we should have done in the first place – check the h.t. voltage. The BU508A was removed and in its place a 100 W bulb was connected between the collector feed and chassis. As we suspected, the h.t. was high. In fact it was 163V instead of 112V and turning the set-h.t. control to minimum reduced it to only 140V. The cause of the trouble was eventually traced to C91 (1 μF, 63V) and once this had been replaced normal power supply operation was restored. Sadly, however, when a new line output transistor was fitted we were confronted with a thin white line across the screen. The field output chip IC6 (TDA2270 in this particular version) had failed.

No sound and set stuck on channel 8: Sound was restored by replacing a faulty TDA1908A chip and fuse. But the set was still stuck on channel 8 with the balance control and remote control unit having no effect. Checks around the M104B1 chip IC1 showed that its 5V supply at pin 14 was missing. The cause was the 5.1V zener diode 5D1 which was short-circuit. Then we just had to reassemble it all!

Fidelity AVS2000

Stuck on channel 8 and no audio: The middle fuse (1 A) of the three attached to the side of the record deck had blown. This fuses the bridge rectifier on the audio panel. Lifting R379, the feed to the 12V regulator, proved that the fault was due to the bridge or the audio output chips. The bridge was innocent, but both IC4 and IC5 were short-circuit. Two new TDA1908As and a new fuse restored normal operation. Stuck on channel 8 was of course due to there being no 12V supply to the M104B1 chip.

Sound OK but no raster: There was field collapse. The TDA2270 field output chip IC4 was the cause of the problem.

Dead TV section: Because of their size these entertainment centres are not our favourite pieces of equipment. This one came into the workshop with a dead TV section. As all the power supplies were in order attention was turned to the line timebase. We found that there was no drive to the line output transistor as the TDA8180 timebase generator chip was faulty.

Fidelity CTV140 (ZX4010 chassis)

About 10 minutes after switch-on the sound and vision would suddenly disappear: A scope check around IC2 (TDA4503) revealed that in the fault condition the line drive frequency increased to over 1 MHz. The culprit was C31 (2.7 nF) in the line generator circuit – it's connected between pin 23 of IC2 and chassis.

Dead set, several faults: This set appeared to be dead but a check on the 114 V line showed that it was very much alive, with some 130 V at the collector of the line output transistor. There was no voltage at the collector of the driver transistor, however. This is fed from a 25 V rail, and we soon found that the 0.68 Ω safety resistor in series with the rectifier was open-circuit. A meter check from the rectifier to chassis gave a reading of only a few ohms: a check on the driver transistor then showed that it was short-circuit. When this was replaced we were rewarded with a nice picture which after a few seconds very slowly drifted off tune. Several things were replaced before we discovered that the isolation diode connected to the slider of no. 5 tuning preset was leaky. Replacing this gave stable tuning.

Dead set, fault in chopper circuit: Check whether the voltage at pin 9 of the TDA4600 chopper control chip is pulsing up and down. If it is, suspect a short on the secondary side of the chopper transformer. In most of the sets that have passed through our workshop with this fault D21 has been short-circuit. Replace with a BY299 or an RGP30.

No contrast: This set had an obscure name on it but appeared to be the Fidelity CTV140. The fault was no contrast. We found that R113 (100 kΩ) in the beam limiter network was open-circuit. Also check R111 (100 kΩ) and R106 (10 kΩ). The Fidelity manual shows a single resistor R106, with a value of 110 kΩ, in this position.

Full brightness with flyback lines: There was a note attached to the manual: we'd had the fault before, caused by failure of R57 (10 Ω) in the supply to the TDA3330 colour decoder chip. Not this time, however. On this occasion it was caused by the fact that the 4.7 μF, 350 V RGB output stages' 160 V supply reservoir capacitor C100 had dried out.

Fidelity CTV14R (ZX2000 chassis)

Dead set, h.t. present: The line output transformer had been replaced a few months ago, so we discounted that. Checks revealed that h.t. was present at R828 and R901 but not at the collector of the line output transistor. A careful check along the print then revealed a hairline crack which couldn't be seen with the naked eye. Repairing this restored the set to normal working order. We've since had several cases of print breaks like this.

Tripping after warm-up: The cause was C412 (100 pF, 8 kV) flashing internally. This capacitor decouples the input to the focus unit.

Intermittent loss of signals and channel 3 reception: The ML923 tuning chip was faulty. After replacing it, however, the signals would still sometimes fail to appear, though channel 3 selection was OK. As the supplies were correct a new tuner unit was fitted. This finally put matters right.

Fidelity F14 (ZX4000 chassis)

Teletext line problem over top third of the raster: The picture was otherwise OK. We'd had another set with this problem (Hitachi NP81CQ chassis) caused by a failure in the blanking circuit. Then this Fidelity F14 came in with identical symptoms. So straight to the blanking circuit where C63 was found to be open-circuit. Just to help, it's carefully disguised on the chassis layout as a wire link, near the back of the set. These are the only two blanking faults I've had in years!

Channel storage failure: If it's not possible to store channels or there's a problem with the tuning, you should check the following; R514 should be 2.7 kΩ (old value 1.8 kΩ); R315–9 inclusive should all be 1.8 kΩ (R319 was 1.5 kΩ); ZD1 (TAA550) 33 V regulator may be faulty (leaky); a 0.1 μF capacitor (C303 in later versions) should be connected in parallel with ZD1 on the print side of the panel; finally add a 100 Ω resistor and a 10 μF, 50 V electrolytic in parallel with ZD1 (negative terminal of the electrolytic to chassis). Check for a 25 V square wave at pin 2 of IC201; if this is absent IC201, TR304 or TR305 could be faulty.

Unreliable start-up: Sometimes this set didn't come on at all. On inspection we found that there was a dry-joint at the centre (collector) leg of the BU508A chopper transistor TR2. This transistor is mounted at the edge of the main PCB. The printed tracks at this point don't totally enclose the holes for the transistor leads, due to the need to keep solder clear of the cabinet guides when the board is slid in. The other two legs had apparently been resoldered by hand during production. There was still a variable start-up delay at each switch-on after we'd repaired the dry-joint. Eventually R84 (15 kΩ, 5 W) in the start-up circuit was found to have gone very high in value. Replacing it restored the set to normal operation.

No chopper output: This set was dead with d.c. to the chopper but no output. The h.t. rectifier D21 was short-circuit. We used a BY299 as a replacement. This brought the set back to life but on scene changes, or when the brightness or contrast control was adjusted, the verticals became corrugated. The cause was D21's reservoir capacitor C87 (100 μF, 250 V).

Just noise on screen, no station search function: The station search and memory buttons are at the back of the set, below the aerial socket. They are part of the back cover moulding. When pressed down they contact circuit switches on the main board. As someone with a heavy hand had been involved the buttons were stuck down hard on the switches. We were able to bend them back gently, away from the board and the switches. Fortunately there was still plenty of spring so that the search tuning system could be operated if required.

Fidelity ZX3000 chassis

Sound distortion: The problem was sound distortion after the set had been on for a few minutes. The speaker was very poor – it had a 'soggy' cone with a rip in it – but the TDA8190 sound output chip was the real culprit.

Line output transformer arcing: The line output transformer in one of these sets was arcing from near the e.h.t. outlet to the adjacent support bracket. Judicious use of silicone rubber sealant put an end to the arcing.

Chopper transistor TR3 shorted and both fuses open- circuit: The mains fuse F1 (T2A) was blackened. When these items had been replaced we still had no results as R97 (4.7 Ω, 4 W) in the feed to the line output stage was open-circuit. Our experience with these sets has been that if any one of the diodes in the bridge rectifier circuit is faulty it's advisable to replace all four with the larger version of the BY127. It's tricky but it can be done.

Green cast on picture: It looked as if the tube had a slight heater-cathode leak. A check on the tube showed that it was OK, so attention was turned to the RGB output stages on the c.r.t. base panel. Comparison voltage checks in the green and red stages brought us to TR10's base bias resistor R214 which had risen in value from 100 kΩ to 150 kΩ. A replacement restored correct colour.

Line tearing, noise from inside set: There was severe line tearing and stressful noises came from within the set. The cause of this situation was the 33 μF, 250 V h.t. reservoir capacitor C100 which was open-circuit. Interesting that the fault condition varied with different brightness levels.

No sound and no picture: A common fault. Advancing the first anode preset usually produces a blank raster and a heavy heart. The first step should be to check all supply lines, which will usually be found to be blameless. Next hook up the scope to check that a video output is present at pin 12 of the TDA3541 i.f. chip IC2. This output goes via various filters to the TDA3562A colour decoder chip IC7 (chrominance input at pin 4, luminance input at pin 8). Some sets have a scart socket, however, and in this there's a TEA1014 switching chip (IC5) between IC2 and IC7. Check that there's a video input at pin 3 of this chip and a video output at pin 12. The TEA1014 chip is often the cause of the trouble. In an emergency and if the customer doesn't use the scart facility you can remove this chip and link the video/audio inputs and outputs across – for audio the input/output pins are 8 and 6 respectively. If everything is OK up to this point have a look at the sandcastle pulse output at pin 15 of the TDA8180 timebase generator chip IC4 (in later versions of the chassis a TDA2578 chip is used in this position, with the sandcastle pulse output at pin 17). For proper operation the sandcastle pulse waveform must be absolutely correct in terms of amplitude and shape. If it's distorted in some way, as it often is, disconnect pin 15 before condemning the TDA8180 chip – the TDA3562A decoder chip can load the pulse. We've had these two chips (TDA8180 and TDA3562A) fail as a pair.

Field expansion: We have on several occasions traced the cause of field expansion at the top of the screen to loss of capacitance in C59. Its value depends on screen size: 1000 μF with 14 and 20 inch sets, 2200 μF with 22 inch sets.

Excessive width with a number of dark bands down the left- hand side of the raster: This usually suggests a lack of damping across the line linearity coil. On several occasions we've found that the cause has been a dry joint on the upper leg of the damping resistor (R103 – 220 Ω, 1 W) rather than the resistor being open-circuit.

Set took an hour to come on: A clue to the cause of the problem was faint pulsing from the power supply which had been noted by the observant customer. A check at pin 9 of the TDA4600 power supply control chip IC8 showed that the 13.5 V start-up supply was low. It built up from 4 V to 10 V, but this was not enough to start the set. The two 10 kΩ resistors R82/3 and rectifier D7 in the start circuit were all OK.

As the voltage at pin 9 varied it seemed likely that the associated 100 μF, 25 V reservoir capacitor C87 was open-circuit. When it was removed we found that it had a discharge from its positive leg.

Poor or intermittent start-up: Check whether C89 (100 μF, 25 V) in the TDA4600 power supply circuit has fallen in value.

Picture tearing: The cause was the 33 μF, 250 V capacitor on the right-hand side of the chopper transformer.

Finlux

FINLUX 1000 SERIES
FINLUX 3000 CHASSIS
FINLUX 3021F
FINLUX 5000 SERIES CHASSIS
FINLUX 9000 CHASSIS

Finlux 1000 series

Dead set, no power supply output: Check the supply to the TDA4600 chopper control chip. There can be dry-joints on thermistor PTCu2 or an open-circuit in the parallel resistor network Ru3/4/6/7.

No functions, loss of memory: If a teletext set has no +vol, +step, FT or memory functions, check that when the dealer fitted the text panel he didn't accidentally cut through the print to pin 28 of the micro-computer chip.

Many symptoms: One of these sets produced the longest list of symptoms imaginable. As you watched the picture would go all blue, all green, off altogether, show false line lock or line tearing. A voltage check at Da1 proved that the 12 V line was low at 9.5 V. The cause was a defective BD241 transistor in position Ta8. Da1 also has a reputation for going high resistance.

Blown 1.25 A fuse: The cause was a dry-joint at one end of LU3. Check this, especially in cases where a replacement fuse appears to cure the fault.

No sound or picture but display OK: Checks showed that the line drive was missing. This was because of loss of the supply to the line oscillator. Transistor Ta2 (BC547), which is part of a voltage regulator circuit on the signals panel, had failed.

Intermittent colour (early series): Two types of colour decoder were used in these sets. This was an early version, with a TDA3652A colour decoder chip. The fault was intermittent colour. Tapping the panel, as we do, gave rise to suspicion that there were dry-joints. But after a little probing we found that the reference oscillator's trimmer Ct12 (6–30 pF) was noisy. To set this up after replacement, simply link pins 24 and 25 and 1 and 5 of the i.c. together and adjust Ct12 for near stationary colour. Then remove the links.

Band of torn lines rolls down the screen: This set would start off all right. But after about half an hour a band of torn lines would roll down the screen from the top to the bottom at a rate of about two per second. The cause was Ch11 (47 µF, 63 V).

Dead set: The BY299 diode Du20, which is the rectifier for the 17.5 V supply on the secondary side of the chopper transformer, was dry-jointed at both ends. Resoldering it brought the set back to life.

Finlux 3000 chassis

Snow on screen and no sound: Channel numbers were accepted and stored, but with no results. Ti4 on the signals panel controls the tuning voltage, and a quick check showed that its collector supply was missing. There are two feed resistors from the 30 V line Ri37 and Ri39. The latter was open-circuit. Fitting a new 330 Ω resistor put matters right.

Loss of tuning: Although failure of the SDA3202 chip in the tuner circuit is a common problem, the result being loss of tuning, it's worth noting that a wet finger applied to pin 18 of this chip will usually bias Ti4 enough to display some signal being received. This will prove that the tuner is working. Not very technical, maybe, but it's effective.

Picture shifted massively to right: If the picture is shifted to the right to the extent that the line scanning starts in the middle of the screen, check Rz19 in the line output stage. Its value is 27 kΩ but you will find it open-circuit. It forms part of the pulse feedback system.

Weird display: There was a bar down the screen and the picture was split and superimposed on another picture that was twice the correct width. When teletext was tried the characters were about 3 inches wide. Scope checks in the line output stage showed that the flyback pulse signal, from the collector of the line output transistor to the base of transistor TZ1, was missing. The cause of the trouble was RZ21 (220 kΩ) which was open-circuit.

Finlux 3021F

Intermittent loss of signals: In addition the set would sometimes go dead. Its cause was traced to Du21 in the power supply. This BY299 diode provides the 7 V supply but was going open-circuit intermittently.

Field collapse: Checks showed that all the components in the field output stage were OK. The chip had been replaced, but the voltage at the input pins was wrong. After some time had been spent getting nowhere we swapped the whole signal panel with one from a good set. As the fault was still present we suspected a fault on the power/line output panel. Wrong again. The cause of the fault was on the control panel beneath the tube, where the microcontroller and memory chips reside. The SDA2526 memory chip had failed!

Colour intensity varies: With some pictures the colour intensity seemed to be greater in different parts of the screen. The cause of the problem was hum on the 12 V supply, the culprit being the 47 Ω resistor Ra34 which is connected in series with the earth pin of the 7812 12 V regulator chip ICa2. A check on the 12 V rail with a voltmeter produced a reading of 12.4 V: when the scope was brought into operation we found that about 350 mV of hum was present. After replacing the resistor the voltage was correct at 12 V and was as smooth as silk.

Excessive width: There were also EW correction troubles. Bright pictures produced tearing at the top and bottom of the screen. We started by checking the supplies, which were all correct and smooth. Moving to the EW correction circuitry, we found that the waveform at pin 19 of ICH2 was incorrect. This pin is linked to pin 8 of the line output transformer, the beam sensing point. The voltage here was varying much more than it should have done. Checks along this line brought me to CH29 (0.1 μF) which was open-circuit. It's on the signals panel. A replacement restored perfect geometry.

Concave picture: The usual cause is D28 on the power supply/line output panel. But in this case the TIP41A EW diode modulator driver transistor TK4 was open-circuit. We're sure that the cause of its failure was the loose mounting screw, which meant that it wasn't held in proper contact with the heatsink.

Finlux 5000 series chassis

Picture intermittently pink with flyback lines: Also the set sometimes blanked out completely. We found that wire link JE38 was dry-jointed. It's partially covered by the plastic at the front of the chassis.

BUZ91 chopper transistor failure: If the BUZ91 chopper transistor Tu1 keeps failing after about 5 or 10 minutes, check the BYV36C diode Du7 in the snubber network. The fault occurs when this diode is leaky.

Dead set: Check the BUZ91 chopper transistor Tu1 is short-circuit and that both fuses, a 1.6 A Wickman fuse on the main board and a 2.5 A fuse on the front panel, have blown. Before replacing these items check whether rectifier diode Du1 (BY500), which produces the 70 V h.t. supply, is short-circuit. Also check the voltages at pins 2 and 3 of ICu2 (TDA4605) with Tu1 out of circuit. There should be 1.2–2.5 V at pin 2. If the reading is 0 V, replace Ru24 and Ru26 (both 150 kΩ). There should be 1.8 V at pin 3. If the reading is high, replace Ru22 and Ru23 (both 374 kΩ).

Produced snow, then went dead: The dead set bit was no problem: there's auto switch-off when the set isn't receiving a signal. The cause of the tuning fault was rather more difficult to trace. The tuner is controlled, at pins 5 and 6, by an I2C bus. Its tuning supply, at pin 3, should be variable from 0 V to 24 V but was stuck at 30 V. The on-screen programming worked and showed the correct numbers, and scope checks at the tuner's I2C pins showed plenty of activity which altered when the channel was changed. A new tuner was tried, but there was still no tuning. The cause of the problem turned out to be the SDA2586 memory chip ICa2, which is an 8-pin pre-programmed EPROM, part number 4400267056. It's available from Nokia or CPC.

Finlux 9000 chassis

Failure to start from cold: You sometimes get failure to start from cold with these sets, especially in cold weather. We've come across two causes of this trouble. Either Ru17 (270 kΩ) has been intermittent or Cu6 has dried up.

Low brightness: Curiously, when the tube's first anode voltage was increased the set developed a Hanover-bar effect, then the colour-difference signals disappeared. After a great deal of head-scratching and unnecessary replacement of components the cause of the fault was found to be in the RGB output section, of all places. We disconnected the emitters of the RGB output transistors Tb5, Tb8 and Tb11 in turn. When the red channel transistor Tb5 was disconnected in this way the brightness and chroma faults cleared. After this it was a straightforward case of checking the components in the red channel. This led us to Tb6 which was short-circuit collector-to-emitter. Normal operation was restored when a new BF881 had been fitted. The faulty transistor must, in addition to affecting the red channel, have been loading the colour decoder section chip which then shut down.

Lack of contrast at switch-on: A common fault with these excellent receivers. The level can be stored in the memory, but the switch-on level is only about 70 per cent at maximum. This isn't actually a fault: it's caused by the design of the circuit. When the normalize button on the set or the remote control unit is pressed the contrast will come up to the desired setting. The problem is that the tube is losing emission: with a high contrast setting a good picture can be displayed. A variable d.c. voltage to the colour decoder chip sets the contrast level. Adjusting the values of the resistors in the potential divider network that produces the control voltage will provide a higher contrast level at switch-on. It can then be adjusted in the normal way. After some experimentation we found that changing the value of Rb29 on the video panel from 56 kΩ to 40.5 kΩ produced the desired effect. With a lower value the contrast wouldn't turn down while a higher value had little effect. Two resistors in parallel were used to produce this value. The modification is not official.

Goldstar

Goldstar CIT2168F (PC04A chassis)

Set stuck in standby: If the power switch was held in, the timebases would start up and a snowy raster would be produced. But there would be no sound and no manual or remote control. A check at pins 31 and 32 of the microcontroller chip IC701 showed that the 10 MHz oscillator signal was not present. So, with no engine the set wouldn't run! A new crystal enabled the microcontroller chip to do its stuff, and the set worked normally. Holding the power switch in had provided the h.t. relay's driver transistor with bias via the switch's pulse contacts. So the power supply had been able to feed the line output stage. But there was obviously no action from the microcontroller chip.

No sound: We found that there was no volume control pulse-width modulated output from the microcontroller chip, so we assumed that this chip was faulty and replaced it. Wrong! In fact the sound was being muted because the microcontroller chip thought there wasn't a locked signal. The culprit turned out to be the TDA1940 sync separator/line generator chip IC401, which wasn't producing an output when coincidence was detected.

Blank screen with normal sound: This was the result of failure of the TDA1170N field timebase chip IC301. In addition R320 (10 Ω, 1 W) in the supply to IC301 had burnt out. The correct current through R320 is 170 mA. To cure slight fold-over at the top of the picture we replaced the 27 V supply's reservoir capacitor C423 and the bootstrap capacitor C302, they are both 100 μF, 35 V components. Note that great care should be taken when removing the back from these sets to prevent damage to the loudspeaker terminals. The only plug and socket available to disconnect the loudspeakers is placed inaccessibly under the c.r.t. rim, and the chassis will not slide out until a front screw is removed.

Goldstar CIT2190F

No sound but a good picture: As we could get hum from the audio output chip we ruled this item out and changed the TDA120T sound demodulator chip. Still no sound. We then got the circuit out. Voltage checks didn't reveal much so we looked for a muting circuit and found that pin 7 of IC401 produced a muting output. When we desoldered this pin and switched on there was full sound. A new chip in the IC401 position put matters right.

No line drive: As there was no line drive this set didn't produce any e.h.t. The cause of the problem was D401 (1N4003) in the supply to the line driver stage. It was open-circuit, removing Q401's collector voltage. A replacement restored normal operation.

Field timebase chip failure: If the TDA1170N field timebase chip IC301 has failed, C302 (100 μF, 35 V) must also be replaced. Otherwise IC301 could fail again.

Goodmans

GOODMANS 1410 (THOMSON TX805 CHASSIS)
GOODMANS 2043T (GOLDSTAR PC04A)
GOODMANS 2180
GOODMANS 2575

Goodmans 1410 (Thomson TX805 chassis)

Dead set: Two of these portables were completely dead when they arrived. In both cases there was no power supply start-up because one of the three series-connected 68 kΩ resistors was open-circuit. One of them, RP44, is rated at only 0.25 W. It hides beneath two large, green power resistors. The other two 68 kΩ resistors are rated at 0.5 W. We replaced them all with 1 W resistors to ensure a long and happy life.

Dead, short-circuit line output transistor: The S2000AF line output transistor (TP10) was short-circuit. Checks showed that in addition RP28 (2.7 Ω, 5 W) was open-circuit and DP11 (BA157) short-circuit. Fortunately the set continued to work during a long soak test after fitting the replacements.

Line tearing, plus a squeak from the chopper transformer: This would develop when this set had been in operation for about half an hour. The cause was eventually traced to CP90 (22 μF, 50 V) in the power supply circuit.

Goodmans 2043T (Goldstar PC04A)

Dead set: The power supply was working, providing the correct h.t. voltage at pin 17 of the chopper transformer, but the set remained lifeless because there was no h.t. at the B+ terminal of the line output transformer. Diode D8035 was open-circuit.

Picture was dark with high and uncontrollable saturation: When a channel was selected it would appear and then almost immediately drift off tune. A check on the 12 V supply, at C429, produced a reading of only 8 V. The two series-connected smoothing resistors R437 and R427 had risen in value. Our circuit diagram gave the values as 3.3 Ω, 1 W (safety type), but those in the set were marked 1.5 Ω. So we fitted replacements of this value, clearing all the symptoms.

Dead set: Checks on the small electrolytics in the chopper power supply revealed that C805P (100 μF, 16 V), which is mounted in front of the chopper transistor, had dried out.

Goodmans 2180

Dead set: This set uses a Nikkai chassis. The power supply didn't work and checks soon showed that R801 (270 kΩ) in the start-up section was open-circuit. A replacement restored normal operation.

Dead set: If the set is dead check R105 (150 kΩ), R106 (150 kΩ), R107 (270 kΩ) and R108 (270 kΩ). These resistors are near the CNX82 optocoupler in the power supply.

Sound but no picture: The h.t. voltage was correct, but further checks showed that there was no voltage at the collector of the line driver transistor – the primary winding of the transformer was open-circuit. A replacement cured the problem. Spares for this set seem to be available only from Comet. This could be a stock fault.

Goodmans 2575

No red output: This 25 inch Nicam set produced no red output from the TEA5181A RGB output chip, which is mounted on the tube's base panel. We soon traced the cause to the 68 kΩ resistor R26, which was open-circuit. We replaced the corresponding resistors in the G and B channels as well, R28 and R29 respectively, as their values had drifted somewhat.

Dead set: There were only a few volts at the collector of the line output transistor. When the h.t. supply was disconnected from the line output transformer the power supply worked normally. Checks in the line output stage showed that a BY448 diode in the EW correction circuit had shorted and burnt the surrounding print. A print repair job and a new diode got the set working again but the raster had bowed sides, indicating loss of EW correction. A new TDA4950 chip (IC18) put that right. You'll find the BY448 diode alongside the large EW coil.

Dead set, with smoke: This set was dead, with smoke coming from C134 (10 nF, 1.6 kV) because of a dry-joint at one of its connections. We were able to clean up the PCB and replace C134. Then we found that D45 (BY228) had also suffered and was short-circuit. Once this had been replaced there was sound and a picture, but the width was excessive and there was no EW control. As the voltages around the TDA4550 pincushion correction chip IC18 were nothing like what they should be, we replaced it. This finally restored the set to its correct working condition.

Grundig

GRUNDIG CUC120 CHASSIS
GRUNDIG CUC220 CHASSIS
GRUNDIG CUC2201 CHASSIS
GRUNDIG CUC2401 CHASSIS
GRUNDIG CUC2410 CHASSIS
GRUNDIG CUC2800 CHASSIS
GRUNDIG CUC3400 CHASSIS
GRUNDIG CUC720 CHASSIS
GRUNDIG CINEMA 9050
GRUNDIG GSC100 CHASSIS

Grundig CUC120 chassis

Intermittent no sound or vision: The set-e.h.t. control R647 can become intermittent with these sets, causing symptoms such as intermittent no sound or vision, varying brightness, or short field output chip life. The symptoms depend on whether the h.t. goes up or down.

Intermittent blank screen, no sound (muted): This is usually caused by a dry-joint inside the tuner-i.f. module. The connection involved is at pin 12, where the 12 V supply enters.

Field collapse: There was field collapse and R2761 (6.8 Ω) was burnt. Naturally the TDA1170 chip was suspected but C2762 (1000 μF, 35 V) was short-circuit.

Intermittent shutdown or refusal to start: This is not an isolated case. What happens is that L631 in the chopper transistor's base circuit develops a dry-joint, though it looks good. The problem is caused by its leadout wires being too thin for the size of hole provided (shades of the Pye i.f. module, remember?). Quite often the TDA4600 control chip hands its notice in as well when the customer is one of those who believes that a good thump on the cabinet cures all . . . So check L631 if you find a dead TDA4600. This is the nearest to a good old stock fault you'll find with these sets.

Took 5–10 minutes to start up: This set took 5–10 minutes to start up, during which time all voltages read low. Once the switch-mode power supply got going you couldn't get the fault to return by applying freezer: the set had to be switched off for several hours before the fault reappeared. We removed and tested the bridge rectifier's reservoir capacitor C626 and it gave a satisfactory indication. But fitting a replacement cured the fault. When C626 was retested we found that it had gone open-circuit.

Luminance level intermittently drops: By substitution we proved that the cause of the problem was on the decoder panel. The luminance delay line and the TDA3561 colour decoder chip were replaced but the fault was still present. It was difficult to take any measurements or use the scope because any such action made the fault disappear. We eventually found that the cause was the luminance coupling capacitor C2527 (0.1 μF).

Very severe patterning varied with brightness: This was caused by ripple on the 19 V rail. C662 (470 μF, 25 V) was completely open-circuit.

Blank raster: If all the supplies are present and correct and the tube's heaters are alight, check whether the TDA1770 field timebase chip IC2775 on the deflection panel is faulty. The blank screen is caused by the c.r.t. protection system coming into operation when no field scan is being generated.

Lack of height: The cause was traced to R2761 which had risen in value to 39 Ω. Its correct value is 6.8 Ω – safety type.

The vision disappeared when the set had been on for a few minutes: This left a blank raster on the screen. We found that the signal at pin 24 of the tuner/i.f. module disappeared – the supplies remained constant. Scope checks inside the module revealed that the BA338 transistor T2251 was going open-circuit intermittently.

Grundig CUC220 chassis

Field fault: When this set (A7400) was very cold there was lack of field scan at the bottom of the screen – the top was OK. There was no fold-over at the bottom. The cause of the fault was C2768 on the deflection

PCB being rather less than the specified 100 μF. Two different deflection PCBs were used in these chassis. One has a TDA1770 field output circuit, the other one using a TDA2655B chip. We've had field distortion with the TDA1770 circuit on a number of occasions, the causes having been the chip itself, D2761 (SKE4G1/04) or R2761 (6.8 Ω, 0.75 W).

Distorted sound: We found that the supply to the audio chip was only around 5 V, though the +G line was correct at 15 V in the power supply. A check with the circuit diagram showed that there's a relay switch in the supply line. The contacts were making poor contact, something we've not come across before in these sets.

Excessively large picture, singing noise: The owner complained that a singing noise came from the back of this set. He said that it had started suddenly and that the picture had gone pink at the time but later reverted to normal colouring. When we checked we found that the power supply choke was indeed singing, but what the customer hadn't mentioned was that the picture was excessively large, with the h.t. at 140 V instead of 119 V. Our first suspect was the 119 V supply reservoir capacitor C657 (100 μF) which can cause problems like this when it falls in value, but a replacement made no difference. Checks around the TDA4600 chopper control chip failed to reveal anything that was obviously amiss, so we went back to our original theory of a low-value capacitor. C647 (1 μF, 63 V) looked a likely suspect as it's close to a hot wirewound resistor. Replacing this did the trick. Incidentally the chassis incorporates auto grey-scale correction: we suspect that this is why the screen went pink when the power supply's outputs rose.

Intermittent start-up: The chassis uses a standard Siemens type power supply with a TDA4600 chopper control chip. The main problem was that C631 (100 μF, 40 V) in the BU208A chopper transistor's base drive circuit had dried out. Replacing this item cured the intermittent start-up, but there was some ringing on verticals. This was cured by replacing C633 (220 μF), C647 (1 μF) and C642 (100 μF) which were all in poor condition. It's also good practice to check the value of R646 (270 kΩ).

No vision, sound or channel display: Checks on the power supply outputs proved to be fruitful: the 15 V supply was very low, even off load. It comes from a standard 7815 regulator, which is one of three mounted on the finned heatsink above the c.r.t. base opening in the main panel. The input to the regulator was OK, but there was a leak between its output and chassis.

Field collapse: After checking everything two or three times we discovered that R2779 (18 kΩ) was open-circuit – but you have to check it out of circuit.

No luminance or chrominance: There was sound and a raster. The supply to the TDA3560 colour decoder chip was OK and the waveforms in this area were all correct. We eventually found that the cause of the fault was D2536 (1N4148) in the beam limiter circuit. Check it out of circuit for low reverse resistance.

Squeal and rumble at switch-on: This fault could possibly apply to any set that uses a TDA4600 type power supply. The set would squeal and rumble at switch-on, with the LED display flashing like a Space Invaders game. But the set would come on if it was repeatedly switched off then on again. A check on the h.t. output from the power supply when the fault was present showed that it was varying. Although the set could be made to work, we suspected that there was a fault in the power supply. Replacing C633 (220 μF, 25 V), which smooths the TDA4600 chip's 12.5 V supply, and the chip itself cured the fault. We also replaced C631 (100 μF, 40 V) which couples the drive to the BU208A chopper transistor as it looked tired.

Set won't tune in: If the tuning voltage is at 30 V and the set won't tune in, the usual cause of the trouble is the U264B prescaler chip in the tuner/i.f. block. It's now no longer available, but a U644B can be used as a replacement. To find this item, look under the small screening can. Pin 9 is the output at the edge connector.

Shuts down when warm: This set was OK when cold. Once it had warmed up, however, it would shut down with a jumbled LED display. Voltage checks showed that the 5 V supply was low at 3.1 V. The culprit was D671 (SKE4F).

Grundig CUC2201 chassis

Set would trip for 20 minutes before coming on: This portable colour set would sometimes trip for about 20 minutes before coming on, after which it would run without further trouble. At other times it would work for days on end without the fault occurring. There were no dry-joints, and heating/freezing suspect components had no effect whatsoever. The cause of the trouble turned out to be the set-h.t. control R637 (1 kΩ).

Intermittently tripping off/on: Tripping off/on, sometimes very intermittently, is usually caused by a faulty set-h.t. control (R637, 1 kΩ). We renew this item as a routine measure whenever one of these sets comes into the workshop.

Dead set, two power supply outputs present: There was no line output stage operation or +C supply. Checks showed that the drive waveform at the base of the line output transistor was present and correct, and a small spark could be drawn from the e.h.t. pulse output of the transformer. But there was no voltage at the collector of the line output transistor because, curiously, the transformer's primary winding was open-circuit.

Grundig CUC2401 chassis

Sound mutes on recorded tape play: Check which type of ABL module is fitted. If it's type 29504–107–31, change it for a 29504–007–28 or disconnect the link from the TDA8185 chip to pin 9 of the ABL module. Removing the link disconnects the mute, so the self-seek tuning won't stop when a station is found etc.

Changes to line output circuit: Early versions of this chassis used a BU508D line output transistor without the internal resistor between its base and emitter – many replacement BU508Ds do have this resistor. To use a transistor with the internal resistor the following changes have to be made to the base circuit: change R521 from 100 Ω to 82 Ω; add a BZX85C5V6 zener diode across R521 with its cathode to pin 1 of the TDA8140 chip; and finally change R523 from 0.15 Ω to 0.12 Ω. If the BU508D is getting much too hot, check with a scope that it's being driven at line rate. According to the nice man at Grundig, when the TDA8140 line drive processor chip fails it can sometimes decide to drive the BU508D at twice the line rate – not something one would normally expect!

Intermittent no sound or picture: Suspect the set-e.h.t. control, especially if it's made by Preh.

Intermittent tripping at switch-on: This set had perfect picture and sound but would sometimes start tripping when switched on. No dry-joints or obviously faulty components could be seen in the power supply. We finally found the cause of the problem accidentally when attempting to reduce the h.t. setting. The 1 kΩ potentiometer R637 was faulty.

No sound: The audio stage was working and the dealer had already replaced the TBA130 chip on the i.f. board. By injecting a 6 MHz signal we discovered that the TBA130 wasn't working, although volume commands were present on the I2C bus line, the mute (koin) line was normal and the chip's supply was present. The internally generated supplies were missing, however, as C2276 (470 μF) was short-circuit.

Dead set, no output from PSU at all: The dealer who brought the set in had changed the TDA3640 chopper control chip IC655 and the associated electrolytics without success. A scope check at pin 18 of IC655 showed that the voltage here was varying between 7 V and 10 V every second or so. This indicated that the power supply was trying to start, but as the running supply at pin 2 didn't rise the chip would shut down again. The drive to the chopper transistor pulsed in sympathy with the start-up supply. Although the scope couldn't detect anything at the overcurrent sensing pin 7, everything pointed to the chip sensing a fault. Resistance checks at the outputs of the chopper transformer showed that nothing was amiss here and we were beginning to suspect the transformer itself. On a hunch we checked D666, which is one of the diodes in the chopper transistor's snubber network. It was short-circuit.

Very intermittent power supply failure: This was caused by a defective set-h.t. control. It's R637, 1 kΩ. In a further set with the same fault we couldn't find anything wrong with R637 but replacing it provided a complete cure.

Intermittent loss of picture: When the set was switched on it worked all right, so we left it on soak test with the first anode control set high. After an hour or two we were rewarded with a field collapse display. So we tiptoed across and measured the 26 V supply at pin 8 of the TDA2655B field timebase chip. It was a mere 4 V, though there was ample voltage at the anode side of the relevant rectifier D2758. We replaced this item, using a BY298, and also the reservoir capacitor C2758 (1000 μF, 35 V).

Set would switch to standby: The customer reported that the set would occasionally switch to standby. This went on for several months. All he had to do was to switch the set back on again, and it could be several weeks before there was a repeat performance. Now, however, the set tripped off only minutes after being switched on. We replaced the tripler and ran the set for two days as a check. It remained on. As a final check we refitted the original tripler, which brought the fault back.

No picture, faint line on screen: There was no picture. While trying the set out in a dark corner of the workshop we noticed that a faint line down the screen would put in an appearance every few seconds. Line collapse was the trouble, because the line output transformer was open-circuit. H.t. was present, but wasn't reaching the collector of the line output transistor. While looking for a replacement transformer we found that the cost of a pattern one was about half that of the original.

Grundig CUC2410 chassis

No picture, RGB stages not driven: This set had no picture as the RGB output stages weren't being driven. The anti-screen-burn circuit was coming into action although the field hadn't collapsed. A check on the field output waveform revealed that it was about half the normal size, which was reasonable as the supply was only 10 V. R525 (0.33 Ω) had gone high resistance.

No results, on/off switch sticky: The complaint was of no results. When we called we found that the on/off switch was sticky. Next day we were back again because of the same complaint. This time we discovered that the set would come on if the switch was operated twice. We were puzzling over this when there was an almighty crack and sparks came from the side of the tripler. Back in the workshop we fitted a new tripler but there was still a power supply fault. This was eventually cured by fitting a new TDA4600 chopper control chip.

Dead set; power supply would fail to start: After a cold check on all relevant resistors and fitting a new TDA4600 chopper control chip, we eventually traced the cause of the fault to C633 (100 μF, 25 V) which was open-circuit.

Tripler arc-over damage repair: A tripler arc-over had caused quite a lot of damage. The dealer had replaced the microcontroller, colour decoder and i.f. chips and a picture was now present, but there was no sound, either from the speakers or the scart socket. The reason for this was that the Koin line (inter-station mute) was low all the time because the TDA2595 chip was faulty. A new TDA2595 completed the repair.

Sound and text, no channel picture: When the brightness was turned up fullly only a blank raster was displayed. We found that there was no video input at pin 8 of the TDA3566 colour decoder chip because the BC548C video buffer/amplifier transistor Tr2523 was short-circuit collector-to-emitter. A replacement restored the picture.

Grundig CUC2800 chassis

Dead set, power supply working: The voltages on the unswitched side of relay RL652 were correct. There were no voltages on the switched side, however, and further checks showed that the relay was being switched. Investigation showed that although the relay was apparently working the contacts didn't make. A replacement put matters right.

Tripler failure: After replacing it the set got going but there were other problems: the sound came from only one speaker, and the brightness couldn't be adjusted. They were caused by a defective microcontroller chip. It's type SDA2011-A005. Make sure that you get the right one for the set.

Low, distorted sound: We traced the cause to the U2829B dual sound intercarrier chip. It's available from Willow Vale.

No sound: The usual cause is failure of the TDA6200 sound processing chip. On this occasion, however, the U2829B sound demodulator chip was responsible. The construction of this chassis does not make servicing easy. There is some access to the print side of the i.f. board, however. A scope check in the coupling circuit between the two chips will show which one is the culprit.

Grundig CUC3400 chassis

Intermittent tripping from cold: This particular set, a T55–340, first came into the workshop a few months ago because of intermittent tripping from cold. The line drive signal going into the TDA8140 line processor chip was OK, but it was distorted at the output. A replacement TDA8140 cured this for a while. When the set came back the fault was fortunately permanent, so we were able to make quicker progress. We found that the power supply was tripping, although it was all right with a dummy load. If the line hold was adjusted so that the frequency was too low the power supply started up and a raster appeared. For the line deflection/switch-mode power supply (it's a variant on the Ipsalo theme) to work correctly the oscillator in the power supply must operate at line frequency. It didn't, at least not quite. The timing capacitor C653 (4.7 nF) had changed value.

Wouldn't start: This set, Model P37–342, wouldn't start. There was h.t. at the chopper transistor and a slight ticking noise from the transformer. All the voltages at the control chip seemed to be about right, and there were no obvious shorts. We were about to put the set on the shelf and order a new chip when we thought that it might be a good idea to check the chopper transistor's drive waveform. A scope check showed that its frequency was very low. C653 (4.7 nF), which is connected to pin 15 of the i.c., sets the frequency. When it was replaced the set started up normally.

Dead set, buzzing Ipsalo transformer: This set was dead, with a buzzing noise coming from the Ipsalo transformer. The +M supply was correct at 11 V when the set was switched to standby, but when it was brought on this voltage fell instead of increasing to 20 V. The difference between standby and on is that in the latter condition drive is applied to the line output transistor. As there didn't appear to be any shorts across the supply lines, and the line drive waveform was correct, it seemed likely that the chopper and line output sections of the Ipsalo circuit were not synchronized. The basic frequency of the power supply is set by C653, with fine adjustment by means of the feedback applied to pin 12 of the TDA3640 chopper control chip. Replacing C653 had no effect, so the components in the network connected to pin 12 were checked. C630 (0.68 μF) was leaky.

Colour disappears intermittently: At the same time the right-hand side of the picture would be blanked out. A scope check on the sandcastle pulse showed that it was misshapen when the fault was present. The TDA2579 timebase generator chip was faulty.

Poor focus or focusing that can't be adjusted: This is usually a straightforward fault with Grundig TV sets that have the focus control mounted on the tripler – in most cases the control itself is the cause of the trouble. If the set is fitted with one of the above chassis, however, and the focus is so bad that the picture can hardly be seen – take care! Remove C433 (1000 μF). If it's marked 16 V fit a 25 V type instead. The 16 V type can go short-circuit, as a result of which d.c. flows through the field scan coils. But instead of the symptom being a picture shift upwards or downwards the result is a full, defocused raster that's dim and impure. While you have the set working the deflection coils are getting hot and there's a risk of cracking the neck of the c.r.t. One to watch out for!

Dead, chopper transistor shorted: Dead with the 800 mAT d.c. fuse S1624 blown and the BUT11A chopper transistor T661 short-circuit is generally an easy fault to deal with in these sets. The cause is usually dry-joints on the Ipsalo (combined chopper/line output) transformer TR665 or faulty snubber network diodes (D663, D664 and D666). It wasn't on this occasion, however. The power supply worked fine when the set was in the standby mode, but when it was brought on the chopper transistor went short-circuit. This was a later production set, in which BYV16 diodes are used in positions D663/4 while D666 is a BYD33M (they are all BA159s in the original version), but we nevertheless replaced them. A replacement TDA3640 chopper control chip and 4.7 nF capacitor (C653) also made no difference. Another change in later sets is that C667 is 2.2 μF instead of 0.1 μF. The 2.2 μF capacitor had dried up.

Would not change channels from cold: From cold this set wouldn't change channels. The number would change, but there would be no tuning. The cause was a dry-joint at pin 15 of the tuner/i.f. unit's socket – the tuning voltage supply pin.

Blue raster with flyback lines: The cause of the trouble was the TDA3505 video processor chip IC9531 on the chroma panel.

Dead set: The fuses were all OK and there was 300 V at the collector of the chopper transistor T661. We decided to carry out some resistance checks on the diodes in the combined chopper/line output stage and found that D691 (BYV28/100) was short-circuit. It's the 20 V supply rectifier on the secondary side of the circuit.

No raster but sound and e.h.t. present: Check for first anode (G2) voltage at the relevant pin of the c.r.t. base socket (thank you, Grundig, for marking the pins on the PCB). If the voltage is missing, suspect an internal leak in the socket. Check it by replacement.

Grundig CUC720 chassis

Faulty teletext panel: When text was selected the TV picture still showed – the same as when mix is selected. We found that the RGB switch line – pin 15 of the module – was at only 0.3 V when it should have been at 1.3 V. As a result the decoder module didn't switch over. Checks around the SAA5050 chip showed that the PO (picture on) line was going low all right but the BL output was less than normal. Resistance checks were then carried out around transistors 2876 and 2881. We found that C2881 (1 nF) was almost short-circuit, the reading was 12 Ω!

On/off switch faulty: The job card said 'on/off switch faulty'. It wasn't, of course, though the power supply was dead. There was 300 V at the collector of the chopper transistor and needle drive pulses were present, but there was no output to speak of about 2 V on the 150 V line. Voltage checks around the TDA4600 chopper control chip showed that the 4.3 V reference voltage at pin 1 was missing. C642 (100 μF) was open-circuit.

Black band at bottom of screen, heads elongated at the top: There was a black band at the bottom of the screen while heads were elongated at the top. To help matters, the fault was intermittent. One thing that does help, however, is that you can plug the timebase (and other) subpanels into the back of the mother board to work on them. After checking several capacitors in the field timebase I replaced the 100 kΩ field linearity potentiometer R2757. It was going open-circuit intermittently.

Sound disappeared: At switch-on the sound appeared for a second then disappeared – the picture was perfect. Operating the front hi-fi switch would also produce sound for a second. The cause of the trouble was the standby relay in the power supply. It has several internal make-and-break contacts that are prone to failure, usually producing the dead set symptom.

Intermittently dead: We tested it for several days before the fault showed up. Then, at switch-on, the LED display flashed '88' briefly. After that it went off and the set remained dead. On investigation we found that in the fault condition the power supply was running but its outputs were all very low – the h.t. was only 75 V instead of 152 V. Replacing several components seemed to cure the fault, but the set came bouncing back. Eventually we found that the culprit was the h.t. preset control R647. When we touched it with a screwdriver the h.t. rose to 200 V. Its wiper was clearly making intermittent contact.

Grundig cinema 9050

Intermittent no sound or vision: When the fault occurred the mains voltage didn't reach the TV power supply. Relay 601 wasn't pulling in because T6021 on the 'omniplatte' wasn't being turned on. At the top left-hand side of the set (viewed from the front) there's a microswitch that should operate when the screen is up. It was faulty.

Intermittent sound or blank raster: These projection sets use an early version of the CUC series chassis for the signals etc. and suffer from the usual power supply and i.f./tuner faults. They also seem to be prone to a problem that doesn't appear to affect non-projection sets, i.e. cracks in the main chassis, to the left of the tuner/i.f. can, where the tracks go round a large hole in the PCB. Faults so far have been either intermittent sound or an intermittent blank raster.

Washed-out picture: We decided to carry out checks on the RGB board. The supplies and the beam-limiter signals were OK but the data switching signal at pin 7 was suspect, the voltage here being 0.6 V. This point is connected to pin 31 of the abstimm (tuning) module. When this pin was disconnected the picture was restored. The 0.6 V was coming from the TMS3743 chip on the tuning board, a replacement solving the problem.

Screen drive motor problem: In this model the screen rises up out of the cabinet like Reginald Dixon at the Tower Ballroom! Usually when you switch the set on the channel display appears, the screen rises then the line timebase starts up. With this set, however, the channel display appeared, the screen began to rise, then the display went to standby and the screen went down again. If you held the mains switch down the screen would move up jerkily with the channel display constantly going from 1 to standby. The set has to be out of standby for the screen to rise. A check showed that the out-of-standby signal from the tuning board was pulsing when it should have been steady. The tuning board is powered by a mains transformer, rectifier and regulator. There was 1 V of 50 Hz ripple on the output from this arrangement. The regulator is on the Omniplatte (convergence etc.) board. As the rectifier D6013 is a full-wave type there shouldn't have been a 50 Hz ripple. Resistance checks on the transformer solved the problem: one half was open-circuit because of a dry-joint where the leads are soldered to the mains fuse board.

E.h.t. splitter had arced over: When this had been replaced there was sound but no picture. The cause of this was soon tracked down to a faulty TMS3743 chip on the tuning module: pin 5 was stuck at 2 V when it should have been at 0 V for a picture. A new chip produced a picture, but with unlocked colour running through it. The TDA3510 colour decoder chip had died. After replacing this the picture was OK and it was time to try teletext. Oh dear! Just a blank screen with an unlocked 'P100' to see. Substitution checks proved that the cause of this was the SAA5020 chip. All that was needed to complete the repair was to clean the mirrors and set up the convergence. Remember convergence?!

Convergence problem: The unusual thing was that the convergence error constantly changed. A check on the supplies to the convergence circuitry showed that the −12 V line was high, with ripple. The cause of the fault was IC6078, which is type UA79L12.

Grundig GSC100 chassis

Fold-over at bottom, cramping at top: The +D supply to the field timebase module was slightly low at 17 V instead of 18.5 V. This is derived from the line output transformer, so we checked the e.h.t. by measuring the voltage at pin C of the transformer. This was also low and adjustment of the width control HS had no effect. The width control is mounted on the e.h.t. control module and checks here revealed that R2522 (2.2 kΩ) which is in series with the control had gone high in value.

Field linearity problems: A check revealed that the supply to the field timebase panel was low at about 11 V, but changing the TDA1170 chip made no difference. Voltage checks in the line output stage showed that several of the supplies were a bit low, although not significantly. When the field module was powered by a separate 18 V supply the scanning was correct, so the problem was a supply one. It eventually transpired that fuse Si627 had gone high resistance, although not high enough to show on our initial meter check. This fuse is between rectifier diode Di627 and its reservoir capacitor C628 and was thus limiting the peak current into C628.

Persistently blew the mains fuse: This 14 inch set persistently blew the mains fuse for no apparent reason. It was a difficult one to prove: we never found any signs of an h.t. short-circuit to cause the fuse blowing. On a damp day, however, we saw the e.h.t. from the cap connector jump across to the degaussing coil. So this was the cause of the persistent failure of the 3.15 AT fuse!

E.h.t. but no picture: There was also no sound, no channel LED alight and R607 sprang open after a few minutes. The cause of the trouble was the SKE2G l.t. rectifier Di511. If in doubt replace this diode. It works very hard and can become intermittent.

Lack of height plus bad fold-over: A check on the +D supply input at pin 6 of the field timebase module showed that the voltage here was only 13.1 V instead of 18.6 V. This supply is obtained by rectifying line flyback pulses. Di627 is the rectifier diode and C628 (2200 µF, 40 V) the reservoir capacitor. C628 was obviously suspect but a replacement made no difference. Neither did a new diode. Yet the voltage across C628 was low at 13.6 V. There's little else here apart from the 1AT fuse Si627. A check showed that it was dropping 0.5 V along its length. So we replaced the fuse. This cleared the fault and the voltage across C628 rose to 18 V. A check on the old fuse showed that it had a resistance of 1 Ω. As a check we fitted a 1 Ω resistor across the fuseholder. Although the voltage dropped slightly the set still worked all right. Just to make sure that we weren't dreaming we refitted the old fuse. This brought back the original fault. Certainly an odd situation. The customer paid rather more than the cost of his new fuse.

Field collapse: First check that the 18 V supply is present at pin 6 of the field output module. If this is OK check diode Di447 – this may save you trying a replacement TDA1170 chip. We've found that a 1N4007 is a suitable diode.

Hinari

Hinari CT16

No sound: We found that Q802 went open-circuit intermittently when warm.

Dead: When we tried to power the set there was a short, timid 'sssh' noise then the set shut down. Checks in the line output stage showed that the boost diode D552 was short-circuit. It's an ERD29–06 and is situated behind the line output transformer. A replacement brought the set back to life, but the diode was very warm. We decided to use a BY229 fitted to a mica kit bolted to a very convenient hole in the chassis above the line output transistor, with fly leads for connection to the PCB. The set then worked normally, with the diode nice and cool.

Hinari CT5

Field collapse: An open-circuit resistor was the cause of field collapse. The offending item was R422, $10\,\Omega$, off pin 5 of the line output transformer. It seems to supply Q310 with collector voltage/bias, although from our photocopied circuit it's not easy to tell!

Dead set: The usual cause is failure of the remote/standby supply transformer. This set was no exception but after fitting a replacement we found that it was stuck on channel 8 and that no analogue or tuning functions worked. The relevant circuitry is mounted on the front panel, which is a real pain to remove and refit and just about impossible to work on. The cause of the trouble was several open-circuits between the front panel and the small analogue function panel to which it is, or should be, hard wired.

Intermittent field bounce: Replace the 2.2 μF, 35 V tantalum capacitor C901 in the field feedback circuit. This capacitor can also be the cause of other field faults such as cramping or a ragged picture.

Dead set: This set sounded, as Les Lawry-Johns once so perfectly described it, like a bee sprayed with freezer. It was dead as the line output stage wasn't running although the power supply was. No overload then. The cause of the problem was no line drive as there was no supply to the line driver stage due to a break in the print in the front left- hand corner of the chassis, where it fits in the runners. As a result the feed resistor R401 didn't receive its supply.

Cross-over distortion: The symptom with this set was cross-over distortion, i.e. cramping across the centre of the screen. Checks on the field output transistors Q301/Q310 and the associated biasing components failed to reveal anything amiss, although the d.c. conditions were slightly high. The cause of the trouble was the 2.2 μF tantalum field linearity feedback capacitor C307.

Intermittently distorted sound: On investigation we found that one of the audio output transistors, Q602, was leaky. A replacement cured the fault but after a few minutes it was too hot to touch. As we had another of these sets in for a replacement line output transformer (a very common fault) we checked with this one and found that Q602 should run cool. While carrying out further checks we noticed that the legs of two resistors were shorting together. A look at the circuit diagram showed that this shorted out the 470 μF audio output coupling capacitor C608. No wonder Q602 was so hot!

Hinari TVA1

Would not come out of standby: After the initial start-up this set was very intermittent/temperamental about coming out of standby. Just about everything in the power supply seemed to be sensitive to heating/freezing, including the relay. The fault cleared when a new STR5412 chip was fitted.

Intermittent failure to come out of standby: The main PCBs used in the TVA1 and CT15 are almost identical, the differences between them being mainly in the front control panel (selector board). Both sets have remote control: the TVA1 also has a built-in LED clock display and a timer. The Sentra Model GX9000 uses the same main PCB. A common problem with these sets is intermittent failure to come out of the standby mode. The basic circuit diagram shows the mains on/off switch as being on the power panel (mains fuse and filter board). It's actually on the front panel (selector board), feeding a small mains transformer, rectifier and 5 V regulator. These supply the SAA1290 tuning and analogue control chip IC001 and a pair of transistors (Q010 and Q011) that form a latch to drive relay T003 which in turn supplies 240 V a.c. to the power supply on the main PCB. If the set fails to come on, first check the mains input fuse F801 on the small mains input board and for 240 V a.c. at points 83 and 84 on the main board (brown and blue wires). Listen to hear whether the mains on/off relay clicks when the standby button is pressed. If it doesn't operate, check the 5 V supply on the front panel PCB, transistors Q010/011, the relay contacts and the relay itself. If the mains supply is present at the main board, check R801 and the bridge rectifier D801. If 310 V d.c. is present at pin 1 of the STR5412 regulator chip, check the start-up resistors R802/3 (both 100 kΩ, 0.5 W). Replace them if they have gone high in value. If they measure OK, try shunting the pair with a single 470 kΩ resistor. Where the fault has been intermittent this will probably restore normal operation, but the STR5412 should be replaced and the start-up circuit should be restored to its original state, i.e. fit a new pair of 100 kΩ, 0.5 W resistors in positions R802/3 and discard the 470 kΩ resistor.

Loss of memory: A common fault which is cured by replacing the M58655P chip. The front panel is rather inaccessible.

Drifting: Look for a faulty 33 V stabilizer on the inaccessible front panel.

Intermittent loss of sound: The set is very similar to the Hinari CT4, but as the TVA1 has a headphone socket a 1:1 isolating transformer is required. This item was the cause of the fault. One of the transformer's pins had been wrapped but never soldered.

Hitachi

Hitachi C14P216 (G7P MK 2)

Grey raster: There was normal sound but only a grey raster with flyback lines and a white line from the top to the bottom of the screen, about half way between the left-hand side of the screen and screen centre. Fortunately when we opened the set up we noticed that C711 (47 μF, 200 V), which is near the line output transformer, was bulging and had leaked. It's the reservoir capacitor for the h.t. supply to the RGB output transistors. A new capacitor and board clean-up cured the fault.

Low sound and buzzing: We wrongly assumed that the cause would be the TA8691 chip, which is usually responsible for buzzing and sound problems with these sets. On this occasion, however, the culprit turned out to be C410 (0.01 μF) which is connected between crystal MF402 and chassis. It was leaky.

Went off after 20–30 minutes: After removing the back we did the technical tap bit. This revealed a perfect dry-joint at capacitor C711, the reservoir capacitor for one of the supplies derived from the line output transformer. Resoldering it cured the problem.

No results: When we switched the set on we noticed that there was momentary arcing at the spark gap connected to the tube's first anode. This was followed by power supply shutdown. So we wound up the mains input gently, using a variac, whilst monitoring the h.t. voltage. It rose to well in excess of the correct 103 V. Further checks showed that R909 (39 kΩ) in the chopper circuit had risen in value to over 90 kΩ. It's in series with the h.t. preset VR901. A replacement restored normal power supply operation and after resoldering a dry-joint at C711 in the 200 V supply we were rewarded with first-class results.

Dead set, standby LED on: A check on the 82 kΩ resistors R902 and R903 in the start-up network showed that they had increased in value to several hundred k-ohms. Fitting replacements got the set going again.

Occasional failure to start: This set had been in the workshop twice during the past month, the complaint being that it would occasionally fail to start up, only the standby LED indicating that there was some life in it. It hadn't gone wrong during soak tests. The third time it came in the symptom was permanent. R903 and R902 (both 82 kΩ) in the bias feed to the chopper transistor were open-circuit. Note that in the fault condition the mains bridge rectifier's reservoir capacitor is fully charged, at 300 V. Use a 1 kΩ resistor to discharge it before commencing work. This will avoid accidental destruction of the power supply, turning a simple job into a nightmare.

Hitachi C14P218 (G7P MK 2 chassis)

Screen goes very bright, then the set trips: A capacitor near the line output transformer becomes dry-jointed because a part of the cabinet back pushes against it, eventually forcing it from the panel. The capacitor concerned is C711 (47 µF), which is the reservoir capacitor for the h.t. supply to the RGB output transistors. We cut the offending portion off the inside of the cabinet back – it doesn't seem to have any purpose.

Stuck in standby: A check at the collector of the BUT11AF chopper transistor Q903 produced a reading of some 300 V, so obviously the mains bridge rectifier diodes etc. were OK. Two series-connected 82 kΩ resistors, R902 and R903, provide a start-up bias for Q903. They are at the front, right-hand corner of the chassis and were both open-circuit.

Stuck in standby: As we didn't have a manual for the set we carried out a few quick resistance checks. R902 (82 kΩ) was open-circuit, a replacement putting matters right.

Dead set: Chopper transistor base resistors were the cause. There are two 82 kΩ, 0.5 W resistors (R902 and R903) that provide base bias for the BUT11AF series chopper transistor Q903. It is quite common to find that these are either open-circuit or high in value, the result being a dead set.

Dim raster: It was just possible to see a dim raster when this set was first brought in. The contrast, brightness and colour can be adjusted only with the remote control unit: because there was an open-circuit print land within this unit, the picture control action could be reduced using the minus button but the plus button had no effect. Repairing the RC unit and carrying out adjustments restored normal operation. Don't ask us why the controls had been turned down in the first place.

Hitachi C2114T (G7PS chassis)

Loss of teletext operation: The cause of loss of teletext operation with one of these sets was traced to failure of crystal X2231 in the teletext circuit. The part number is E516045.

Lost its sound after a few hours: This set also wouldn't respond to either front panel or remote control commands. We found that the TNP47C1237N microcontroller chip responded to heating and cooling. A replacement cleared the fault.

Hitachi C2118T (G7PS chassis)

No signals: The tuning display could be called up, and the bar moved across the screen when search tuning was started. But there was no variable voltage at pin TU of the tuner. A wet finger test between the 12 V supply and the tuning voltage input proved that the tuner was working, as some sort of picture appeared – albeit with hum from the finger. When we traced back to the source of the tuning voltage we found that R044 (12 kΩ), which provides the link to the h.t. line, was open-circuit.

Tripped at switch-on: At switch-on this set tripped back to standby and the e.h.t. sounded rather too healthy – on test it peaked at about 40 kV. Checks in the power supply revealed that R909 (39 kΩ), which is in series with the set-h.t. control, had risen to 45 kΩ. Luckily there was no other damage.

Dead set: This set was dead though there was 320 V on the primary side of the power supply. Experience has shown that R903 and R902 (both 82 kΩ) can be the cause of various faults, so these were replaced. The set was now in standby and no function would get it started. Further checks in the power supply brought us to the 130 V zener diode ZD903, which was short-circuit. A replacement restored normal operation. Which item went first?

HT rises, then set goes to standby: Try replacing C906 (4.7 μF, 250 V). The 39 kΩ resistor previously mentioned (R909) is the usual that it varied with each flash. The culprit, but this capacitor has started to produce similar symptoms.

Intermittent field bounce: If the fault persists after the MC7809 9 V regulator (IC703) has been checked for dry-joints, replace it. This device does get rather hot and suffers as the months go by.

Hitachi C2564TN (G10Q chassis)

Dead except for a flashing standby light: A faint tripping sound came from the power supply. There was 330 V across the reservoir capacitor

C907 but the voltage didn't reach the chopper transistor Q901, a f.e.t. type, because link FB910 was dry-jointed. The offending joint was not immediately apparent as it's hidden beneath a blob of glue. Link FB911 is also worth resoldering as it can cause similar problems.

Tripping: If one of these sets is tripping, check fuse F902 (630 mAF) on the line output panel. Failure of this fuse will result in the protection circuit operating. Check for dry-joints in the line output stage, the power supply and at the joint which earths the heatsink of IC800 on the tube's base panel when F902 has failed.

Partial field collapse: When this set was powered there was partial field collapse, a very bright screen, then tripping. It's quite a common fault. You'll find that the 1 Ω safety resistor R609 is open-circuit because IC601 is faulty. Replace them. IC601 is the TDA8178 field output chip. R609 is in the rectifier circuit that produces the 27 V supply for IC601.

Field collapse: The TDA8178 field output chip was blameless: replacing the 100 μF, 63 V flyback boost capacitor C604 restored the raster.

Dead apart from tripping squeal: We found that the EW loading coil L650 was dry-jointed, F902 was open-circuit and D704 was leaky. When these points had been attended to we were rewarded with start-up, e.h.t. rustle, then a bang and a puff of smoke from the TDA2009 audio output chip IC100. A new chip completed the repair.

Standby LED pulsing: A check on the standby supply showed cause of the trouble was the CNX82 optocoupler OPTO1, part number T548009, which relays the power on/off command to the isolated side of the power supply. The LED in OPTO1 was open-circuit.

Hitachi C2574TN

Faint purple band through picture: This was noticeable only on very dark scenes. The cause was ringing in the chroma delay line (DL501) – scope checks showed that the band was actually a reflection of the burst signal. The cure was a new delay line, part number 2164051.

Failure of EW output chip: A problem that has cropped up with several of these sets is sporadic and random failure of the EW output chip

IC751 and EW modulator diode D706. The fault is most likely to occur when the set is switched on from cold. The cause is a dry-joint at R610 on the signals panel.

Tripping: It's advisable to check the field and sound output stages in these sets as they can be the cause of tripping. In this case the sound output chip IC4451 was short-circuit. The chip usually fails because the customer has shorted the speaker leads, so advise the owner on this point when installing the set – otherwise it will be back!

Tripped, audio fault: The owner had been told by the shop at which he had bought this set that he could connect the external speakers to it and to the external amplifier at the same time. But when a loud piece of music came on the set tripped out and went dead. We soon found that the main audio output chip IC4451 (type TDA7263M) was short-circuit, so a replacement was fitted. The set then started but couldn't be tuned in. Checks showed that the correct voltages were reaching the tuner and the if strip, but there was still only noise. We then checked the supply to IC4451 and found that it was missing. It comes via Q956, which is used for on/off switching and was open-circuit. A new BD438 restored the supply and brought the set into full operation. A look at the circuit diagram showed that the switched 25 V supply also feeds a voltage regulator.

Hitachi CPT1444 (NP84CQ chassis)

Dead set, but power supply working: Check the feed resistors R710 (2.7 kΩ) and R713 (2.2 kΩ) in the supply to the line driver transistor Q702. The chances are that only R710 will be open-circuit, but replace both resistors and stand them clear of the PCB to improve the cooling.

Dead set, no supply to line driver stage: Feed resistor R710 was open-circuit.

No picture or sound but LED channel display lit: The LED channel display lit and the channel numbers could be changed. Before ordering a manual we decided to check the fuses and found that FS901 (1.6 AT) behind the on/off switch was open-circuit. No contributory cause could be found. Note that if the fuse next to it, FS902, goes open-circuit there will be no channel display and no tuning, just a snowy raster.

Dead set, fuse open circuit: The 800 mA fuse FS903, which is on a little panel between the tuner and the power supply heatsink, was open-circuit. We usually find that the cause of this is a short-circuit 2SD1453 line output transistor (Q703). Unfortunately there was also a hole in the line output transformer while Q902 (BU806), Q903 (BF422), the 36 V zener diode ZD901 and the LM317T regulator IC901 in the power supply were all short-circuit. When we got the set going we found that the tube's emission was low.

No audio output: We found that the audio output stage was working when we touched the input with a finger. Most of the sound circuitry is contained within the big TDA4503 multi-function chip IC203, however. We decided not to replace it immediately, as we've done in the past for many different faults only to find that the chip was not the cause. After various checks we applied 12 V to pin 11, the d.c. volume control pin. This brought the sound up, but turned out to be a red herring. The sound was restored by fitting a new TDA4503 chip.

Hitachi CPT1455 (NP84CQ-2 chassis)

Line tear, pulling and drift: There were also vertical shadows across the screen, similar to a misadjusted line linearity coil. The line output transformer was tried without success, also the various associated smoothing capacitors. We eventually found that the trouble was due to C214 (470 µF) which decouples pin 7 of IC201.

Vertical strations: After repairing a dead power supply we got a picture with a few vertical striations on the left-hand side. A scope check at the tube's first anode produced a tell-tale pulse and ring display. Adding extra capacitance did nothing to help so the earth print had to be open-circuit. There was an almost invisible crack on the c.r.t. base panel. It had been caused by the drag of the leads to chassis while I had the chassis out. I've been bothered by this sort of thing several times recently, on new sets of various makes, especially on very large panels.

STR4211 power supply chip short-circuit: When this happens the 2.7 V protection zener diode also fails – make sure you check it when you have a faulty power supply. When the defective parts had been replaced there was a nice picture but the sound was distorted. The collector of Q421 was at only 40 V because the feed resistor R423 (56 Ω) had gone high in value. You get a similar problem with the Amstrad CTV2000. When the h.t. goes high R313 suffers giving very distorted sound.

'Faulty on/off switch': The set was dead, as these usually are when they come in. The STR4211 chip in the power supply had failed. A replacement was fitted along with the precautionary bits (see Hitachi data and sheets) and the set was switched on. The sound was poor, as if the loudspeaker was off centre with coil rubbing. It was also extremely weak. The 56 Ω resistor that feeds the two output transistors was open-circuit and one of the transistors was short-circuit. Nothing unusual so far. The faulty items were replaced along with the 1N4148 diode. This produced clean sound, in a way, until the volume control was at about midsetting. At higher settings there was no increase in volume and back came the distortion. The driver transistor was OK and all the voltages were about right. In despair we hooked up the scope and fed a sinewave into the TDA4503 chip. At above the half-way setting of the volume control the output became distorted. A new chip cured the fault. It didn't cure the customer when he read his bill for 'a switch'.

Dead set, power supply noises: We found that the protection diode ZD953 was short-circuit. When this had been replaced a check showed that the h.t. varied between about 80 V and the point where ZD953 again went short-circuit. The cause was traced to C908 (4.7 μF, 160 V).

Sporadic brightness variations: We found that the 200 V h.t. supply was unstable because C761 (4.7 μF, 250 V) was defective.

Hitachi CPT1471

Mk 1 set tripping, with arcing noises: The complaint was of tripping and going dead, accompanied by ticking or arcing noises. This was caused by the usual power supply problems. A kit is available, including the STR6020 power supply chip, the field chip feed/supply transistor etc. After fitting these items the set gave us new symptoms. Although the power supply was now working well the field scan was accompanied by text lines and the flyback blanking was very poor. The field generator and driver stages, also the line oscillator and blanking circuits, are all within the LA7801 chip. Replacing this restored normal operation. It presumably died during the power supply's failure.

Partial field collapse: The cause was dry-joints at the legs of the STA441C field output chip.

Intermittently go dead: This set would intermittently go dead, sometimes a few minutes after switching on or at other times after many hours' use.

Just prior to the occurrence of the fault the set would make a sound that was not unlike arcing in the line output transformer. But the transformer checked out OK when tested. A slight increase in the h.t. voltage occurred just before the shut down. Replacing the STR6020 chopper chip IC901, using plenty of heatsink compound, cured the fault.

Hitachi CPT1474 (NP84CQ4 chassis)

Very intermittent loss of sound: The fault was sensitive to movement of the board around the audio output area. Everything here was soldered but this was no good. Use of a scope then showed loss of signal from the output transformer, the input being OK. The problem wasn't dry-joints on the lead-out wires but an open-circuit secondary winding inside the housing. A new transformer put an end to the trouble.

Bright raster, no sound at switch-on. After 3 seconds the set tripped and screamed: When we next switched the set on we turned down the first anode control. This time the set remained on but still had an overbright raster and no sound. A check at the collectors of the RGB output transistors showed that the voltages were only 40 V, although there was 200 V at the line output transformer. It seemed that the RGB output transistors were conducting heavily, so the drives from the TDA3565 colour decoder chip IC501 were disconnected. Up came the collector voltages and out boomed the sound. A new TDA3565 cured the problem.

No picture, no sound, no field scan: This was a Model CPT1676. The 12 V supply was found to be missing in places on the chassis. The cause was an open-circuit $0.47\,\Omega$ safety resistor, R228. We've had the same problem on the CPT1474 where the culprit was R222 (same value).

Vertical rolling: The cause of vertical rolling was traced to C906 ($0.2\,\mu\text{F}$) in the field oscillator circuit. It measured OK when checked with a capacitance meter, but a replacement cured the fault.

HT instability: The HT voltage was also low. Replacing C910 and C919 (both $10\,\mu\text{F}$) cured the trouble.

Hitachi CPT1476 (NP84CQ4 chassis)

Teletext lines over top third of picture and field bounce: We decided to check at the field flyback blanking pulse source, which is pin 7 of the field output chip IC601. Sure enough there was a dry-joint here. Pins 6 and 5 also looked as though they could do with attention. Resoldering put matters right. Field bounce and similar trouble with these sets has been traced to dry-joints around this chip. It appears that the pins push up through the print due to heat expansion. It's best to melt and resolder all seven pins at the same time.

No colour: We found that the voltage at the colour control pin 5 of IC501 was only 0.7 V. It should vary between about 1.8 V for minimum colour and 3.8 V for maximum colour. Correct voltage readings were obtained at the slider of the colour control, the cause of the fault being the decoupling capacitor C507 (2.2 μF) which was leaky.

Field collapse: Careful examination of the field output chip's terminations revealed a number of suspect looking joints. Resoldering them restored the field scan.

Programming the Hitachi CPT1476R series

Introduction

Microcomputer chips are now widely used to control various functions in TV sets. The extent of the control operations varies from just volume and tuning in simpler sets to control of the signal processing and deflection circuitry via master and slave devices and a digital bus in all singing, all dancing models. Unfortunately when problems occur with microcomputer chips the symptoms can be many, depending on what they control. The days when faults like no sound or field collapse were nearly always due to a simple cause in the relevant output stage are passing. Nowadays we may have to deal with new fault conditions where the microcomputer chip or a slave device is the cause. All manner of weird and wonderful faults can be introduced by microcomputer control.

Hitachi has used microcomputer control for some years now. Generally, as with other manufacturers, the systems have been very reliable. There is, however, a particular problem you get with the CPT1476R/1646R and CPT2196/2198/2578 ranges. When the problem occurs, the microcomputer chip tells its separate memory i.c. to open its memory. The LED display changes from the programme number to the letters 'CH'. As a result the customer thinks 'what's up?' and starts to press the buttons on the remote control unit in an attempt to restore the programme number display. This doesn't have much effect as far as the viewer can see, but most of the numbered buttons that have been pressed will have set or reset a particular bit in the microcomputer control system's memory. Eventually the customer gets bored with trying to get rid of the erroneous display and then switches the set off. When the set is next switched on the display is back to normal and the customer thinks no more about it. Until, that is, the set starts to do odd things because of the corrupted information in its memory, things like going into standby after a couple of minutes, not going into the teletext mode when asked, not tuning in, remaining permanently in the video or RGB mode, or lighting random characters or dots in the LED display.

Reprogramming

The only remedy is to reprogramme the memory restoring its contents to their correct states. To do this you have to dismantle the remote control unit and add a programming switch between two pins of the chip. With the CPT1476R and CPT1646R pins 15 and 23 of the SAA1250 chip IC1 should be soldered to a pushbutton switch. With the larger-screen sets the chip is an M50467: wire pins 12 and 20 to the switch. Figure 1 shows the switching arrangements.

When the switch is pressed the LED display should show 'CH'. Press the switch again and the display should change to 'OP'. Now press number 1 on the handset producing 'P1' in the display. Press the standby button on the handset and the set will go into the standby mode and enable the memory's contents to be altered.

Press the store button on the set itself and hold it in. While doing this press button number 9 on the handset. The set should now come on with channel number 9. If there's a dot in the display this must be removed before programming can commence. Other channels should also be checked to see if a dot is illuminated. To remove the dot, select

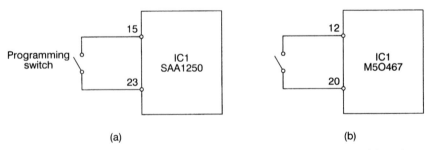

(a)   (b)

Figure 1 *Adding a programming switch to the remote control handset, (a) with the small-screen models, (b) with the larger-screen models*

a channel which doesn't have an illuminated pot and press the store button. With the store button still pressed, use the handset to select the channel number with the unwanted dot. The display should then show two flashing bars followed by the channel number which should now be dot-free. If the offending dot is still present, repeat the procedure until it's extinguished.

The programming procedure depends on the model.

Small-screen models

With the small-screen Models CPT1476R/1646R, start by pressing the programming switch three times. The display should then be as shown in Figure 2(a). The handset numbers correspond with particular segments in the display: pressing the relevant number will light or extinguish, i.e. set or reset, a particular segment of the display and memory location in the i.c., see Figure 2(b). Press the relevant buttons until the display is as shown in Figure 2(c), i.e. with byte 1 segments 1 and 2 of the bit display are lit and segments 3, 4, 5, 6, 7 and 8 are out.

Once this first step is correct, pressing the programming switch again should advance to byte number 2 as shown in Figure 2(d): press the appropriate buttons to light segments 1 and 2 of the byte 2 display.

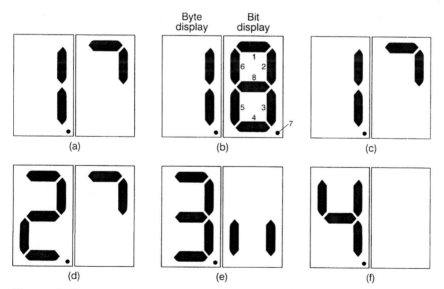

Figure 2 *Programming displays for Models CPT1476/1646*

Press the programming switch again to advance to byte 3, then press the appropriate buttons to obtain the display shown in Figure 2(e): with byte 3 only segments 3 and 5 should be lit.

Once this is correct, press the programming button to advance to the last byte, number 4. With this byte the programming is correct when no bits are lit, as shown in Figure 2(f).

When all the bytes have been set correctly press the handset's standby button to fill the memory. The set will then go to standby. Programming is now complete and the set can be turned on and tuned in.

Large-screen models

The procedure with the large-screen Models CPT2196/2198/2578 is as follows. When the programming switch has been pressed three times the display should be as shown in Figure 3(a). The handset numbers correspond with the same bit segments as with the small-screen models, see Figure 2(b). Thus with byte 1 bits 1, 4, 5 and 7 should be lit/set.

Press the programming switch again for byte 2 and set the bits as shown in Figure 3(b). Press the programming switch once more for byte 3 and set the bits to obtain the display shown in Figure 3(c). Press the programming switch the final time for byte 4 and set the bits as shown in Figure 3(d).

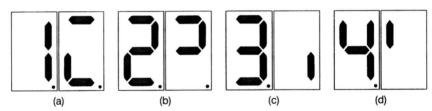

(a)   (b)   (c)   (d)

Figure 3 *Programming displays for Models CPT2196/2198/2578*

When everything has been set correctly press the handset's standby button to fill the memory. Then turn the set on and tune in the channels.

Recommendation

Once these procedures have been carried out the set should be in good working order. It's probably advisable to check the memory's contents whenever one of these sets comes in for repair of any sort of fault, as the memory could have been corrupted without any obvious symptoms being shown – until a later date, when a call-out might be needed.

Hitachi CPT1646 (NP84CQ chassis)

Tuning problem: This was a new set. When we tried to tune in the stations the raster would suddenly flash white and the video vanish as the correct tuning point was approached. The field blanking period would then roll down the screen and a moment later a normal picture would return. This effect pulsed on and off about once a second. Reducing the brightness and contrast control settings was not very helpful and it was impossible to store any station in order to make a detailed study of the a.g.c., a.f.c. etc. While attempting these latter two checks we got hum bars with wavy verticals and weak contrast between the bars. This effect disappeared as quickly as it came and didn't return. Lengthy checks were made on the decoupling and we discovered that reducing the first anode voltage from 470 V to 440 V markedly improved matters, though the picture was then too dark. Flicking the contrast suddenly to fully up seemed to provoke the fault, while on low contrast scenes reducing the contrast and brightness way down cleared it. These things suggested a fault in the beam limiter circuit. Replacing D702 and D703 improved matters noticeably but didn't provide a complete cure. C718 (10 µF, 50 V) measured OK on the Avo but fitting a replacement finally cured the problem.

No picture, no sound, no field scan: The 12 V supply was found to be missing in places. The cause was an open-circuit 0.47 Ω safety resistor, R228. We've had the same problem with Model CPT1474 where the culprit was R222 (same value).

Dead set, standby light on and channel 1 shown: This set was dead with number 1 and the standby LEDs alight in the display. If the programme up or down button was pressed the number changed but the standby LEDs remained lit. It seemed that the microcomputer control chip thought that the set was in standby and on at the same time: very strange! We followed the standby line from the chip. This brought us to the BC548B standby switching transistor Q1455, where the voltages showed that something was definitely amiss. The transistor was open-circuit base-to-emitter. A replacement restored normal operation.

Hitachi CPT1646R (NP84CQ chassis)

Colour was pulsing: The pulsing effect was very faint, but the set belonged to a sports club and apparently the fault was quite noticeable with football. When a colour bar pattern was fed to the set a very faint shimmer was visible in the green and blue bars. It was more noticeable with the colour turned up and the contrast turned down. Several checks were made around the colour decoder chip but we were unable to arrive at a diagnosis. The chip was changed of course but the fault persisted. So more checks were made and various components were replaced, all to no avail. Then, by chance, we noticed that when fine tuning the station the colour seemed to pulsate in sympathy with the flashing LED display. The problem disappeared when the supply to the display was disconnected. But rather than replacing the display we decided to fit a new SAA1293 remote control decoder chip (IC1501). This provided a permanent cure. We've had a few faulty SAA1293 chips.

Dead set, e.h.t. died: When we switched it on the e.h.t. rustled up then very quickly died away. An LM317T three-terminal adjustable regulator is used in this chassis, with a large bypass resistor. A check on the LM317T showed that all pins were at about 150 V. We tried powering the set via a variac and found that at about 70 per cent of the full mains input everything worked all right. So no other damage had been done. A new LM317T was tried, on the assumption that the original one was short-circuit, but the fault condition remained the same. As the voltage at the adjustment pin couldn't be reduced using the set-h.t. control, the cause of the fault was likely to be in this area. Tracing back from the adjustment pin brought us to two resistors, R907 (56 kΩ) and R908 (22 kΩ). R908 turned out to be open-circuit.

Wouldn't search/tune stations: Here's a good one for the unwary. On being called out to retune a customer's TV set and VCR we discovered that the TV set wouldn't search/tune stations. The front panel buttons were found to be misbehaving, which seemed to indicate a system control problem. After checking on the price of a replacement SAA1293H remote control decoder chip we decided to look elsewhere before committing the customer to great expense. Our notes on this model suggested that memory corruption could be the cause of the trouble. To solve this one see the entry under 'Reprogramming the Hitachi CPT1476R Series' above. When we carried out this procedure things returned to normal.

Picture roll: On test we found that the e.h.t. trip thyristor would operate and remove the line drive. So we disconnected the trip and ran the set up using a variac. This showed that the h.t. was high at 113 V instead of 103 V. The h.t. could be set up correctly, but the control was at the end of its range. Further checks showed that R908 in the regulator circuit had risen in value from 22 kΩ to 27.5 kΩ. Replacing this, readjusting the h.t. and reconnecting the trip completed the repair. The fault could also occur with Models CPT1444 and CPT1446.

Hitachi CPT2050 (Salora J chassis)

Erratic height variations: This multi-standard Salora set had erratic height variations with simultaneous changes from pink to green faces. Pin 12 of the TDA2653 field chip provides a 50/60 Hz control voltage for system switching. At 50 Hz field frequency the voltage should be 0.2 V. It was fluctuating wildly, which confused the decoder. Fitting a new chip cured the height problem and restored pink cheeks.

Intermittent green faces: This chassis uses a TDA3562A colour decoder chip, with the PAL switching controlled by the sandcastle pulses. As the pulses didn't change when the fault was present we suspected the TDA3562A, but a replacement made no difference. Neither did replacement of the components connected to pin 2. While we were studying the fault we noticed that the field height increased when the ident dropped out. This was the clue that led us to the cause of the fault. The TDA2653A field timebase chip detects the frequency of the sync pulses, either 50 Hz or 60 Hz, putting out a voltage at pin 12 to adjust the height and switch the colour decoder chip to NTSC operation as appropriate. For PAL operation the voltage at this pin must be zero. When it rises above 12 V the set changes to NTSC operation. In this particular set the voltage at pin 12 varied intermittently. Fitting a new chip restored correct operation.

Set displayed only red chrominance: There was no G, B or Y picture content. Because the set was not set up correctly the symptom was not what you usually get (no picture) when the top-end resistor in the potential divider that supplies the tube's first anode goes open-circuit. A new 470 kΩ resistor plus setting up restored a super picture. A new volume control was also required: the original one was so noisy that the sound went up as you turned the control down past half way.

Hitachi CPT2078 (NP83CQ MK 2 chassis)

Works OK for 10 seconds to 10 minutes: This set would switch on and work normally for a period that varied from 10 seconds to about

10 minutes. Then the seven segment display would go out, the picture and sound would disappear and the three LEDs on the front would light up. A check in the fault condition showed that there was no 5 V supply at pin 20 of the MAB8422 microcontroller chip. This supply comes from a bridge rectifier and voltage regulator on the CITAC panel. The 7805 regulator had only 3 V at its input. Substitution revealed that the bridge rectifier diodes were breaking down, though they read OK when cold.

Blank screen, no sound and the green video LED lit up: The cause of the trouble was that the 12 V rail was low at 6 V. We found that R716, which is connected to pin 3 of the line output transformer as a surge limiter, read 72 Ω instead of 2 Ω as specified in the manual. It's the first time we've come across a change of value with a safety resistor.

No sound or picture: Not caused by failure of the line output transformer on this occasion. The set was stuck in the AV mode. When the AV switching link from the CITAC panel was disconnected the signals were restored. We found that the 12 V zener diode ZD1438 was leaky.

Strange picture, no colour: The verticals were bent and smeared. The teletext panel was not to blame, but the picture was restored to normal when the AV panel was bypassed. The cause of the trouble was the HEF4066 chip IC1201.

Hitachi CPT2158

Turn-on required repeated pressing of on/off switch: This set led us a dance over some months. The initial complaint was that repeated pressing of the on/off switch was required before the set would come on. But no fault could be found during its repeated visits to the shop. Then on one occasion we found that the 2SD1877 line output transistor had failed. In fact it would fail after about four weeks, and the customer continued to complain about the need for continued on/off operation. Many components were replaced and dry-joints resoldered. We ran the set daily for three weeks in the shop, but the 2SD1877 failed only at the customer's house where the set was being run from a three-way adaptor, along with an electric fire and a table lamp. With thoughts of the G11, we felt that this could be the cause. We sold the customer a four-way trailing socket, but the set continued to come back. A chance discovery led us to the cause of the fault. If the set was left in a very cold place it would trip at switch-on, with a large e.h.t. surge. Since the crowbar Q706 is fired by excessive line pulses at pin 1 of the LOPT, it seemed likely that the power supply's output varied with temperature. We connected a 100 W bulb across the power supply's output as a dummy load and found that the 152 V output could be made to rise to 250 V if C917 was cooled. This capacitor is part of the negative feedback conFol loop: when the feedback falls, the output rises to compensate. In the interests of temperature stability we decided to replace C917 with a 4.7 μF tantalum capacitor. Note that it's shown the wrong way round in the manual: D905's anode should go to the negative side of the capacitor.

Blank raster, no sound or picture: All the supplies were OK. The cause of the fault was the TDA4505 chip. Make sure you get the right one: some can give no sound or no field sync.

Odd sound fault: This set, for which we didn't have a service manual, had a very odd sound fault. The sound was low, with an audio tone that whistled away in the background. When the volume up/down buttons were pressed, the tone would change to a series of bleeps that varied in tone and frequency. If you have ever listened to a computer game on a tape recorder, this is exactly what the set sounded like. A squirt of freezer on IC402 (AN5836), which we assumed was involved with volume control, slightly improved matters. So a replacement was ordered. When it was fitted, the fault remained the same. Disconnecting the L and R inputs made no difference. Neither did disconnecting the inputs to the audio output chip. This suggested that the fault was still to do with IC402. But voltage checks seemed to be inconclusive. Pin 2 was at 4.9 V, which seemed about right if the device was supplied from a 5 V line. As we'd gone as far as we could, we turned out the drawer of Hitachi manuals to look for another set that uses the AN5836 chip. We found it in Model C25-P238, and the manual provided us with the information we needed. The chip is a stereo tone control and preamplifier device which should have a 12 V supply at pin 2. When we traced the print back we came to R926, which had 15 V at one side and 4.9 V at the other. R926 is connected to two transistors, Q903 and Q904, which are presumably part of a regulator circuit. Q904 had no base voltage, because the 12 V zener diode ZD902 was short-circuit. Replacing this item cured the fault.

Hitachi CPT2174 (G6P chassis)

Dead set, squealing power supply: There was a short-circuit reading across the line output transistor. On this occasion the line output transformer was the cause.

Intermittent loss of colour: Application of slight pressure to any component in the vicinity of the chroma delay line DL501 made the colour come and go, so we replaced this item, to no avail. We eventually discovered that pin 30 of the HA51338SP timebase generator/colour decoder chip IC501 had never been soldered – the pin was making barely perceptible contact with the print. The amazing thing is that the set had given no previous trouble.

Varying sound level: Sometimes the sound would go up to full volume. It behaved impeccably on the bench for the best part of a week, however. Tapping and freezing made no difference, and no dry-joints were revealed by a visual inspection. A few days later, 10 minutes after switching on, the fault was very active. A check on the stabilized 12 V supply then showed that it was varying between 12 V and 13.4 V, the volume rising and falling in sympathy. Zener diode ZD791 (HZ11LC2) in Q791's base circuit was the cause of the trouble. It was hidden beneath a large blob of glue. After fitting a replacement a soak test confirmed that the set was now OK.

Came on with a white raster but tripped after 30 seconds: We found that R771 was open-circuit and D763 leaky. These are the surge limiter and rectifier respectively in the h.t. supply to the RGB output transistors. A BY500 works all right in position D763.

Dead set, purr from power supply: The cause of the trouble was eventually traced to the line driver transformer. Its body is used as an earth return, and one of the lugs was dry-jointed. Resoldering cured the fault.

Hitachi CPT2176 (G6P chassis)

Tuning information can't be stored: You can tune the set in but when you try to store the station by pressing the preset button the picture goes off. With the memory chips used in modern sets a negative supply is usually obtained from the line output transformer. With this Hitachi model the supply is −30 V and the memory chip is IC1102. We've found that R772 goes open-circuit: it's a 1.2 kΩ, 0.5 W resistor off the line output transformer.

Dead set, low whine from PSU: We found that the voltage at the collector of the line output transistor Q781 was low. The cause was C909 (220 μF, 160 V) which smooths the output from the chopper power supply.

Dead set, line timebase whistle: We'd experienced this fault many times before and had no hesitation about replacing the line output transformer which once again restored normal operation.

Intermittent failure to switch-on: The power LED was off and there was a gentle squeal from the power supply. The cause of the problem was dry-joints on the line driver transformer. Not on the operational pins but on the lugs that secure the transformer's metal frame to the PCB lands – they are used as an earthing link between two sections of the circuit. We've also encountered this with other makes like Tatung, Finlux and ITT.

Set was tripping: An unusual fault with this chassis – the power supply usually either works or it doesn't. Surprisingly the h.t. voltage was constant at 110 V. What was actually happening was that the line drive was pulsing. It comes from the M51338SP colour decoder/timebase generator chip IC501, and checks here showed that the start-up voltage at pin 24 was pulsing. This comes via D913 and Q904 which are on the secondary side of the power supply. Q904's base is in turn under the control of the standby switching transistor Q905. We found that this transistor was being switched on periodically, in turn switching Q904 off so that the set couldn't start up. When Q905's base was disconnected the set started up and everything worked correctly, except that there was of course no standby operation. So we moved across to the microcontroller chip IC1101 and connected an oscilloscope, on a very low-speed setting, to pin 30, the power-control output that goes to the base of Q905. The voltage here could be seen to rise to 5 V, stay there for a second or so, then return to zero. It seemed logical to suppose that the cause of the fault lay in the microcontroller chip's start-up circuit or reset line. The latter is connected to pin 16 and is controlled by transistor Q1102. At switch-on the 5 V rail rises and C1101 (2.2 µF) in this transistor's base circuit charges. The sequence is that Q1102 then switches on, C1101 discharges, Q1102 switches of and IC1101 resets. What seemed to be happening was that the reset line was being held low for too long. This curious fault was cleared when we replaced C1101.

No station memory: Before condemning the memory chip IC1102 we decided to check its supplies and found that the −30 V rail was low at only −2 V. Tracing the source back brought us to R772 (1.2 kΩ) which had risen in value to over 900 kΩ. A replacement put matters right.

Hitachi CPT2188 (Salora K chassis)

Dead set, distorted waveform at the base of the line output transformer: There was also an almost identical waveform at its collector – a check showed that its base-collector resistance was only 85 Ω. We didn't have the original type (2SD1577) in stock so we decided to try a BU508A instead. The 2SD1577 has an insulated body, so its heatsink is soldered to chassis. The BU508A was therefore mounted on the healsink using a conventional mica insulator, spacer and bolt system. A long test run showed that the transistor ran cool.

Line tearing and screen would go bright blue intermittently: There was line tearing and the screen would intermittently go bright blue. Was it a single fault or two separate ones? At first I suspected faulty decoupling. The 12 V line goes to the sync/timebase generator and the colour decoder chips, but it was rock steady with an insignificant line-frequency content. What else could be common to the two symptoms? The sync/timebase generator's validate pin was the answer. It's pin 13 of the TDA2579 chip ICB500. With a 50 Hz signal there should be 11 V here, falling to 7.5 V with a 60 Hz signal and 0 V with no signal at all. A check showed that the voltage was stuck at zero even though the set was tuned in, with sound and a picture present. This line also goes to the auto grey-scale correction circuit in the colour decoder chip ICB200, hence the video fault. A new TDA2579 chip cured the trouble.

Intermittent failure to start: The owner overcame the problem by leaving the set in standby all the time. Eventually the set refused to start from standby, so he had to bring it in for repair. We traced the cause of the trouble to the fact that C604 (1000 μF, 25 V) was open-circuit.

Hitachi CPT2198 (G8Q chassis)

Intermittent failure to switch on from cold: Replacing TH902 in the start-up circuit restored normal operation.

No life in line output stage: The sound was OK and the channel indicator was lit but there was no life in the line output stage. A check at the base of the line output transistor showed that there was no drive here. We had to check back to the TDA2579 timebase generator chip whose line drive output at pin 11 was missing. Replacing this i.c. restored normal operation.

Complete failure: The cause was cracked print at the chopper transformer.

Hitachi CPT2218 (NP81CQ chassis)

Teletext would suddenly come on: The customer complained that after about half an hour a teletext display would suddenly invade the screen of her set, obliterating the picture. This would happen on any channel, and after a moment or two the set would go into standby. If left for, say, a quarter of an hour everything became normal. The problem was said to occur only at night. Repeated tests in the workshop, even using the hair dryer on the text and interface panels, showed nothing amiss. So we resorted to drastic measures. We put the set in a large plastic bag with a thermometer visible and blew air from the hair dryer in at the bottom. At 90°F the fault really did happen. Spraying IC2101 (SAA1272) with freezer cured the trouble, and reheating it confirmed that it was the culprit. So we changed it and returned the set. Next day the customer reported that the set was no better. Repeating the bag and hairdryer treatment again showed that she was right. This time IC2101 was blameless and it was the SAA1251 that caused the trouble. Soon after another of these sets arrived. This time the teletext thing crept slowly over the picture from the right, pushing the picture away. After our previous experience we decided to replace the SAA1251 without further ado. It did the trick.

Set would not come out of standby: All the legs of the chopper transformer T901 had hairline cracks at the solder connections – they were visible only when a strong magnifying lens and lamp were used.

Field collapse: We had replaced the HM6251 field output module some months previously. A scope check at pin 4 produced a field-rate square wave. Odd – this could only mean that the set was running without the effect of the field coils' inductance being present. So we checked the 220 μF, 50 V field scan coupling capacitor C610. It had dried out, a replacement restoring normal operation.

Hitachi CPT2478 (G6P chassis)

No text graphics: Teletext operation was OK. Now it's worth pointing out that transistors Q858 (green-on-screen) and Q857 (red-on-screen) on the c.r.t. base panel do give trouble. But not this time. The character generator chip IC1104 was OK, but D1113 and D1114 were both leaky. Replacing them restored the on-screen graphics but a week later the same fault occurred again. We replaced the two diodes and left the set soak testing. After a couple of days the e.h.t. arced over: so this was the cause of the diodes' downfall! We cleaned the final anode cap and so far (three weeks later) the diodes are still OK.

Dead set: Checks in the power supply showed that there was a low resistance between the base and emitter of the chopper transistor Q901, although it proved to be OK when checked separately. An internal short could be measured between pins 5 and 6 of the HM9205 chopper control/drive module CP901. Replacing this item restored normal operation.

Sometimes no text or jumbled text: It didn't take us long to discover that the 12 V input at pin 2 of PL1204 on the text board measured only 10.5 V. R761 had risen in value from the correct $1\,\Omega$ to $5\,\Omega$.

Hitachi CPT2488 (Salora K chassis)

At switch-on picture blanked out except for a narrow strip at the top and bottom: As the set warmed up the blank area shrank, one line at a time, towards screen centre. Within 20 or 30 minutes the blank bar had disappeared, leaving a normal picture. All this was caused by C574 (100 μF, 40 V) in the field timebase area being low in value.

No results: It was reported that the set would shut down when the VCR was put into the search mode, and then only when the VCR was operating in the LP mode. We confirmed that this was the case, the set going completely dead when search was selected in the LP mode. The standby light would remain off. After a few minutes the set could be powered up again by using the on/off switch. Though the symptom was strange, the cause of the trouble was simple. The h.t. was too high! When the set had warmed up, the h.t. rose from the correct 144 V to 160 V. The line output stage was unable to handle the LP mode tracking bars with this high h.t. Checks showed that R952 in the power supply had risen in value from $68\,k\Omega$ to $74\,k\Omega$.

Picture slow, centre blanked: The picture was slow to appear and when it did the centre was blanked for several minutes. The cause was C574 (100 μF) in the field output stage. We wouldn't recommend using freezer to prove the point, as this results in the sudden demise of the field output chip!

Hitachi CPT2508 (G7P MK 2 chassis)

Dead set, 2.5 AT mains fuse F901 blown: The usual cause is the 4.7 nF, 1 kV disc capacitor C919 in the chopper transistor's snubber network. Use the correct type from Hitachi as others are unable to handle the spikes present at the collector of the BUT11AF chopper transistor. If you use the wrong capacitor you'll end up with a blown TDA4601 chip (IC901), BUT11AF transistor (Q901) and the two 150 kΩ and 120 kΩ resistors R931 and R932. For no start-up check the 82 pF capacitor C911 connected between pins 2 and 3 of the TDA4601 chip.

No life: The cause was the fact that the main surge limiter resistor R901 (3.9 Ω, 7 W) was open-circuit while C919 (4.7 nF, 1 kV) and C928 (2.2 nF, 2 kV) were both short-circuit. Both capacitors had split, burnt bodies. C919 is in the snubber network while C928 is connected across the chopper transistor.

Dead set, power supply fault: It didn't take us long to find that the little blue capacitor (C919, 4.7 nF 1 kV) in the chopper transistor's snubber network had split open. In addition the 3.9 Ω, 7 W surge limiter resistor R901 was open-circuit. Strangely neither the chopper transistor nor the mains fuse had been damaged.

Dead set, power supply fault: Replacing the TDA4601 chopper control chip and the usual high-value resistors (R931 and R932) failed to produce a cure. We then found that the chopper transistor Q901 has a 1N4148 diode (D904) connected between its base and emitter. It had gone short-circuit, a replacement restoring the set to life.

No results: There was only 33 V at the collector of the line output transistor. This was found to be the result of the mains rectifier's 150 μF, 400 V reservoir capacitor C909 being open-circuit.

Dead set: There was 320 V at the collector of the chopper transistor Q901 and 7.7 V at pin 9 of the TDA4601 chopper control chip IC901. A scope check at the base of Q901 showed that start-up was attempted but lasted for only a few pulses, indicating an overload. We found that the RU2M 20 V supply rectifier D911 read about 100 kΩ in the wrong direction. Dry-joints were evident at the chopper transformer. As the set has separate speakers, we didn't try the sound until later. We then found that the LA4270 audio output chip had a hole in it.

Hitachi CPT2578 (G8Q chassis)

Blank raster and no sound on some channels: There was normal sound on other channels. Reprogramming normally provides a cure, but after doing this channels 4 and 5 remained blanked out. The problem was solved as follows. Select the blanked channel (in this case 4), press the tune/a.f.c. button, then the volume minus button and the store button. This will produce a normal raster. Repeat the procedure with the other blanked channels. Finally press the tune/a.f.c. button.

Blank raster and no sound: Checks showed that the 12 V output from the 7812 regulator IC932 was low. The chip was all right, however, the cause of the trouble being C933 which smooths its input. A new 2200 μF, 25 V capacitor restored normal operation.

Channel tuning problems: The MDA2062 (blue spot) memory chip IC1502 is usually the cause. If the complaint is no teletext, however, suspect the memory chip first, not the teletext decoder which is very reliable.

No sound: The set's audio section seemed to be OK, and pin 34 of the SAA1293H chip ramped up and down as the volume control was operated. But the voltage at pin 12 of the d.c. volume/tone control chip didn't. Replacing this chip made no difference, however. The TDA2579 timebase generator chip is linked to the chip via D701 (1N4148), presumably to mute the sound on channel change. D701 was found to be leaky, measuring about 200 Ω. We had to replace the TDA2579 chip as well as D701.

Thin wavy lines with diagonal bands of interference: 'Thin wavy lines, with diagonal bands of interference,' the lady said. 'Have you got a satellite receiver,' we asked? 'Yes' came the answer. 'Oh, we know what that is, madam, we'll put that right in a jiffy.' Did we? No! When we looked at the set the symptom was just like a VideoCrypt patterning problem. Only it was just the same when the satellite receiver was switched off. We had to take the receiver back to the workshop, where we eventually found that the cause of the trouble was l.t. ripple. C932 (470 μF, 16 V) was responsible. The symptom was more easy to see in the text mode.

Hitachi CPT2656 (Salora K chassis)

Dead with the supply line voltages low: Every one of these sets that has come in for repair has had exactly the same fault: dead with the supply line voltages very low and a slight chirp from the power supply at switch-off. The cause has every time been slight reverse leakage in the BY228 diode D508 in the line output stage. We use a DG3P diode as a replacement.

Finlandia badge – bottom part of picture missing: This set wore a Finlandia badge but we were able to match it up with one of our Hitachi service manuals. Over a period of eight months it has been back to the workshop on several occasions, but each time it failed to display any fault during a soak test. The customer's complaint was that the bottom part of the picture was missing. Despite replacing many components in the field output stage the set kept on coming back. It was difficult to know what to do as we'd not seen the fault. On its latest visit, however, the fault put in an appearance: after about an hour the bottom of the picture began to cramp up while the top widened out. By feeding a cross-hatch pattern signal to the set we could see that the actual symptom was change in linearity. A slight touch on preset RTB573 (470 Ω) cured the fault. So a replacement linearity potentiometer was fitted and the set was handed back to the customer with confidence.

Would not come out of standby: This set was OK in standby, but died when any attempt was made to switch it on. Checks in the line output stage showed that DB506 (BY228), one of the diodes in the EW modulator circuit, was leaky.

Hitachi CPT2658 (Salora L chassis)

Severe herringbone patterning: The symptoms occurred both with off-air and scart input signals. There were striations down the side of the picture, and both the line phase and frequency were wrong. In cases like this it's usually a good idea to tie the faults down to one part of the chassis. But it was a bit difficult here because separate faults seemed to be present in the i.f./video, line oscillator and line output departments. The only thing in common of course is the 12 V supply. A check was made for ripple but none could be detected. We decided to replace CB616, the 1000 μF reservoir capacitor for the 17 V supply that feeds the 12 V regulator, and as if by magic all the symptoms cleared.

Intermittent loss of picture: There was a very intermittent fault with this set, loss of the TV picture with the screen going very dark. No amount of kicking, tapping, freezing or frying would bring it on. Luckily it eventually put in an appearance for a decent length of time and we were able to pounce. The cause seemed to lie right at the beginning, in the control system, not on the main panel. In the text mode there should be a steady 4 V at pin 9 of IC13. In the mixed mode this voltage should pulse and it should be zero in the TV mode. When the TV picture vanished the 4 V was present. As the 5 V supply was present we suspected the chip and consulted our helpful Hitachi distributor. He pointed out that the chip is expensive and that they'd had a few odd video faults that a new chip didn't cure. He suggested replacing the 17.73 MHz crystal XTG2 – and proved to be dead right!

Appeared to be stuck in video mode: Sure enough there was a blank raster and no sound. The set wasn't actually stuck in the video mode, however, because a video signal fed in via the scart socket produced a good picture on channel 0 only, proving that the video switching stages were working correctly. A quick check on the tuning voltage line suggested that the SAA1293 remote control chip was also probably doing its job correctly. As all the voltages around the CD4053 switching chip ICB102 were in order in the video and TV modes we made our way to the TDA4505 chip which contains the i.f. circuitry. A replacement restored good pictures in the TV and video modes but a new problem was present. There was grossly excessive height and the operation of the height control had become very one-ended. We consulted Hitachi technical who told me of a modification to carry out when this chip is replaced. It can take a good half hour as it involves 14 components. The changes are as follows: change RB117 to 1.2 MΩ; RB155 to 18 kΩ; CB149 to 2.2 nF; CB126 to 33 nF; RB129 to 1.8 kΩ; CB127 to 4.7 μF; RB134 to 2.2 kΩ; RB135 to 470 kΩ; RB122 and RB123 to 220 kΩ; RB110 to 1.5 MΩ; CB130 to 33 pF; and RB128 and RB587 to 3.3 kΩ. When this had been done we switched on and found that the field scanning was now OK and well within tolerance. But yet another problem was present. There was no colour and the line timebase was twitching. Very slight adjustment of the line hold control produced colour, but it kept dropping out and was very unstable. We stared at the circuit and noticed that two of the above modifications were in the line hold circuit. The original values for CB126 and RB129 were 68 nF and 1 kΩ respectively. When these were fitted the set finally performed correctly, with good pictures, colour and field scanning.

Dead set, no channel indicators: There was 300 V across C707 but nothing at the collector of the S2000AF line output transistor T701 because R707 (22 Ω, 5 W) was open-circuit. A check then showed that T701 was short-circuit collector-to-emitter. Replacing R707 and T701 brought the set back to life: check the orientation of T701 when fitting it on the BB as it's very easy to insert it the wrong way round, with disastrous results.

No sound or a blank raster: The Salora L chassis is used in this model. We've had two of them in recently with the TDA4505 chip faulty, the symptom being no sound or a blank raster. Depending on the make of the chip, adjustments to the line and field hold presets may be necessary after fitting a replacement. Before condemning this chip, try disconnecting the text panel (if fitted). If the picture then returns, replace the wire links (about eight in all) that connect the upper and lower tracks on the panel and resolder the other components that link the two print patterns. If the text panel is still faulty, the DPU2540 chip is usually the culprit.

Dozen squiggly lines on display: The display on this set's screen consisted of a couple of dozen squiggly lines. There was an accompanying smell of burning. Our first conclusion was that the scan coils had failed, which can happen with this model. The culprit turned out to be C507, however. It had gone short-circuit.

Hitachi G6P chassis

Power supply shut down: The fault here was that the set started up but a couple of seconds later the power supply shut down. We found that the crowbar thyristor Q902 was being fired. After disconnecting its gate we wound the mains supply up with a variac. As the set operated perfectly it seemed that the fault was in the protection circuit itself. After checking various components we found that the 24 V zener diode ZD752 in the e.h.t. sensing circuit was leaky, a replacement curing the problem.

Failure of line output transformer: This has been a frequent fault with these sets. With this one, however, the symptom was red streaks that appeared at random, starting at the left-hand side of the screen. They would then merge into a vertical red band at the left. This gradually broadened, with a ragged edge, filling the whole screen. The red raster would then brighten to the point at which flyback lines became visible. As a lot of work, with no success, had been carried out on the c.r.t. base and around the colour decoder chip, we decided to look elsewhere. The circuit diagram was consulted, and this gave us an idea. Maybe the graphics generator chip was at fault. A gentle puff of freezer on IC1104 (M50450–023P) proved the point.

No response to remote control: The on-board controls worked all right. The handset emitted the right control pulses, and a very convincing pulse train was present at pin 15 of the control microcomputer chip. It took us some time to discover that L1201 in the i.r. preamplifier circuit was open-circuit. Normal operation was restored when L1201 had been replaced – but there was no discernible difference in the data bursts at pin 15 of the micro!

Snow-filled raster: This set refused to do anything other than produce a snow-filled raster. Operation of the remote and manual controls had no effect, and there were no on-screen graphics. So we carried out checks around the M50432551SP microcontroller chip IC1101. There were normal scan pulses when commands were present, a 4 MHz clock signal was present at pins 33 and 34 and there was a ripple-free supply at pin 42. But there was no movement at any of the data output pins. So there was nothing for it but to replace the chip. Fortunately the set was happy with its new brain and burst into life – we didn't even have to tune it in.

Fail to start-up from standby: The cause was a dry-joint at R1172 (1 kΩ), which is connected between pin 38 of the microcontroller chip IC1101 and the collector of Q1103. To come out of standby at switch-on, Q1103 momentarily connects pins 2 and 38 of IC1101.

Dead set, fuses OK: There was 320 V at the collector of the chopper transistor, but there was no power supply activity at all. The three 82 kΩ resistors were OK, so resistance checks at the power supply outputs seemed to be the next step. The 110 V supply read 20 Ω to chassis, the line output transformer was faulty.

Hitachi G7P MK II chassis

No results: A quick cold check on all likely high-value resistors showed that R920 (100 kΩ) was open-circuit.

Dead set: When we looked at the power supply several high-value resistors caught my eye – R919, R920 (both 100 kΩ) and R932 (120 kΩ). The one that looked most distressed had risen in value from 100 kΩ to 130 kΩ. The other 100 kΩ resistor, which looked brand new, was open-circuit. R932, which is connected to pin 4 of the TDA4601 chip to provide chopper transistor collector current simulation, had risen in value to 130 kΩ. When these three resistors had been replaced the set was speaking to us and displayed a picture.

Tuning drift: There was tuning drift between channels 23 and 33 and nothing above channel 36 could be tuned in. We first suspected a tuner fault, but there was only some 7 V across the 33 V regulator ZD101. The cause of this was none of the usual suspects (regulator, feed resistors etc.). It was the HD401304R12S chip IC101. A clue was provided by the fact that the on-screen tuning bar swept the whole band.

No picture: There were no shorts, but the power supply produced an inadequate h.t. output of 30 V. The TDA4601 chopper control chip IC901 had failed. We also replaced the associated small electrolytic capacitors C914/5/6/7.

Hitachi G7P chassis

Intermittently dead, blew chopper: After replacing the 2SC3679 a few times we found that the cause of the fault was R932, which is connected to pin 4 of IC901 (TDA4601). It had increased in value to about 220 kΩ. Many sets use this type of power supply and it seems to be quite common for these high-value resistors to cause problems.

Intermittent cut-out after minutes or hours: This set would sometimes start then cut out after a few minutes, or run for hours then stop. Tins of freezer and much time with the hairdryer failed to reveal the cause. We eventually found that the two 100 kΩ resistors R919 and R920 which feed pin 5 of the TDA4601 chopper control chip had gone high in value – R920 read about 600 kΩ and R919 130 kΩ. They are rated at 0.5 W. When both had been replaced the set worked perfectly.

Set wouldn't store stations: D706 and R723 in the line output stage are the cause. They were only dry-jointed this time, but it's best to replace them.

Hitachi G8Q chassis

Dead set, mains fuse not blown: This set was dead – no surprise here. Praying that the mains fuse hadn't been blasted, we removed the back. Thankfully the fuse was OK, so a full power supply rebuild wouldn't be necessary. As the h.t. was present and correct we checked the supply to the timebase generator chip IC701. The 12 V supply was present but line drive wasn't being generated. A check was then made on the standby line from the microcomputer panel. The voltage here is normally between 8 and 12 V when the set is running, 0 V in standby. It was found to be 3.5 V. This led us to suspect a fault on the microcomputer panel, and sure enough when the standby line was disconnected the set started up. Before condemning the micro-computer chip we decided to check a few other components. This proved to be a fruitful approach as the 1N4148 diode D1504 in the standby circuit was faulty.

No tuning: We were unable to tune in any stations, obtaining only a blank raster with most programme numbers while the buttons at the front of the set operated incorrect functions. Replacing the tuner unit and the microcomputer chip got us nowhere. A call to Hitachi produced the advice that the memory chip IC502 should be reprogrammed. This is an involved process and the instructions that have to be followed are not listed in the manual. Details for some models were included in the entry 'Reprogramming the Hitachi CPT1476R Series' above. This procedure does get one out of the trouble.

No tuning: The complaint with this set was no tuning. It employs pulse-width modulated tuning, the relevant output from the SAA1293 chip being integrated to provide the tuning voltage. As there was no tuning voltage at the tuner we checked back to the source. The pulse-width modulated output from the chip was present at the base of Q1506 and its mark–space ratio altered as the tuning was varied. This transistor's collector voltage also varied, but only slightly – not enough to span the required tuning range. The path from the collector of this transistor to the tuner is via several resistors, first on the tuning panel, then on the control panel and finally on the main board. We eventually found that one of these resistors, R1534, had gone high in value, a replacement (39 kΩ) restoring normal operation.

Bright white raster only: We found that the 4.7 μF reservoir capacitor for the supply to the RGB output transistors, C720, was open-circuit.

Intermittent start-up fault: When cold the set would sometimes come on only in standby. But if the mains switch was held the set would eventually come on correctly. The cause of the fault was traced to the start-up thermistor TH902.

Warning – heavy charge retained in reservoir capacitor: In this chassis the mains bridge rectifier's reservoir capacitor C906 (220 μF, 385 V) retains a heavy charge for several days. If the power supply isn't working it charges to the usual 300 V or so. With normal operation this drops to around 170 V. Before cold testing or carrying out any resoldering C906 should be discharged, using a 1 kΩ resistor, otherwise you may soon regret not having done so. If you are lucky, you'll receive only a nasty shock. But if, for example, the pins of Q902 (SGSIF344) are momentarily short-circuited a small spark can occur. This may damage the UC3844 chopper control chip IC901. If power is then applied without replacing this chip fuse FS901 may blow together with Q902, ZD901 (BZX79–C27) and D905 (BYD33D).

No picture: When the first anode control was turned up we found that there was field collapse. Checks around the TDA2579A timebase generator chip showed that there was no field drive at pin 1 although there was a linear sawtooth waveform at pin 3. Just about every component that could have caused the loss of field drive was checked, but no faults were found. What now? On switching the set back on again we were amazed to find that there was a full picture. No amount of tapping or heating/freezing would make it misbehave. Perhaps a poor joint had unwittingly been repaired? We then noticed that the raster had some pincushion distortion. Not a lot, but it was there. The presets altered the raster geometry, but not by enough. Again every possible component was tried, even the TDA2031A correction chip. After soak testing the set for two days to make sure that the field fault had been cleared we returned the set. The customer didn't seem to be too bothered about the pincushion distortion but said that it hadn't been there before. It was one of those sets you can't help feeling you'll see again.

No teletext: That's what we thought too, but on closer examination very weakly contrasted teletext was present in the mix mode. The on-screen channel display was similarly affected. We tried a new teletext panel but this made no difference. Over to the video section, where the text signals are inserted at pins 12, 14 and 16 of the TDA3562A colour decoder chip. A new TDA3562A cured the fault.

Dead power supply: Q901 (BUZ71A), Q902 (SGSIF344) and a couple of diodes were short-circuit while R910 ($0.5\,\Omega$) was open-circuit. A colleague showed me an Hitachi service bulletin that recommends replacement of TH902, Q901, D905, D907, IC901, ZD902, Q902 and R910. Whenever the power supply fails it's essential to replace TH902. CHS supply a repair kit (KIT HIT 1) that contains all the components required. It's not cheap, so make sure that you allow for this in your estimate.

Lots of smoke, then dead: On investigation we found that there was charred print at one leg of C715, which is one of the capacitors in the EW diode modulator circuit. We scraped away the burnt remains of the print, then soldered the capacitor's leg to a good section of print. This solved the problem.

Tripping: Try loading the 145 V h.t. line with a 100 W bulb. We've found that these sets trip when the line output stage isn't working. The most common reason for this is failure of the TDA2579A timebase generator chip IC701. If the set still won't come on, lift R713 which is connected to pin 16 (trip input) of IC701. You will probably find that the set springs to life, but with a snowy screen and no activity around the microcontroller chip IC1501. In this situation the first suspect is the 4 MHz crystal X1501. The cause of the fault could, however, be IC1501 or the MDA2062 EEPROM chip IC1502 we've had all these items fail at one time or another. If the set is fitted with the PCF84C type of microcontroller chip, the replacement comes with a modification sheet – it requires a different supply voltage.

Dead, no power supply: Cold checks revealed that D907 (BYV10–40) was short-circuit. It's connected to the gate of the f.e.t. chopper transistor Q901 to prevent the drive from the UC3844 control chip IC901 going negative. As we've not had this diode fail before we decided to replace Q901, Q902 (which is in series with Q901) and IC901 as well.

Takes several seconds to come on: The set would eventually produce a picture with severe line tearing and ragged edges. The cause of this was C933 (2200 μF, 25 V) which was open-circuit. It's the reservoir capacitor in the supply to the 12 V regulator IC932.

No sound or vision: When the setting of the first anode preset was turned up a blank raster appeared. So we jumped to the incorrect conclusion that there was a programming error. We fitted a switch to the remote control chip's pins, but no programming sequence came up on the display. On closer inspection the cause of the fault was found to be a dry-jointed connection to the sleeved wire that links the 8 V supply to the text board and the 5 V regulator IC001 (supply to the tuner).

Hitachi NP81CQ chassis

Lack of line and field sync pulses: We started by replacing the relevant chip – IC701, type LA7801 – which incorporates both timebase generators as well as the sync separator circuit. We then replaced every capacitor within a 3 inch radius of the chip but to no avail. If it wasn't the chip and it wasn't the capacitors, it had to be a resistor. Sure enough R723 (22 kΩ) was open-circuit.

Teletext display on the top one-third of the screen: The trouble was a brilliant display of 13 beady teletext lines covering the top one-third of an otherwise perfect picture. There was no non-linearity and no lack of height. Now we all know that the text information is neatly gathered up into a little package of unused lines at the top of the raster, so we deduced that the problem wasn't flyback suppression but one of slow flyback at the end of the field. It must have accelerated sharply after the 13 lines, since four were clearly seen in the proper place. But then odd things happen with modern technology. Every single component in the field timebase circuit was substituted or tested as appropriate, without success. Better brains in higher places agreed that the problem was slow flyback, but couldn't help us. So we contacted Hitachi who asked whether we'd checked the flyback suppression. We truthfully said that we had: field and line in that order were OK on the scope, going into the little flyback suppression panel at the front of the chassis, and since the scope was still at line speed we got things at line speed out of the panel. Hitachi were adamant, however. Must be no field pulses coming out, and not to fret about the slow flyback which is normal with this chassis. So we did what they said and replaced Q2201 (BC548B) and zener diode ZD2201 (BZX79C5V1). And that was that: everything now fine! Question: do all sets show 13 lines over the top third of the raster when the field flyback suppression is taken out?

Teletext display would invade screen: The customer complained that after about half an hour a teletext display would suddenly invade the screen of her set, obliterating the picture. This would happen on any channel, and after a moment or two the set would go into standby. If left for, say, a quarter of an hour everything became normal. The problem was said to occur only at night. Repeated tests in the workshop, even using the hairdryer on the text and interface panels, showed nothing amiss. So we resorted to drastic measures. We put the set in a large plastic bag with a thermometer visible and blew air from the hairdryer in at the bottom. At 90°F the fault really did happen. Spraying IC2101 (SAA1272) with freezer cured the trouble, and reheating it confirmed that it was the culprit. So we changed it and returned the set. Next day the customer reported that the set was no better. Repeating the bag and hairdryer treatment again showed that she was right. This time IC2101 was blameless and it was the SAA1251 that caused the trouble. Soon after another of these sets arrived. This time the teletext thing crept slowly over the picture from the right, pushing the picture away. After our previous experience we decided to replace the SAA1251 without further ado. It did the trick. I don't trust this device!

Intermittent colour: Sometimes the colour would flash on in parts of the picture. We noted that where there was colour it was of correct frequency and phase. Quite some time was wasted making checks around the colour decoder chip. Not until the dual-trace scope was hooked up to the display were the incoming chroma signal and the burst gating pulse both revealed. The gating pulse didn't coincide with the centre of the 10 cycles of burst on the back porch of the sync pulse. In fact the line pulse wasn't being delayed. Replacing L506 completely cured the problem – this little choke, which in this case had its green plastic cover missing, is used to delay the burst gating pulse. We were lucky to have a similar set to hand, enabling us to make comparative checks.

Intermittent loss of colour: The action of the colour control was also very coarse. This was caused by failure of C718, a 2200 μF, 16 V electrolytic which is the reservoir capacitor for the 12 V supply derived from the line output transformer.

Intermitent excessive brightness: The fault was so intermittent that it sometimes didn't show up for several weeks. We eventually found that the 5.1 V zener diode ZD802 on the tube base was the cause – it would sometimes 'zener' at 3 V, thus overdriving the tube.

Huanyu

Huanyu 37C-2

Line collapse: Caused by dry-joints on transformer T782 in the scan circuit.

Power supply whined, low h.t.: The h.t. line read only about 2.7 V. When the base and emitter of the line output transistor had been shorted across and a dummy load had been connected across the power supply's h.t. output the voltage was restored to the correct 115 V. Checks in the line output stage led us to the transformer, which had a short on the secondary side.

Would intermittently die, no colour: There were two problems with this Chinese 14 inch portable. It would intermittently die, and there was no colour. The cause of the intermittent failure was traced to a poor soldered joint at the collector of the line output transistor. When we investigated the no colour fault we found that there were no line pulses at pin 4 of the D7193AP colour decoder chip. R318 (12 kΩ), R778 or R779 (both 1.5 kΩ) could have been the cause, but the chip itself was the culprit. It was short-circuit internally. We fitted a TA7193AP as a replacement and this worked all right. It seems safe to assume that all the D prefixed chips in this chassis can be replaced with TA series ones.

Huanyu 37C-3

No line drive: As there was no load the power supply was whining. The line output transistor had 110 V at its collector but there was no line drive. Drive was present at the secondary winding of the driver transformer. It was lost due to a print fault between the transformer and coil L781 in the output transistor's base circuit. The print looked to be intact and had to be bridged to provide a cure.

Dead set; stock fault: This set seems to be a Chinese copy of an Hitachi model with which we are familiar. It suffered from the same stock fault. Someone who claimed to have an electronics background brought it in, saying that he didn't have the time to do the repair. After we'd removed a 2N3055 and fitted the correct 2SD898B line output transistor we replaced the 27 V zener diode ZD907 in the power supply. All was then well.

C907 +/− markings transposed: Take care when replacing C907 (2.2 μF, 160 V) in the power supply: in the set we had in for repair the − and + markings on the PCB were transposed. We just managed to switch the set off before our replacement exploded. The correct way round is with the positive terminal connected to the cathode of D905.

Line collapse: We hadn't seen one of these Chinese portables before so we were relieved when the customer handed us a dog-eared piece of paper with the comment 'here's the instructions'. It turned out to be the circuit diagram. Although the circuit wasn't essential in this case, it does save time poking around identifying components. The problem was line collapse − a bright, thin, vertical white line. A resistor in the horizontal centring circuit was badly burnt. It proved to be open- circuit as it had been carrying the full line scan current. The root cause of the problem was the line linearity coil. Repairing and resoldering the coil connections and replacing the 220 Ω, 1 W resistor (R783) restored the line scan.

No picture, although PSU OK: The power supply lines were all at the correct voltages and the line output stage was working. The cause of the fault was the 100 μH coil L509, which is in the 12 V feed to the colour decoder chip IC501. It had gone open-circuit.

Intermittent sound: This is a fairly common fault with these sets. The loudspeaker is the cause: the flexible braid connection to the push-on connector is poorly soldered. Resoldering often provides a cure.

Dead set, blackened mains fuse: As no shorts could be detected we fitted a new fuse and powered the set via a variac. It worked fine, so we left it on soak test for a couple of hours then switched it off and on a few times. It continued to work normally. We then left it switched off for an hour. When it was switched on again from cold the fuse exploded. The cause of this fuse blowing was eventually traced to a faulty thermistor in the degaussing circuit.

No e.h.t. and tube heaters out: The cause of the trouble was traced to dry-joints at the brown plastic coated inductor that's alongside the scan-coil plug.

Chopper transformer whined: Otherwise the set was dead. A check showed that the full h.t. reached the collector of the line output transistor. We eventually found that L782, which is in series with the line output transistor's emitter connection, was dry-jointed.

No luminance: This was put right by cleaning the service switch on the PCB.

ITT

ITT CVC1175 chassis

Picture collapse after 5 minutes: Picture brightness went low, with lack of width and field fold-over. C716 (10 μF, 350 V) on the chopper module was leaky – it gets hot.

Appeared to be faulty line output transformer: This dead set had been checked at the airbase where its owner worked. The technician had diagnosed a faulty line output transformer. The trouble was in fact due to the 115 V line reservoir capacitor C716 (10 μF, 350 V). It looked OK until it was unsoldered, when it was found to be leaking electrolyte.

Low power supply output: The output from the power supply was low at about 50 V. When the feed to the line output stage was disconnected and a 100 W bulb was connected as a dummy load it lit up and the h.t. rose to around 100 V. From this you might suspect that there was a fault in the line output stage, but the actual cause of the problem was the 10 μF, 350 V h.t. reservoir capacitor C716 which had dried up.

ITT CVC1200 chassis

Low h.t. when first switched on: For low h.t. when the set is first switched on, check whether C701 (10 μF, 350 V) has dried up.

Dead set, Si651 (1 A) fuse blown: We found that the chopper transistor T713 was short-circuit but after fitting a replacement we could still measure a very low reading between the fuse and the power supply 'earth'. C701 (4.7 μF) was short-circuit.

Low h.t.: For low h.t. at 90 V instead of 145 V check whether the pulse feedback transformer Tr712 in the switch-mode power supply is open-circuit. You get the same effect when the line output stage isn't working, for example if the line driver stage is inoperative because the h.t. feed resistor R744 is open-circuit.

Dead set, on-board fuse blown: It's common to get a dead set with the on-board fuse blown because C701 (4.7 μF, 350 V) is short-circuit.

Chopper power supply problem: We've never been completely at home with discrete component chopper power supply circuits. In our experience they either work normally or self-destruct. This case fell into the latter category. At switch on the BU508A chopper transistor would go short-circuit, blowing fuse Si651 (F1 A). By using a variac we were able to prove that the filter capacitor C701 (47 μF, 350 V) was faulty. Nothing unusual about that. But it took rather longer that it should have done because a replacement had only recently been fitted. The set worked when C701 and the chopper transistor had been replaced but the output voltages were low. We cured this by replacing the various small electrolytics and the ZPD8.2 zener diode D721 in the control section on the isolated side of the circuit. After setting up the 117 V rail we were rewarded with normal pictures and sound. This lasted for only half an hour, after which T712 (BC328) in the chopper driver stage went short-circuit. It had presumably been weakened by the earlier problems with the BU508A chopper transistor.

ITT CVC1210 chassis

Very intermittent sound and vision: After a few minutes the fault disappeared and no amount of tapping would bring it back. The cause was eventually traced to glue on pin 32 of the tuner/i.f. module (CMR800/3). This glue is used in some quantity to secure C1075 (2200 μF) in position and had spilt on to the contacts.

Blank raster, no picture: We replaced the line output transformer because of insulation breakdown (a hole in the side). This resulted in a blank raster with no picture. After much component checking and replacement we found that C542 (33 nF, 250 V) in the earthy side of the e.h.t. circuit was open-circuit. This was upsetting the operation of the beam limiter circuit.

Dead set: You can tell from the chopper transformer's squeak whether the mains side of the supply is working. It was. So we disconnected the scan coil plug and checked the power supply by loading it with a bulb across the interlock pins of the socket. This proved that the cause of the fault was in the line output stage, where the output transistor was short-circuit. When this had been replaced there was some sort of waveform at its collector, and after a short time the faintest glimmer of line collapse appeared on the screen. A few further checks revealed that the EW modulator driver transistor was short-circuit and R503 (100 Ω) open-circuit. The cause of all this was the scan coil coupling capacitor C511 (330 nF, 400 V). When it was removed one leg remained in the board.

ITT compact 80R 90° chassis

Electronic fuse was operating: The set ran all right with a 15 kΩ resistor connected between test points 701 and 702. The cause of the trouble was that the 20 V feed to the drive circuit on the primary side of the power supply was low at about 13 V. C703 (100 μF, 40 V) was to blame.

Excessive width: Neither the width nor the EW pincushion presets had any effect. Replacing the BD135 transistor Tr563 and slight adjustment of the presets (R556 and R565) restored a normal raster.

Intermittent switching to standby: This is a fairly common fault with these sets. The cure is to change C703 and C713 to 10 μF and connect a 1.8 Ω, 0.25 W resistor in series with D712 (lift the cathode and add it here).

ITT Digi-3 110° chassis

'Set kept going off, now dead': The line output transformer, a common by a field technician a month or so previously. When the set came into the workshop we found that the chopper transistor was short-circuit. This was replaced and the set was put on soak test. After a while the picture and sound disappeared and a loud humming noise came from the set. The new chopper transistor had failed. We eventually found that the culprit was C714, which decouples the start/run l.t. supply to the non-isolated section of the power supply. It was going open-circuit intermittently.

No sound or picture: The standby light was flashing on and off. Another faulty line output transformer? Not this time. Our Konig LOPT tester gave it a green light. But the set had all the symptoms of a short-circuit somewhere in the line output stage. Tests on the supplies derived from the LOPT showed that the diodes were all OK and that there were no short-circuits to chassis. It was a different matter when we checked the resistance across the field output chip, which was short-circuit. Beware of chips that operate with split supplies! We had a TDA2170 in stock, but Chas Hyde can supply an alternative. Replace the field flyback boost capacitor C402 as well.

No sound or raster: We disconnected the scan coil plug and used a 100 W bulb across C795 as a dummy load for the power supply. This proved that the power supply was OK. The line output transistor and transformer, which are the most usual causes of the fault, were then replaced. But no luck. Disconnecting in turn the rectifier diodes fed from the line output transformer brought me to D547 in the 13 V supply. It was short-circuit. The TDA2170 field timebase chip IC401 was also short-circuit.

ITT Digi-3 chassis

Locking out remote control signals intermittently: There's a modification to prevent this. Connect a 10 kΩ resistor between pin 6 of IC1404 and chassis – this i.c. is on the IFB286 control unit.

Set changes channel by itself: Use the above modification.

Dead set: Check whether D795 is short-circuit.

Incorrect or patchy colours: Suspect the degaussing posistor R701.

Shaking picture, as though the line phase is shifting: Try a dash of freezer on IC1402 on the IFB286 module.

Faulty colour: It looked as though the reference oscillator was just off frequency. Substitution proved that the fault was in the digiboard, but changing the video codec and PAL processor chips made no difference. Replacing the clock generator chip IC610 put matters right.

Squegging on vertical lines: Replace the video codec chip IC650.

Line fault at switch-on and paterning: This set had a line fault at switch-on. But after only a minute or two there was also a patterning effect – almost as though the a.f.c. was pulling the tuning very slightly upwards to the point where the line sync starts to break up. We expected to find a problem with the deflection processor chip. But not so. Closer investigation led me to the signal unit (h.f. module) where we found that C209 (47 µF) and C234 (10 µF) were both in trouble. Replacing them provided a complete cure.

Electronic fuse operated: A whistle could be heard and the channel indicator lit for a second or two, then the electronic fuse operated. We disconnected the scan coil plug, connected a 60 W bulb between the 145 V line (pin 4 of the scan coil plug) and chassis and switched on again. This confirmed that the power supply was working correctly. Checks in the line output stage then showed that there was a short-circuit across the 13 V supply. Diode D547 (BA158), the TDA2170 field output chip IC401 and C548 (1000 µF, 16 V) were all short-circuit.

Crackling on sound when hypo-sound selected: The cause of the trouble was the APU2400E chip IC670 on the digiboard.

ITT monoprint B chassis

Set wouldn't come on from cold: If the switch was operated a number of times a small raster appeared with a loud noise that sounded a bit like line output transformer arcing coming from within the set. The cause was C707 (2.2 µF) in the power supply. Once the set had warmed up it worked OK.

Difficult to power up: Having been left off overnight this set wouldn't power up without multiple pressings of the on/off switch. The TEA2165 chopper control chip IC701 (TEA2162 in some versions) was shutting down because of inadequate negative bias at pins 4 and 5. C704 (47 µF) was the cause, being unsatisfactory at low temperatures. You could see the ripple across it change when freezer was applied. Three months later the symptom reappeared. This time the mains bridge rectifier's reservoir capacitor C701 (4.7 µF) was found to have 100 V peak-to-peak across it when cold, a replacement curing the fault.

Dead set and intermittent start-up: This set originally came in because it was dead. The mains rectifier's 5.1 Ω surge limiter resistor R652 was open-circuit, the 115 V supply rectifier D733 (BYW36) was short-circuit and the line output transformer was faulty. When these items had been replaced we were presented with another fault. The cause of intermittent start-up and tripping when cold was traced to C707 (2.2 μF, 63 V) in the power supply. It's the reservoir capacitor for an 8.5 V feed to the chopper control chip IC701. When we checked it with the component tester we found that it lost all capacitance when cold.

Switch-on problem: Initially the set wouldn't come on although the supplies, including the e.h.t., were present and the heaters were alight – the standby LED was on. After a few minutes the set would work, after which it worked normally. No reset pulse could be measured at pin 12 of the MDA2061 memory chip IC1403, while the 8 V supply at the cathode of D1429 was low at 7.4 V. Replacing the supply's 470 μF, 16 V reservoir capacitor C722 brought it back up to 8.6 V, after which the reset pulse was present at every switch-on.

Intermittently goes into standby: We found that as the temperature rose the TDD1605S 5 V regulator chip IC405 became leaky between its input pin and chassis.

No picture or sound, e.h.t. present: The tube's heaters were alight. The audio output was OK when a screwdriver was applied to pin 1 of the scart socket. This indicated that the set was in the AV mode. Pin 32 (AV select) of the microcontroller chip operates the AV/TV switching via the BC238 transistor T1415. This was short-circuit base-to-emitter.

ITT TX3326 (monoprint B chassis)

No teletext, although green 'text on' light would show: If text was selected via the remote control handset the channels wouldn't change. The cause of the trouble was the DPU2541 chip on the text panel.

Sizzling noise at switch-on plus a thump from the loudspeaker: Then the set went dead. This all happened within a couple of seconds. We found that pin G of the chopper transformer was badly dry-jointed – to the extent that the board had carbonized. A thorough clean-up and resoldering job cured the problem.

Wouldn't stay on: The set wouldn't stay on long enough to produce a raster. A quick check in the short time available revealed that when the set came on the h.t. was too high, at 156 V. It should have been around 115 V. So the set was detecting an overvoltage condition and shutting down. In these receivers the set-h.t. potentiometer isn't in the power supply area: you'll find it near the line oscillator. As we suspected, adjusting it made no difference to the fault condition. What had happened was that the 300 kΩ resistor in series with it, R613, had gone high in value. When this had been replaced the h.t. could be set up and the receiver performed normally.

JVC

JVC AV21F1EK (JX chassis)

Would come on for 10 seconds, no raster: The power supply protection shutdown circuit would kick in. As we could find no shorts in the power supply or the line output stage we uplifted the set for workshop attention. With the set on the bench the preliminary findings were first confirmed then plug R on the power supply panel was disconnected to isolate the 127 V feed to the line output stage. A 60 W bulb was fitted as a dummy load instead. On test we found that the h.t. supply rose to 123 V before the power supply shut down. When plug Q was removed, disconnecting the 27 V feed to the audio PCB, the power supply came on and stayed on. When the dummy load was disconnected there was a normal picture. A new TA8216AH audio power amplifier chip (IC801) cleared the fault.

Appeared to be a hum bar: This set suffered from an unusual effect that looked like a hum bar. Scope checks showed that there was indeed ripple on the h.t. line, but the cause proved to be rather elusive until L2001 in the chopper transistor's base drive circuit was checked. It was open-circuit (part number CELC005–2R5).

Loss of the picture: There was just snow. There were also no on-screen menus. The cause of the fault turned out to be dry-joints at the 5 V regulator IC522.

No sync, intermittent: About 5 minutes after switching the set on from cold the picture would shake from left to right, sometimes violently. If text was selected the display was all right. If mix was selected there would be no line lock between the text and the picture. This would last for about 3 minutes, after which the fault would clear. No amount of freezing or heating would then instigate the fault. Switching the set off overnight would bring the fault back, however. JVC thought that the problem might be caused by the 6 MHz crystal on the text panel – the sync pulses come from the text processor chip in these sets. So we started to check around in this area and found that when the fault was present there was noise on the sync pulses in the video input waveform. This comes from the i.f. unit, where a tantalum capacitor (C02) that decouples the i.f. a.g.c. was found to be the culprit. Replacing it cleared the fault completely.

Drifts off tune intermittently: We traced the cause to dry-joints at all the pins of the 5 V regulator IC522.

No teletext: When teletext was selected the screen would blank for a second then a normal picture would return. The cause was IC004, type PCF84 C81P/064, on the text PCB.

JVC AV25F1EK (JX chassis)

Series of intermittent faults: The customer said that the remote control sometimes locked up and didn't appear to work, the signals would disappear briefly and then come back again, among other things. I removed the back cover and tapped around the tuner section. This instigated the fault condition, which was caused by dry-joints at the 7805 regulator chip IC522. A good solder-up here cured all the intermittencies, including the apparent remote control problem.

Power supply would stop after 1 second: The power supply would work for a second or so then shut down. Voltage checks showed that while the power supply worked the h.t. voltage was excessive. The 2SC1815YG transistor Q031 was the cause: it was open-circuit.

Dead set: A fault that now occurs on a regular basis is a dead set because of a dry-joint at the chopper transformer pin that's connected to the collector of the chopper transistor. It seems that the rivet itself is not soldered to the print land. Resoldering provides a complete cure. Also check and resolder the pins of the TEA2261 control chip IC001 and the CNX82A optocoupler IC002.

JVC AV25S1EK (MX II chassis)

Dim picture with green tint: The cause was no voltage at IC201's contrast control pin because capacitor C1206 had become leaky. A new 3.3 μF capacitor restored good pictures.

No line drive: The standby light would change colour but there was no line drive. Voltage checks showed that there was no supply to the line driver transformer because the standby switching transistor Q902 was not being switched on. Its base bias resistor R903 (2.2 kΩ) had risen slightly in value.

Corrupt values for geometry: If the values for height and vertical geometry become corrupt it is best to replace the 24C01A/P memory chip IC1707 that's connected to the bus line. Reprogramming it may provide only a temporary cure.

Loewe

Loewe MS56 (C8001 chassis)

No sound and vision: Check whether the line driver transistor's supply resistor R534 (3.3 Ω) is open-circuit. If the h.t. is high at around 230 V (should be 142 V) check C638 (1 μF).

Short in line-output transformer: The dead set symptom with this set was caused by a short in the line output transformer. When a replacement had been fitted and the h.t. had been set up (142 V) there was an EW fault. This was cleared by replacing R583 (22 Ω) and the TDA4950 EW drive chip IC1581.

Severe venetian blinds: This set's picture was marred by very severe venetian blinds. We found that the 1 kΩ LZ AMP preset P323 had gone open-circuit.

Logik

Logik 4090 (Ferguson TX90 chassis)

Blew 1.6 A fuse at switch-on: Resistance checks around the line output stage revealed a short-circuit between the collector of the line output transistor and chassis. When tested the transistor seemed to be OK, so a new output transformer was fitted. This restored normal working.

No colour with faint blue cast: This was the result of C157 (22 nF) having developed a leak – when measured it read about 25 kΩ. It's connected to pin 18 of the μPC1365C colour decoder chip IC103, being part of the a.p.c. detector circuit. In the past we've also had D103 (BAV20) cause loss of colour with these sets. It's connected to pins 19 and 23 of the colour decoder chip, in the line pulse feed.

Corrugated ripple on verticals: This could be reduced by lowering the brightness or contrast level, i.e. the beam current. The cause of the fault was C189 (22 μF, 50 V) which when checked measured only 0.5 μF.

Logik 4298 (Ferguson TX100 chassis)

D15 (BY299) short-circuit in PSU: We had two of these sets in on the same day, both with the same fault. In each case D15 (BY299) in the power supply had gone short-circuit, giving the dead set symptom. The first repair was straightforward, but after we'd replaced the faulty diode in the second set we switched on to find that the field scan was badly distorted. The TDA3651 field output chip had to be replaced.

Fast tripping: Go straight for the line output transformer. It's quite a common fault now.

Random channel changing and going to standby: This was cured by resoldering L20 in the power supply. It had become dry-jointed.

Dead set: We found that the BY299 h.t. rectifier diode D15 was short-circuit.

Dead set, no short-circuit h.t.: For a change there was no short-circuit reading across the h.t. line. We found that D28 (BY299), which is in series with the h.t. feed to the line output transformer, was short-circuit. When this was replaced there was a concave raster. R711 (2.7 Ω, fusible) on the EW correction PCB had failed. It's in series with the emitter of the Darlington driver transistor TR73.

Luxor

Luxor SX9 chassis

Wouldn't give real-time clock on ITV: We scratched our heads over this one, as did Luxor's UK technical department. The Swedish head office knew the answer right away, however. The customer had programmed an unused page number into the text memory. The number was used on each channel except ITV, which meant that the text decoder was searching for something that didn't exist. This was confusing to the clock.

Remote control intermittent: We put the set on soak test for a long time and confirmed that the remote control system stopped working on several occasions. Another handset was tried, and we found that by switching the set off and on normal operation was restored. One time while the set was on soak test we noticed the width jump in and out. A check around the switch-mode transformer revealed that there were several dry-joints, particularly at pin 1. Repairing these cleared both the width and the remote control problem. Incidentally if you get one of these sets that switches on and goes straight to standby, check the diodes on the secondary side of the switch-mode transformer. If these are OK the problem is almost always due to shorted turns in the line output transformer.

Line driver transistor failure: This is becoming a common problem with these sets. It's a special type, BD419, available from NCS. If the problem is encountered it's best to change DH01 (1N4448), CH03 (68 nF), RH02 (150 Ω), RH03 (6.8 Ω), CH02 (22 μF) and, in stubborn cases, the line driver transformer LH01. It's also possible that RN19 (1 Ω) in the power supply has failed. It's also vital to check the soldering to the large copper heatsinks, as these form earth continuity paths.

The most important one is the print land in the area of the line oscillator chip. When servicing these sets it pays to check the soldering on the chopper transformer, particularly at pin 1, the condition of the capacitors in the power supply and the condition of the mains switch. If you have difficulty getting a BD419 a TIP48 is an unofficial alternative.

Matsui

MATSUI 1420
MATSUI 1420A
MATSUI 1422
MATSUI 1424 (TATUNG 190)
MATSUI 1436
MATSUI 1436X
MATSUI 1436XA
MATSUI 1440A
MATSUI 1455
MATSUI 1460
MATSUI 1466
MATSUI 1480
MATSUI 1480A
MATSUI 1481B
MATSUI 14R1 (GRUNDIG G1000 CHASSIS)
MATSUI 2091
MATSUI 2092T
MATSUI 209T
MATSUI 2160
MATSUI 2180TT
MATSUI 2190
MATSUI 2580
MATSUI 2890

Matsui 1420

R503 goes open-circuit, no start-up: A stock fault that's beginning to appear on these very popular portables is that R503 (82 kΩ) goes high or open-circuit. As a result the power regulator (Q501, 2SC3158) fails to start up.

Failure to start when cold: This was traced to C508 in the power supply. It's a 4.7 μF, 50 V electrolytic.

Intermittent failure to start: Visual checks revealed several dry-joints at the pins of the chopper transformer T501. Resoldering these cleared the trouble.

Field cramp: If field cramp develops with one of these sets after it has run for 2–3 hours, replace C412 (1 µF, 25 V) in the linearity feedback circuit. It's one of those nasty tantalum capacitors.

Dead set: This set came in dead and we found that the SR2M overvoltage protection diode D508 was short-circuit. Checks in the power supply failed to reveal any defects and when a replacement SR2M had been fitted the set ran for days. Then the diode again went short-circuit. The power supply was not of the STR type but had a transistor on a subpanel. This was blameless, however. The cause of the problem was an intermittent line driver transformer. When the line drive ceased the h.t. rose and D508 failed.

Green screen: It nearly always pays to carry out a visual examination. Sure enough R807 on the c.r.t. base panel had been badly overheating. When removed and checked this 8.2 kΩ resistor was found to have risen in value to 450 kΩ. Somewhat different! A new resistor restored normal operation.

Matsui 1420A

Dead set, power supply trying to work: The h.t. was very low and R434 (10 Ω, 7 W), which is in series with the emitter of the line output transistor, was very hot. When the h.t. feed to the line output transformer was disconnected (pin 4) the h.t. rose to its normal 103 V. A new line output transformer restored normal operation.

Intermittent field scan: Check for cracks near the line output transformer, where the securing lugs go through – especially near R437 and D409. This is becoming a common problem.

No sound/picture, e.h.t. OK: When the setting of the first anode preset was advanced we found that there was field collapse. The 12 V regulator that provides the supply for IC401 was red hot. Cause of the trouble was C436 (470 µF, 16 V) which was leaky.

STR50103 chopper chip etc. faulty: We've had a number of these sets with the STR50103 chopper chip, the SR2M overvoltage avalanche diode and the 5.6 Ω surge limiter resistor all faulty. Be careful when you apply the mains supply after replacing these items. If the line output stage appears to be in distress, switch off at once or you'll get a repeat performance. The faulty component may be the line output transformer, which often has a pinhole in it, or the scan coils. The latter are not readily available on their own: it's wise therefore to save the deflection yokes from scrapped sets.

No line drive: There was no line drive with this portable. The 2SC2271 line driver transistor Q401 was open-circuit and the line driver transformer T401, though measuring correctly from the d.c. point of view, failed to transform the drive to the secondary side.

Matsui 1422

Whistle came from power supply: We thought it must be the transformer. A little pressure on the transformer with a screwdriver stopped the whistle but, just as we were about to remove it, we noticed a small 4700 pF, 1 kV capacitor with a split in it next to the transformer. This capacitor (C613) normally shuts the set down when it splits, but on this occasion it decided to oscillate instead. Still, cheaper than a replacement transformer.

No display and no tuning: It's worth checking D403 on the front panel and the associated circuitry.

Dead set: This portable was dead. Fortunately the cause was very obvious. C617 (4700 pF, 1 kV) in the chopper circuit had split in half. When a replacement had been fitted the set worked normally.

No picture, just a blank raster: L108 in the i.f. strip had gone open-circuit.

Dead set or a buzzing power supply: The cause is almost always C617 (4.7 nF, 1 kV), a blue disc ceramic capacitor. Also check C612 which lives next door. Faulty ones will have started to crack, but this isn't always visible with the component in situ. It's easy to mistake this fault for an overload in the line output stage. The surge limiter resistor will fail if the set is left on.

Dead set, h.t. high: Because the h.t. was high, the line output transistor Q305 had gone short-circuit. The basic cause of the trouble was C609 (47 μF).

Lack of height: There was only about an inch of scan. The cause was traced to C308 (2.2 μF) which is part of the field linearity feedback

No sound: The cause proved to be C150 which was short-circuit. It's at the input to the audio output chip.

No power: On inspection we found that C617 had burnt out, something that's quite common with this chassis. I replaced it and switched on, but the set was still dead. Unusually, Q604 was short-circuit. The set was restored to life when a replacement had been fitted.

Field collapse: Some quick checks showed that the field output stage's 56 V supply, which is derived from the line output transformer, was missing because fusible resistor R306 (10 Ω) was open-circuit. Replacing this item restored a full picture.

Snowy raster with no signals and no channel display: The cause was loss of the l.t. supply because F602, which is in series with D610, was open-circuit. We have to say that this is a horrid bit of construction. The fuse has wires soldered to both end caps. One end is soldered to the board, the other to the anode of D610, the whole lot being encased in a plastic sleeve. There's no room for a proper fuseholder.

Matsui 1424 (Tatung 190 chassis)

Failure of power supply to start: This can be a stock fault. R802 and/or R803 (both 16 kΩ) has gone high in value or open-circuit.

Sound lost when channel changed: Once this set had warmed up the sound would be lost when the channel was changed. The cause of the fault was traced to the HD401220 microcontroller chip (part number 1983156).

Line frequency was wrong: Adjustment of the set's control would produce only a very unstable lock. The cause of the trouble was R109 (30 kΩ) in the line oscillator circuit. It had increased in value to about 31 kΩ.

Matsui 1436

Bright white raster for 10 minutes: For the first 10 minutes after switch-on this set displayed a bright white raster. The contrast then slowly increased until a nice picture was present. Use of freezer soon took me to the culprint, the TA8691N multi-function chip. A replacement restored normal operation.

No field scan: By using the highly technical tap fault- finding technique we narrowed the cause down to a dry-joint at R602. Resoldering it restored the scan.

Would work for a minute then stop: This set would work for a minute or so then cut out, with h.t. present but no line drive. A voltage check at pin 30 of IC301 showed that the X-ray protection circuit was in operation. The cause was R492 (5.6 kΩ) which had risen in value. It's the lower resistor in the potential divider for the X-ray protection circuit.

Line tearing after warm-up: Heating and freezing IC301 made the fault come and go. A new TA7698AP chip cured the trouble.

No green in display: The green cathode voltage seemed to be about right at 130 V. A bright green raster was produced when we briefly earthed the green cathode. After doing this, the green content had returned to the picture! We eventually found that the 3.3 kΩ flash-over protection resistor R505, which is connected between the cathode and the collector of the green output transistor, had risen in value.

'Loud fizz and went pop': On investigation we found that the degaussing posistor had exploded, taking with it the standby mains transformer, relay RL650 and the degaussing coils. All was well after replacing these items.

Didn't come on, no standby light: This portable didn't come on and there was no light from the red standby indicator. A check showed that the relay wasn't being operated. There's a 160 mA fuse, which had failed, in the circuit that controls the relay. Repair was simply a matter of replacing the fuse.

Field collapse: The supply to the field output stage was OK, and we were beginning to suspect the chip. But we decided to check the service switch to make sure that it was in the correct position – we've been caught out by this before. It didn't work, but we did notice that when the switch was moved the field tried to expand. Investigating this further brought us to Q601 where the voltage was low. Its feed resistor R608, a safety type, was open-circuit. Replacing this resistor and Q601 restored full field scanning.

Low or no line drive: Check the value of R756 (0.5 Ω). This resistor is connected between the base of the line output transistor and the secondary winding on the line driver transformer.

Lifeless, but would come out of standby: The power supply was working, and there was h.t. at the collector of the line output transistor. There was also line drive at the output from the driver transformer. But it didn't reach the base of the output transistor, because of a dry-joint on the link between these two items. The dry-joint was visible only when examined with a magnifying glass. Resoldering it cured the trouble.

Colour in vertical moving bands: This portable had a really strange fault. When it came on the colour would be in vertical bands that moved across the screen from left to right. Fortunately the fault was intermittent – it responded to gentle tapping with a screwdriver. The cause of the trouble was a badly soldered joint at the chassis side of C140.

No sound or picture: The standby switching worked but there was no sound or picture. Checks in the line timebase showed that while line drive was present at the primary winding of the driver transformer T751 it didn't reach the base of the line output transistor. A replacement driver transformer cured the fault – the original one had shorted turns.

Matsui 1436X

Set refused to come on: A look in the power supply showed that R651 and R653 both had burn marks on them. When we replaced them, and also IC650, the set still refused to work although there were no burn-ups. Cold checks showed that C655 was short-circuit and R652 open-circuit. Replacing these items restored normal operation.

Smeary picture at switch-on: As we took the back off the fault started to come and go, so we started a tapping session. This led us to a dry-joint around R513 on the c.r.t. base panel. After soldering all the connections around the base the set was just fine.

Picture would fade after 40 minutes: When this set was switched on a normal, clear picture would appear. But after 10 to 40 minutes the picture would go dark and lose its colour, then the width would decrease with a folded line scan. The cause of all this was the line output transistor, which was leaky. Most modern chopper power supplies tend to trip in this situation. This one seems to ration the h.t. power continuously, however, by reducing the voltage.

Matsui 1436XA

Field collapse: The voltages were all about right, but we found that there was no drive from IC301 to the field output chip IC701. As replacing IC301 made no difference we had to take a closer look. While checking voltages we noticed a poor connection to C706. When this was resoldered the field drive returned.

Field twitch: It was caused by a dodgy preset, VR702, in the field timebase circuit. Replacing it cured the problem.

Intermittent field or line problems: Check the small presets. Their rivets become loose. Crimping the rivets can provide a cure, but replacement is best for long-term reliability.

No colour: The colour began to appear after about an hour, however. Freezer applied to the chroma circuitry showed that C318 (2.2 μF, 50 V) was very heat sensitive. A replacement restored full colour.

No picture: The line output stage was getting its h.t. supply, but the line driver stage wasn't. The line driver transformer T751 was open-circuit.

Dead, no illumination from standby: The small 160 mAT fuse FU211 in the standby supply was open-circuit.

Matsui 1440A

Dead set, chopper chip failed: The usual causes of a dead set are the STR50103A chopper chip, the SR2M protection diode and the 56 Ω surge limiter resistor. If you find that they are all OK, check R502 and R503 (both 330 kΩ) and Q108 (2SB698). In one set Q108 was short-circuit collector-to-emitter and R503 was open-circuit. These components are in the start-up circuit.

Noisy controls: A common fault with these sets is noisy volume, colour, brightness and contrast controls, although most customers only complain about uncontrollable sound. We've found that repeated application of switch cleaner fails to cure the problem. The best thing to do is to remove the whole potentiometer bank and dismantle each one in turn, cleaning the wiper and carbon track with a powerful solvent such as RS 554–153 1.1.1.trichloroethane.

Tripped after a minute: We've had two of these sets with the same problem. They worked perfectly from cold, then tripped out after a minute. In both cases the STR50103A chopper chip IC501 was the culprit. We've also had the R2M overvoltage protection diode D508 cause tripping with this popular portable.

Horizontal banding and patterning: The picture suffered from flickery horizontal banding and patterning. This was being caused by h.f. oscillation in the UPC1378 field output chip. C417, a 15 nF capacitor which is connected between pin 4 of the chip and chassis, was found to be open-circuit. Other models in these ranges use the same circuit and chip, but the component reference and capacitor values differ.

Very bright, very poor picture: A check showed that the 180 V supply to the RGB output stages was low. The culprit was the supply's reservoir capacitor C431 (4.7 μF, 250 V). A replacement restored the voltage and the picture quality.

Matsui 1455

No luminance, bad character display: There was no luminance and the on-screen display characters flared badly to the right, extending to the edge of the screen. A check on the luminance waveform at pin 23 of IC202 showed that all was well here, but the signal was missing at the emitter of the 2SA562 luminance amplifier transistor Q202 which turned out to be leaky. A replacement restored full luminance but the on-screen display fault was still present. The on-screen display signal is fed to the base of Q202 via a 1N4148 diode which also turned out to be leaky, but the fault was still present when a replacement had been fitted. Further checks brought us to the collector of the 2SC1815 buffer transistor Q408 where the voltage was 7 V. The correct figure is 0.13 V. The transistor tested OK when removed but a replacement cured the fault.

Vision would sometimes disappear leaving a blank white screen: A check showed that when this happened it had gone into the AV mode. But when the on-screen display was called up the channel number appeared. A check on the voltage at pin 6 of the microcontroller chip

TC401 showed that this was correct, i.e. low for TV and high for AV. Our next checks were on the AV panel at the rear of the set where we found that the AV switching transistor Q1103 was without its 12 V collector supply. On tracing the source of this supply we came to R121 which had voltage at both sides. The only thing between R121 and Q1103 was the print, which turned out to have a hairline crack.

Field collapse: The output stage seemed to be working normally. Checks around the jungle chip were inconclusive and a replacement made no difference. So it was back to the driver and output stages. The power supply arrangement here is a bit unusual. By making cold checks we eventually found that the $10\,\Omega$ safety resistor R310, which acts as a surge limiter, was open-circuit. This stopped the driver stage working.

Standby switching circuit problem: This set caused its owner some distress. When she switched it to standby it remained on and when she switched it on from standby there was an overbright, pulsating display. A quick check on the h.t. feed to the line output stage showed that it was not being shut down when standby was requested. We traced back to the standby switching circuit and found that Q605 was short-circuit. Replacing this cured the problem, to the relief of the customer.

Dead set, no output from PSU: Check the $180\,k\Omega$ start-up resistors R903 and R904 first. If these are OK it's likely that either C613 ($0.0047\,\mu F$, 1 kV) or C614 ($0.0022\,\mu F$, 1 kV) is faulty. They are both small blue disc capacitors that are connected in parallel with the h.t. rectifier D607 on the secondary side of the chopper transformer. For no luminance, another very common fault with this set, check the luminance emitter-follower Q202 (2SA562TM-Y). If this is OK it's likely that either Q407, Q408 (both type 2SC1815Y) or D417 (1N4148), which are behind the AV board, is faulty. As a quick check, for a picture to be produced Q408's collector voltage must be low at 0.13 V. We've never had to replace the TA7698AP colour decoder/timebase generator chip IC202.

Strange pattern at turn-on: If you turned the brightness and contrast down the pattern disappeared and you could see a rather nasty hum bar. Taking a look at the inside we noticed that the mains bridge rectifier's $100\,\mu F$, 400 V reservoir capacitor C604 had put on weight: when touched, with the set switched off of course, it was extremely hot. As it measured strangely when removed from the circuit we decided to replace the BA10G bridge rectifier BR601 as well. At switch-on there was a perfect picture.

Field collapse: Check whether D306 is short-circuit and R310 (10 Ω) open-circuit. The usual cause of field collapse (a very bright white line) is loss of the 12 V supply to the multi-function chip IC202: zener diode D219 goes short-circuit.

Intermittent sound and picture: Tapping the line output transformer cleared the fault. A screw near C316 bonds the line output transformer's screening plate to chassis. It was loose.

High h.t.: This is a common fault with this set. The cause is C607 (47 μF) in the power supply drying out, the result being weak chopper transistor turn-off drive. Usually the 2SD1426 line output transistor is damaged. So we quoted a price and began work. After replacing the above two items we were left with a string of other faults – and the profit margin rapidly dwindled. Field collapse was caused by loss of the 12 V supply to the TA7698AP chip. We had to replace R310 (10 Ω) near the line output transformer and the 12 V zener diode D219 which was short-circuit. Look around the luminance delay line for this component. For loss of the luminance signal we had to replace Q202 (2SA562), Q408 (2SC1815) and D417 (1N4148) which were all leaky. Finally the colour was wrong. This took us to the tube base where presets VR503 (10 kΩ) and VR504 (500 Ω) were burnt up, R510 (1.5 kΩ) and R511 (4.7 kΩ) were open-circuit and C505 (180 pF) was short-circuit.

No line drive: A quick check showed that the supply to the driver transformer's primary winding was missing. We then noticed that the feed resistor R308 was under stress. The associated decoupling capacitor C311 was leaky.

Auto-sweep tuning not working: It wouldn't stop when a station was found. This was because there was no sync pulse at pin 36 of IC202 when a station was passed, although the pulse was OK at the base of Q404. The cause of the problem was a small crack in the print.

Dead set: The cause of a dead set is usually R603/4 or C613. Not this time! There was 300 V across C604, and no distress in the power supply. We decided to check the transistors in the discrete-component chopper power supply and discovered that Q602 was leaky. As a precaution, the chopper transistor's base drive coupling capacitor C607 was replaced with a high-temperature component.

Wouldn't come out of standby: The reason for this was a low supply voltage at pin 42 of the microcontroller chip – 3.5 V instead of 5 V.

Capacitor C612 (220 μF, 25 V) in the power supply had dried up. For reliability, C607 (47 μF, 25 V) should always be replaced when one of these sets is serviced.

No picture, sound OK: The RGB outputs were being blanked because of a fault in the control circuitry. Q408's collector voltage was high – there was only 0.2 V at its base. But there was 10 V at the collector of the preceding DC coupled transistor Q407. So we replaced Q408, which made no difference. The cause of the trouble was D417 (1N4148) in Q408's collector circuit. It was leaky.

No line sync: Resetting VR203 restored a picture, but it was unlocked and at best there was hooking at the top of the screen. The line-frequency pulses that should have been present at pin 35 of the do-it-all chip IC202 were found to be missing. They come from the line output transformer, via R274 (27 kΩ) which was open-circuit.

Rolling picture after warm-up: On close examination we could see slight line tearing which became worse when the set was warm. Heating the tuner section drastically increased the line tearing, and the picture began to roll. But this was a bit of a red herring. The cause of the fault was the 1N60 diode D103, which is connected to pin 5 (i.f. a.g.c.) of the i.f. chip IC101.

Matsui 1460

Dead set, open-circuit fuses: To start with the set was dead, with open-circuit mains and supply fuses. Checks showed that the STR451 power supply chip was short-circuit between pins 1, 2 and 3, the SR2M avalanche diode was short-circuit and the 2SD869 line output transistor was low resistance between its base and emitter. When these items had been replaced the set was powered up. It came on with normal field scanning but there was reduced width and line fold-over at the left-hand side of the screen. After checking the scan-correction circuit we found that the scan coils were arcing to the tube at the top left. Examination of the scan coils after removing them showed that the insulation had burnt with the result that the line scan coils were shorting to each other. In the past we've had noisy coils and open-circuit tag connections to windings: this is the first time we've had windings shorting to each other and arcing to the tube.

Set dead, channel display OK: A resistance check at the regulator chip's output pin indicated that a short was present. Normally the cause is the

R2M overvoltage protection diode D805, but this time the STR451 regulator itself (Q801) had failed. Both items were replaced.

Field collapse: The 35 V supply to the field output transistors was present and correct, and there was a good sawtooth waveform at the field hold control. A few cold checks soon brought us to R323 (1.8 Ω) which was open-circuit.

Matsui 1466

Set completely dead: We found that the set has a standard TDA4601-type power supply. Checks showed that the BU208 chopper transistor was short-circuit while the fuse was blackened. We replaced these items then checked the 270 kΩ resistor associated with the TDA4601 chip. As usual it had gone high in value. After fitting a replacement we switched on and up came the picture. A nice, easy repair we thought. But the picture then slowly collapsed to a bright line. Checks showed that the 25 V supply to the field timebase was OK but the 9 V supply slowly disappeared due to the regulator breaking down. We replaced it and, with crossed fingers, switched on. This time the picture remained.

No sound/vision, channel indicator OK: We went straight to the regular culprit with this model, R502 (330 kΩ) in the power supply. It goes high in value. Fitting a replacement restored normal operation.

Intermittent failure to switch-on: We decided to replace the STR50103 chip in the power supply and the 330 kΩ start-up resistor. After that it wouldn't switch on at all. As checks on the rest of the components in this area failed to reveal anything amiss we refitted the original power supply chip. The set then worked normally at every switch-on.

No picture/sound, channel change OK: This is a regular problem. We replaced R502 and R503 and got the set working.

Low, distorted sound: The cause was traced to the 10 kΩ bias resistor R353 in the audio output stage. It was open-circuit.

Dead set, channel appears briefly: If you find that the set is dead, with the channel display appearing for a few seconds then going off again, check the start-up resistors R502 and R503 in the power supply. The fault occurs when one or both go open-circuit. They are both 330 kΩ resistors.

Matsui 1480

No sound or raster: This was easy to track down as there was no line drive. Eventually, after changing the line chip, we noticed that there was a sticky film on the nearby print. Cleaning this brought the set to immediate life.

Set had a dark picture: Adjustment of the first anode and preset brightness controls helped, but something was still wrong with the clamping. In addition a slight hum bar could be seen at the left of the picture. Scope checks showed that there was a 50 Hz waveform on the 10V line. This took us to C509 (470 μF, 16 V) which was open-circuit. A replacement produced a surprisingly good picture with the presets returned to their original positions.

No signals and no channel indications, just snow: You'll probably find that the primary winding of transformer T101 is open-circuit.

Matsui 1480A

Picture intermittently blanked off: Check that the teletext board is making good electrical contact via its leads. With the method used you may find that the plastic instead of the bare wire is gripped.

R2M protection diode short-circuit: It's quite common to have to replace the R2M protection diode D508 in the power supply: it goes short-circuit, shutting the set down with only the channel indicators working. We had a merry dance this time, however: the set was still dead after replacing D508 and the STR50103 chopper chip IC501. There should be a 330 V peak-to-peak waveform at pin 3 of IC501 and 103 V d.c. at pin 2. Both were missing. I've had failure of one or other of the 330 kΩ start-up resistors R502 and R503 in the past, but they were both intact. The culprit turned out to be the 2SB698E power switch transistor Q108, whose emitter is connected to pin 2 of IC501. A check produced a low-resistance reading between its emitter and collector. We fitted a BC327 transistor and the set worked fine.

Dark picture: If the first anode preset was turned up there was a good bright picture, but the brightness control didn't work. As we didn't have the right manual we wasted a lot of time before we discovered that C483 (4.7 μF, 50 V) in the beam limiter circuit was the cause of the trouble. It had dried up which is not surprising as it is mounted right next to the field output chip's heatsink. In the model whose manual we were using the beam limiter acts on the contrast. In this model it acts on the brightness.

Intermittent screen blank out: The sound was not affected. Eventually the set's owner got fed up and asked us to take a look. We weren't too keen. We left the set on test in the workshop, and don't know how many times it misbehaved while our backs were turned. After a few hours we saw the fault, which was exactly as described. The screen went blank for about 5 seconds, then the picture returned. Off came the back cover and, as luck would have it, the picture blanked out permanently. When we turned up the first anode control's setting there was a faint negative image. So we used the scope to trace through the luminance signal path, all the time wondering how long it would be before the set decided to work normally again. In fact the signals were OK up to the RGB outputs from the AN5612 chip IC602, but there were no RGB signals at the CRT base panel. The only item in between is the teletext board. We found that the fault could be made to come and go by slightly flexing this PCB, and eventually traced the cause of the trouble to connector CD902. It looked perfectly OK, but the slightest touch would instigate the fault. We cut the wires and soldered them directly to the board. After that no amount of prodding would bring back the fault.

Matsui 1481B

Stuck on one channel, no remote control operation and no response from controls: This means microcontroller chip lock-out. Replacing R158 will cure the fault. But be careful. In the manual it's shown as a 315 mA circuit protector (ICP401). It looks like a small, white, wirewound resistor.

Weak red: If the trouble is weak red, most noticeable with colour bars and teletext, replace C964 (1 μF, 50 V). It's part of the red cut-off network connected to pin 29 of the TA8807N chip on the tube base and tends to go open-circuit.

Barely audible sound: Some tapping soon revealed dry-joints at the 6 MHz filter CF301 in the i.f. section. Resoldering brought the sound back to its full level.

Matsui 14R1 (Grundig G1000 chassis)

No sound or picture: It was soon obvious that a fault in the line output stage was dragging down the output from the power supply. As a short-circuit reading was obtained at the line output transformer we suspected this item, but the short remained when the transformer was isolated. Without a circuit diagram we had to start checking around and eventually found that D301 was short-circuit. A replacement put matters right.

Green screen at switch-on: A quick check on the c.r.t. base panel, around TR901 and TR904, soon revealed the cause of the fault. Replacing TR901 restored the picture. Another of these sets had no sound. There was no l.t. supply to the sound output chip because the fusible 4.7 Ω resistor in the feed was open-circuit. A replacement got the set going with plenty of sound.

No sound output: On investigation we found that the 4.7 Ω safety resistor R550 had gone open-circuit.

Matsui 2091

Dead set, but h.t. present: A dead set with h.t. present at the collector of the line output transistor directed our attention to the line driver stage. The 2SC2230 transistor here had 18 V at its collector but no drive at its base. This comes from the TDA1940 chip IC401 which was without its supply at pin 4. Tracing the print back (we'd no manual) brought us to D407 which was open-circuit with 9 V at its anode but nothing at its cathode.

Set remained dead when brought out of standby: D401 (1N4003) in the supply to the line driver stage was open-circuit.

Goes into standby when the mains switch was replaced: X901 (10 MHz) was the cause of this fault.

Partial field collapse: Check D301 (1N4003). You will usually find that it's open-circuit. Replace it with a good quality 1N4007 diode.

Dead set, power supply working: On closer inspection of the power supply outputs I noticed that there was a dry-joint at D803S. Resoldering this restored the sound and picture. As a precaution some other poor connections, in the line output stage, were resoldered as well.

Set reverts to standby at switch-on: If the on/off switch was held in, the set would stay on but with a snowy raster and no on- screen display. The culprit turned out to be the microcontroller chip's 10 MHz crystal.

Field timebase chip failure: If the TDA1170N field timebase chip IC301 has failed, C302 (100 μF, 35 V) must also be replaced. Otherwise IC301 could fail again.

Matsui 2092T

Slow to start from cold or completely dead: The usual cause is R507 (1 MΩ) in the power supply. For added reliability R506 (820 kΩ) should also be replaced.

Line output transformer tip: A number of different line output transformers that are not interchangeable were used in these sets. This one was fitted with a Philips transformer, part number 043221008P. It was short-circuit, a replacement bringing the set back to life.

Brightness level drops: The brightness level would drop drastically when this set had been on for a while. Tapping around brought me to R195. It was not dry-jointed: instead it was bent over, with the result that it made intermittent contact with a jump link. Straightening it cured the fault.

Faulty line output transformer: It took three different replacement transformers before the set worked. With these sets it's essential that the part number on the replacement transformer is identical to that on the original one. You'll know if you've fitted the wrong transformer type, because there will be reduced width and a picture with no contrast when you switch on. Unfortunately suppliers just send you what's in stock, listed under the model number. You've been warned! After fitting a new transformer, whether of the correct type or not, you may well encounter two more problems. So make a note of this for future estimates. First you might not be able to switch the set to standby using the remote control unit – instead the picture disappears into snow, the set remaining very much alive. Fortunately the cure is simple: replace the standby switching transistors Q504 (BF421) and Q503 (2SD401), one or both of which will be short-circuit. Use the correct types. We are talking about the non-relay version of the set of course. We thought our troubles were over when the correct transformer had been fitted, producing a correctly sized picture (with normal standby operation). Then we noticed that grey to dark areas of the picture were very dark, even with the brightness and contrast control settings at maximum. A dark suit, for example, appeared in black with no detail. When the first anode control's setting was advanced flyback lines appeared, with no improvement. The culprit was the TA8867N colour decoder and timebase generator chip.

Matsui 209T

Very intermittent loss of signals: This was traced to L101 (150 μH) going open-circuit. It feeds the 12 V supply to pin BM on the tuner.

Odd intermittent faults: All sorts of odd intermittent faults can be caused by R630 which, despite its designation, is in fact a 1.25 A fuse. The holder is usually to blame. We find it best to replace both the fuse and its holder.

Flyback lines at top of picture: We've had three of these sets in all with the same problem, flyback lines at the top of the picture, also the six RGB test lines for the auto grey-scale correction visible. Replacing C301 (100 μF, 50 V) and C303 (4.7 μF, 160 V) cured the problem in each case. Both capacitors are in the field output stage.

Matsui 2160

Small pulsating picture: There was a small, pulsating picture, the sound varied up and down and the brightness was low. We get quite a few of these sets so we went straight to the line output transformer-derived 180 V rail where the reading was low at only 73 V. Fitting a new STR58041 chopper chip (IC501) restored normal operation.

Sound but no picture: There was h.t. at the collector of the line output transistor but no e.h.t. because the transistor's base drive was missing. R428 and R427 which provide the supply to the driver transistor Q401 were red hot but the transistor was OK. A check at Q401's base showed that there was no input, only a d.c. voltage of 1 V which turned the transistor on. The μPC1420CA chip IC401 had the correct 12 V at pin 38 and the voltage at the X-ray shutdown pin was correct at 0 V. A d.c. reading of 3 V was obtained at the line drive output pin 26, however. After disconnecting this pin the voltage rose to 12 V. Clearly the chip was defective, a replacement curing the fault.

No sound or picture with normal channel change lights: Check the output from the STR58041 chopper chip IC501. If this is missing fit a replacement.

Power supply had no output: We found that the five-legged STR58041 chip was faulty.

Matsui 2180TT

Dead set: In these circumstances the first thing to check is the wirewound 5.6 Ω resistor R502. You quite often find that it's open-circuit because the STR58041 chip IC501 has failed. This time the resistor was OK, but we replaced the chip as we've known it to go open-circuit. There were still no results. Further checks with the meter revealed that R516 (1.5 Ω) had failed. Replacing this restored normal operation.

Dead set (STR58041 chip dead): The mains bridge rectifier was providing a supply for the STR58041 chip in the power supply but the latter provided no output. A new chip plus mica washer and heatsink compound got the set going.

R512 and R518 in PSU open-circuit: When one of these sets came in we found that R512 (0.47 Ω) and R518 (1 Ω) in the power supply were open-circuit. On replacing them the STR58041 chopper chip IC501 and the 5.6 Ω surge limiter R502 promptly died. After putting this right there was a raster full of snow with no channel number showing. There was no tuning voltage supply because the L5631 30 V regulator IC104 was faulty. Replacing this finally got the set working.

Dead set, STR58041 chopper faulty: This set was dead because the STR58041 chopper chip was faulty. Another repairer had given up and claimed that spares are not available. The STR58041 is available almost anywhere.

Field jitter: We found that there was a dry-joint at the scan plug P570. Resoldering its connections cleared the fault.

Matsui 2190

No sound: R312 (330 Ω) was burning up simply due to one of the TDA2030 audio output chips, in this case IC302.

Flat out brightness: The supply to the RGB output stages comes via R613 which was open-circuit – it's in the line output stage. Two power supply faults. C808 (1 μF, 63 V) was the cause of a fluctuating output. R810 (100 Ω) being open-circuit was the cause of a dead set. It's in the 12 V supply to the TDA4601 chip once the start-up phase has been completed and runs quite hot. The correct replacement should be obtained and fitted.

Dead set: It's always worth checking the secondary supplies with this model. We found that the standby 5 V supply was missing. The special 4 A fuse F802 had gone open-circuit while diode D807 was short-circuit. Replacements restored normal operation: a nice easy fault for a change.

Dead set, power supply working: Checks in the line output stage showed that there was 140 V at the collector of the line output transistor, but there was no line drive at its base. So we moved back to the line driver stage where the 24 V supply was missing. It comes from the emitter of TR803 in the power supply. This transistor acts as a stabilizer, in conjunction with a ZPY24 zener diode (D804) in its base circuit. D804 was short-circuit and TR803 open-circuit. We used a TIP41C to replace the transistor.

Starting problem: The set would trip when it tried to start. If you held down one of the keys on the remote control unit for a few seconds the power supply would begin to motorboat. After holding the handset button down for some time the set would start up. A check on the standby 5 V line showed that it dipped to only 4 V when the set tried to start up. The 8.5 V supply from the chopper transformer also dipped, from 12 V in the standby mode to 5 V, as start-up was tried. There are two 2200 μF electrolytics for this supply. Both were low in value, replacements restoring normal operation. These sets are not made to come on from the standby condition by operation of the mains switch. We find it annoying having to find the handset to switch on. The mains switch is the remote control type, however, and the simple addition of a diode (1N4148 or similar) in position D1511 will provide starting via the mains switch.

Dead, 115 V h.t. supply OK: We found that the 115 V h.t. supply was OK, measured at the cathode of D802. But a 5 V supply was missing. It comes from the M78L05 regulator chip IC804, whose input voltage was OK. A new M78L05 put matters right.

Matsui 2580

Set was tripping out: It needed a new BU508A line output transistor as the old one had never been fastened on to its heatsink. At switch-on the picture appeared without EW correction: the EW loading coil L901 then melted. The cause of this was a dry-joint on one of the EW modulator diodes. A replacement coil was obtained from Mastercare and fitted. The EW correction was now right but the picture needed a slight adjustment to its height. Now this chassis is a digital one that uses the ITT Digi chip set, so no board adjustments are possible. To enter the service mode you hold down the service switch inside the set and press the channel button on the remote control unit. Press the full button to select from the on-screen menu the adjustment required. For height, press the volume up/down button. To store, press normal then standby.

Height variations before warm-up: Check the 24 V supply. A small fluctuation here will result in height alteration. D804 (ZPY24) is usually responsible.

Dead set: A look in the back of one of these sets for the first time can be a bit worrying, as they are full of huge digital chips much like the ones used in some ITT models. The power supply is a fairly basic TDA4601 type, however, and this set was dead. After checking some resistors we found that R808 (270 kΩ) was open-circuit. When this and the chip had been replaced the set worked normally.

Dull red on screen: Red flyback lines were visible in only the darkest areas of the picture. This was traced to IC101 (VCU2133).

Wouldn't come out of standby: Diode D118 was going open-circuit intermittently.

Dead set, standby light on: This set was brought in because it was 'dead'. The standby light was on, however, and a check on the 150 V h.t. rail produced a reading of 78 V. After about 30 seconds the h.t. voltage rose rapidly to 175 V or so. When the handset's on button was held down the 24 V supply came up and the h.t. voltage squegged between about 70 V and 100 V, still with no picture. We disconnected the scan coils and tested the power supply with a light bulb as the load. Still the h.t. wobbled, so it was time for component tests in the power supply. A check on the capacitors soon brought us to C808 (1 μF, 63 V) which was open-circuit. A replacement restored the set to life.

Picture would break up on channel change: It was as if the a.f.c. was locking out. Also line sync would occasionally be lost on channel change when the set had been on for a long time – once the picture had locked it would run perfectly, provided the channel wasn't changed. Fortunately we had another of these sets in at the time, so we tried swapping over the plug-in chips. This proved that the MDA2062 memory chip IC1503 was faulty.

Set is dead or apparently dead, with no standby LED indication: The first thing to check is the 145 V h.t. voltage – at the scan coil interlock. If it's OK or high, check the 8.5V supply at L805. If this is missing, look at the link wire between L805 and R826. It may look OK but is often poorly jointed. With sets that go into standby intermittently, check the fuseholder which may be loose. This can be confirmed by taking a close look at the fuse: if it's loose you will see small burn marks at both ends. Also the fuseholder may not be correctly connected to the PCB – the hole isn't always through the land but to the side of it. The mains input connector can also be to blame, and is often 'sparky' or the print is cracked at the point where the connector is soldered to the panel.

No raster, supplies OK: If the supplies are all present and correct, the tube's heaters are lit, the sound is OK but there's no raster, a quick scope check at the field scan coil plug can save time. There should be a huge sawtooth waveform here. If this is missing, check for 24 V at pin 2 of the TDA8175 field output chip IC401. If this is OK, check at the input (pin 7) for a sawtooth at about 2.5 V peak-to-peak. The TDA8175 chip is the usual cause of the fault, however. As with most sets nowadays, loss of the field scan produces the no raster symptom. In this case a guard circuit operates in the codec chip.

Matsui 2890

Field collapse: The white line was very bright indeed. Had this additional factor registered with us time wouldn't have been wasted looking for a fault in the field output stage. The cause of the fault was in the video output supply, where D406 was short-circuit and the series safety resistor R440 was open-circuit.

Dead set, no channel display: Checks showed that there was no a.c. supply to the bridge rectifier as relay RY101 wasn't being energized by a low signal from pin 30 of the microcontroller chip IC103. This was hardly surprising as IC103's 5 V supply (pin 52) was missing. It comes from the 5 V regulator IC106 which was very hot. A resistance check showed that there was a short-circuit across the 5 V line. The cause turned out to be the 5.6 V zener diode D142, which is on the print side of the PCB, connected between pin 52 of IC103 and chassis. The set worked well once this item had been replaced.

Intermittent going to standby: Here's fault that can catch you out with these sets. The symptom is that the set goes to standby intermittently when warm. The cause is the standby transformer T101, which goes open-circuit intermittently to produce the fault. The part number is 040535009C.

Mitsubishi

MITSUBISHI CT21M1TX
MITSUBISHI CT2227BM
MITSUBISHI CT2531BM (EURO 4 CHASSIS)
MITSUBISHI CT2532TX (EURO 4 CHASSIS)
MITSUBISHI CT25A2STX
MITSUBISHI CT29B2STX

Mitsubishi CT21M1TX

Power supply immediately shut down at switch-on: This sort of trouble is normally caused by a heavy overload, but checks for shorts across the outputs and subsequently disconnecting them individually failed to reveal any. The run supply rectifier D905 and its associated safety resistor R905 were in order and replacement of the TEA2261 power supply control chip made no difference. We eventually found that the opto-isolator PC951 was defective, a replacement restoring normal operation. In retrospect we should have gone for this first as the set had failed during a thunderstorm.

Intermittently dead from cold: We instinctively reached for the freezer can and hair dryer. This soon brought us to C905 (470 μF, 25 V), which is in the start-up supply.

No sound: This was because an l.t. supply to the audio chip was missing. We found that Q952 (2SA950) was open-circuit. A BC640 proved to be a suitable replacement.

Very loud whistle: It was similar to that from a 405-line output transformer. We tried impregnating the chopper transformer T901 with shellac, but this made no difference. Scope checks then showed that the chopper power supply was running at a very low frequency. The slow-start transistor Q902 was the cause. Incidentally all the supplies produced at the secondary side of the chopper transformer were correct when the circuit was operating in this l.f. mode and producing the awful noise.

No audio output: The cause was found on the power supply board, where Q952 (2SA950) was leaky collector-to-emitter. It's the regulator for the 22 V supply to the audio output stage.

Mitsubishi CT2227BM

Would stick in standby: This set would operate normally for many days after which it would go into the standby mode and couldn't be switched by means of the remote control unit. The cause of the trouble was the M50124P ETS chip IC7A0.

No sound and no raster – the mains input circuitry: If this is in order check the l.t. supply at the remote control panel – the safety resistor R7A0 (1.2 Ω) tends to go open-circuit.

Dead set caused by dry joints: We service a lot of these sets and have found them to be very reliable. A dead set call usually means that there are dry-joints at the line driver transformer or that the safety resistor in series with the remote control supply rectifier has failed. Only the correct type should be used.

No sound: Often due to R371 being open-circuit. This resistor is in the feed to the sound supply bridge rectifier. We've found no cause for these resistors to fail.

Field collapse: The 330 μF, 50 V field scan coupling capacitors C412/3 are often the cause of this, but were blameless on this occasion. Voltage checks showed that the supply to pin 12 of the timebase generator chip IC401 was missing. C403 (470 μF, 16 V) was short-circuit.

Mitsubishi CT2531BM (Euro 4 chassis)

Normal sound but no raster: As the tube's heaters were alight we turned up the first anode control and found that the cause of the trouble was field collapse. A quick check showed that the 24 V supply was missing at pin 7 of the AN5521 field output chip IC401. This supply is derived from the line output transformer. We found that the 0.82 Ω safety resistor R563 in the relevant rectifier circuit had failed. Somewhat to our surprise a new resistor restored the full field scan. Normally the resistor goes open-circuit when the chip fails. A long test run proved that all was OK.

Set dead, high pitched whistle: Checks on the line output transistor showed that it was short-circuit all round, and a replacement immediately bit the dust. A new STR59041 chopper chip brought the set to life, but the picture had an odd look to it: we can only describe this by saying that the top half of the screen looked as though it had been scribbled over. After a minute the scribble disappeared, then we realized that we couldn't see the MTV logo in the corner of the screen. The h.t. was found to be at 198 V instead of 155 V. Further checks in the power supply revealed that C908 (10 μF, 100 V) had fallen in value to around 3 μF. Replacing this finally restored normal operation. R903, R910 and R917 were all-dry-jointed as well.

H.t. voltage too high: The h.t. voltage should be 156 V in these sets. In this one it had risen to 190 V. As a result, it was destroying line output transistors. The STR59041 chopper chip IC901 can be troublesome, but in this case C908 (10 μF, 100 V) was responsible. It's the reservoir capacitor for IC901's −41 V supply.

Mitsubishi CT2532TX (Euro 4 chassis)

Dead set: The mains input and rectifiers were OK and there was 320 V across the reservoir capacitor. We decided to check the line output transistor in circuit before delving into the power supply. It gave a short-circuit reading, but was OK when checked out of circuit. There was little else to cause the short which turned out to be in the line output transformer, between pins 2 (h.t. feed) and 7 (chassis).

No tuning storage: A check on the M58630P memory chip IC702 showed that its −31 V supply was missing at pin 2. This comes from the subpower supply, where pin 4 of the chopper transformer was open-circuit. Fortunately the break was at the connecting pin and could be repaired.

Lines on screen during warm-up: Then there was lack of height. In addition a shrill noise came from the power supply. The culprit was C908 (10 μF, 100 V) which is the reservoir capacitor for the −41 V feed to pin 1 of the STR59041 chopper chip IC901.

Mitsubishi CT25A2STX

No sound or vision: When the set was switched on the red standby LED lit, then the green one, then both LEDs went out. Application of heat from a hair dryer in the vicinity of IC901 enabled the set to start up normally. By further heat and freezer testing we discovered that the cause of the trouble was the 4.3 V zener diode D909.

Trips when comes out of standby: While we were carrying out some checks the line output transistor went short-circuit. So we fitted a 60 W bulb as a dummy load instead. When the set was brought out of standby the bulb glowed brightly, with the h.t. well above its correct value. The cause of this was traced to C609, the 47 μF capacitor in the chopper transistor's base circuit. A new capacitor reduced the h.t. voltage to the correct level while a new line output transistor restored the picture.

Wavy dot pattern on picture: This set had an intermittent fault with rather obscure symptoms. A wavy dot pattern was superimposed on the picture. The luminance response was also affected, and at times the picture blanked out completely. Teletext was OK. Tests showed that the luminance signal was present wherever it should be. In the fault condition the feedback pin of the VCJ chip was at 1.5 V instead of 3.5 V. Freezing this chip cured the fault – but replacing it didn't! When we checked the blanking voltage from the teletext board we found that it was at 0.6 V instead of 0 V. The culprit turned out to be the JC501Q transistor Q7705 on the text board.

Line output transistor shorted: If the line output transistor Q522 is short-circuit, the four small electrolytics in the power supply should be replaced using high-temperature (105°C) components. They are as follows: C905 (470 μF, 35 V) C906 (47 μF, 50 V), C909 (2.2 μF, 50 V) and C920 (100 μF, 25 V). Note that the transistor type varies with screen size. With a 21 inch tube it's type 2SD1877 (part number 260P606010), with a 25 inch tube it's type 2SD1878 (part number 260P607010) while with a 29 inch c.r.t. it's type 2SD1879 (part number 260P608010).

Intermittent loss of luminance: One of these sets suffered from intermittent loss of the luminance signal. At first it seemed that a dry-joint was likely to be the cause, but the culprit was actually Q202 (J501).

Mitsubishi CT29B2STX

Line output transformer short: If the line output transistor has gone short-circuit it's likely that the h.t. is too high. In this event replace R961 (150 kΩ), C906 (47 μF, 50 V) and IC950.

North/south distortion: This is a rare fault these days. With the 29 inch tube, the symptom looked pretty bad. There's a 2SA950 30 V switch transistor (Q4009) on the NS correction panel. It was supplying only 2.3 V. A replacement cured the fault.

Line output transistor shorted: If the line output transistor Q522 is short-circuit, the four small electrolytics in the power supply should be replaced using high-temperature (105°C) components. They are as follows: C905 (470 μF, 35 V) C906 (47 μF, 50 V), C909 (2.2 μF, 50 V) and C920 (100 μF, 25 V). Note that the transistor type varies with screen size. With a 21 inch tube it's type 2SD1877 (part number 260P606010), with a 25 inch tube it's type 2SD1878 (part number 260P607010) while with a 29 inch c.r.t. it's type 2SD1879 (part number 260P608010).

Nei

Nei 1451 (Indiana 100 chassis)

BU508 chopper transistor shorted: If one of these sets comes in with the BU508 chopper transistor Q800 short-circuit, before fitting a replacement and switching on check R809. You'll find that this 270 kΩ resistor is open-circuit.

Intermittent failure: The cause was traced to thermistor R802. When we removed it the board beneath was scorched. A good clean-up, a new thermistor and some resoldering in the power supply area cured the problem.

Intermittent standby or dead set: This set would behave perfectly for days, then suffer a bout of unexpectedly going into standby and sometimes becoming completely dead. Our fifth-generation copy of the service data was of little use but the chopper circuit appeared to be a conventional arrangement with a TDA4600 type chip and a BU508A chopper transistor. After the usual checks for dry-joints, we noticed that the 5.6 kΩ, 3 W resistor which feeds the line driver stage was rather tired looking. This was replaced, along with the 100 kΩ resistor connected to pin 5 of the TDA4600 type chip – it was slightly out-of-specification. We then tested the set continuously for several days. As it seemed to be OK, we returned it to the customer. It bounced back two weeks later. Some serious head scratching and lengthy meter and scope checks eventually revealed an interruption in the drive between the chopper control chip and the base of the BU508A transistor. Coupling is provided by the usual electrolytic capacitor and a small RF choke of unknown value. Replacing the latter with an equivalent-looking device robbed from a scrap VCR chassis restored normal, reliable operation.

Network

Network NWC1410R

Tuning drift: We found that there was a dry-joint on the tuner's a.f.c. pin.

Sound output transistors failed: Both the sound output transistors Q652/3 had failed. We replaced them along with, for good measure, the two $1\,\Omega$ resistors R659/660 which were looking distressed. This restored the sound.

Suddenly lost picture and smoked: A look at the main PCB showed that R338 ($10\,\Omega$) was in a sorry state – it was burnt black. This wasn't the main trouble, however: the line output transformer had developed shorted turns. Replacing these two items restored the picture – with no smoke.

Nikkai

NIKKAI BABY 10
NIKKAI TLG1409
NIKKAI TLG99

Nikkai baby 10

Loss of power: This was traced to failure of the protection components F402 and D402. The cause had of course been reversed connections to the battery. Make sure that the user understands the importance of correct connection when you get this sort of thing.

Dead set: These little sets have sold well. The main problem we've had with them has been failure of the potted regulator. This particular one was a bit different, however. Our field engineer found that the 5 A h.t. fuse was open-circuit. As he couldn't measure any shorts and the fuse had simply died (not blown) he replaced it and switched on. At this a piece of print from the bridge rectifier to the h.t. line burnt up pretty spectacularly. So the set was brought into the workshop. We found that there was a short-circuit across the h.t. line. The reverse polarity protection diode was hot favourite but blameless, as was the main reservoir capacitor. The cause of the problem was a short-circuit in the d.c. jack, which had melted. Spares for these sets are available from Willow Vale.

Picture became darker: We've had a few of these little sets in for repair, generally for straightforward faults. This one took us a little longer to pin down. At switch-on the picture was perfect, but after the set had been on for a while the screen gradually became darker until there was no picture. Out came the freezer, and the area of the fault was traced to Q303, the picture mute transistor. Fitting a replacement, 2SC1815 or BC184, cured the fault.

Switched itself off intermittently and field bounce on a change of scene: On test we found that the picture was oversized, with bent verticals. A check on the potted regulator chip IC402 showed that its output and input were the same. So a new one was ordered. When this came we found that it had been improved for the better, having a diecast case to improve the heat transfer to the heatsink and a fixing hole so that it could be screwed down tightly.

Dead set: Both fuses were intact and the relay latched, so attention was turned to the voltage regulator chip IC401 which was found to have a very low output. A quick check with the bench power supply confirmed that IC401 was the cause of the fault.

Dead set: This set was dead but for a change it wasn't the 12 V regulator. R402 (1 Ω, 0.5 W) was open-circuit.

Only one inch of field scan: There was a bright line across the screen. When we checked at the pins of the field output chip with our meter the scan opened up. If the service switch was operated a couple of times you might get full scan or a blanking fault with flyback lines. There was no on-screen display. The cause of the fault turned out to be the AN5512 field output chip. Why it caused loss of the on-screen display we can't say.

Intermittent shutdown: At last a fault other than the 12 V regulator! The set would shut down intermittently. We found that the line output transistor's emitter leg was dry-jointed.

No picture, sound and display OK: A long search finally brought us to R425 (270 kΩ), which provides biasing in the beam limiter network. It was open-circuit.

Nikkai TLG1409

Dead, squealing power supply: This set was dead. A squealing noise came from the chopper power supply and there was a smell of something cooking. The cause of the trouble was C617 (4700 pF, 1 kV) which had split open and was very hot. You'll find it connected across the 109 V h.t. rectifier diode D116, in series with R618.

Field collapse: We found that the 10 Ω safety resistor R306 was open-circuit for no apparent reason.

Slow to come on: When the chopper transistor was sprayed with freezer it would shut off. With a new transistor fitted the symptoms remained the same. Closer inspection showed that there was a slight gap between the transistor and its heatsink. Tightening the nuts and bolts that secure the transistor cured the trouble.

Nikkai TLG99

Dead, previously intermittent low gain: This colour portable had two faults. It arrived in the workshop dead, but prior to that it had had an intermittent low-gain problem. As it has remote control, standby switching is incorporated. This is done by altering the voltage applied to pin 2 of the STK5412 power supply chip IC104. The switch-on signal comes from the front panel via the three transistors Q116/7/8. D.c. continuity checks showed that the print between the collector of Q117 and R106 was open-circuit. The cause of the low-gain problem was an a.g.c. fault. This was again due to open-circuit print – between pin 4 of the tuner and pin 5 of the TDA4501 chip IC101. The damage had been done by insertion into the board of TP1 (the a.g.c. voltage test point), the print being a little thin here.

Dead set faults: These sets always seem to come in dead. Here are various faults we've had. R109 (180 Ω, 0.5 W) goes open-circuit. This resistor's body colouring makes it look as though the value is 1.8 kΩ – we've even had a faulty one measure 1.8 kΩ! Q117 (2SC1573A) often goes open-circuit. It's an npn transistor rated like a video output device. Another regular failure is the remote standby transformer whose primary winding goes open-circuit. The 12 V supply filter resistor R104 (5 Ω, 1 W) can and does go high in value. This usually results in a dead set though in one case the symptom was persistent field collapse because the low 12 V supply upset the TDA4503 chip, removing the field drive.

Very weak, distorted sound: There was very weak, distorted sound. As R153 was open-circuit there was no bias voltage at the base of Q103, the lower transistor in the class A push-pull output stage.

Intermittent no colour: There was also field fold-over and horizontal shift-off. The cause was traced to the wire link between R218 and pin 27 of IC101 – loss of the sandcastle pulses produced the symptoms.

Persistent field collapse: R104 (5 Ω, 1 W) going high in value causes the 12 V supply to go low, upsetting the TDA4503 chip and thereby removing the field drive.

Orion

Orion 14ARX

Set stuck in standby: There was no 103 V h.t. supply, although we found that it did come up initially at switch-on, subsequently dying down. The video/chroma/i.f. chip IC401 had no voltage at pin 40. This comes from the 103 V line via R458, being stabilized by the 9.1 V zener diode D403 which was short-circuit. A replacement restored normal operation.

Loss of field sync (picture rolling): A replacement TA8808N chip (IC401) provided a cure.

Dead set: We were told that this set hadn't had much use. Certainly it was clean inside, but it refused to start. The 330 kΩ start-up resistors were OK, also the STR50103 power supply chip. The clue was provided by R532, a large 2.2 kΩ wirewound resistor, which got very hot after a few seconds. It feeds a diode and a 3.3 μF, 250 V capacitor, C530. This item was dead short, a replacement restoring the set to life. It's no wonder that the capacitor had failed. It is of minuscule proportions and is mounted very close to three wirewound resistors. The original one had obviously been running very hot, as the outer plastic cover was discoloured. We used a high-grade type of larger proportions, rated at 105°.

Would not come out of standby: A dead set or one that fails to be awakened from the standby state would probably have you giving the power supply suspicious looks, especially as it uses an STR50103 chip. Before you change this item, however, take a look at the supply to the 5 V regulator (IC105) at the front of the chassis. It's derived from the mains supply via a half-wave rectifier and a suitably substantial resistor, with C530 (3.3 μF, 250 V) to provide decoupling at the hot end. On a couple of occasions recently we've found this capacitor to be open-circuit or very low in value.

Stuck in standby: For the microcontroller chip to respond to remote control commands it must have a 5 V supply. Only 2.8 V was present. The 5 V regulator that provides this supply obtains its input from the mains circuit, via D531, R531 and R158, the voltage at the junction of these resistors being smoothed by C530. Only 4.8 V was present here, because C530 was open-circuit. This voltage was insufficient to operate the regulator correctly. Incidentally, once the set is running the 5 V supply is derived from the line output transformer. We discovered that if we applied 5 V to the regulator's output the set would start up and continue to run until it was switched off and then asked to start from cold.

These sets, which use a simple resistive dropper to power the standby circuit, tend to come to us stuck in the standby mode. To prove where the fault lies, disconnect Q117. If the set doesn't come on, the cause is probably the STR50103 chip or the series-connected 330 kΩ resistors. If the set now works except for standby – the cause of the fault is usually in the standby 5 V feed. If this falls below 4 V, the LED still lights but the set doesn't respond. C530 (3.3 µF, 250 V) is the usual cause of this problem. It's mounted by the mains switch, in a hot corner. A high-temperature component is a good idea here. This chassis is used in Matsui, Bush etc. sets.

Osaki

Osaki P10R

Field collapse: We found that C428 (0.022 μF) was leaky.

Dead set but no blown fuses: The cause of the fault was no output from the 12 V regulator IC402, a big, square block. We fitted the Nikkai AL2711K and all was well during a long test run. The set is similar to the Nikkai Baby 10.

Field collapse with no sound: Replace zener diode ZD302 (12 V, 1.3 W) and the two 51 Ω, 2 W parallel connected resistors R730 and R731.

Osume

Osume CTV1484R

Loss of channel selection and sound control: This remote controlled cash'n'carry weirdo turned out to be an Alba set in disguise. The trouble was loss of channel selection and uncontrollable, maximum sound when the set had been in operation for a few minutes. Heat and freeze probing suggested that a 455 kHz crystal on the front remote control/tuning board was faulty. A suitable transplant was obtained from a defunct Sharp remote control panel. Fitting it restored normal operation.

Sound but no e.h.t.: There was only about 1 V of drive at the base of the line outpul transistor Q111. The line driver transistor Q110 had a healthy input but not much came out at the other end. The unexpected cause was the line driver transformer EM115 whose primary winding read high at 80 Ω. A check on the replacement produced a reading of 50 Ω.

Tripped on switch-on: A lot of time was wasted because we assumed that the chopper power supply was tripping due to an overload. After drawing a blank in the 120 V department and then a check on the 12 V supply, it finally dawned on us that there was no standby voltage at C128. In this circuit the result is a cyclic on/off effect. The cause of the trouble was the standby transformer EM112 which was open-circuit. Note: all tripping is not the same!

Panasonic

PANASONIC ALPHA 1 CHASSIS
PANASONIC ALPHA 2 CHASSIS
PANASONIC ALPHA 2 W CHASSIS
PANASONIC ALPHA 3 CHASSIS
PANASONIC EURO 1 CHASSIS
PANASONIC EURO 2 CHASSIS
PANASONIC TC1485 (Z3 CHASSIS)
PANASONIC TC1785 (Z3 CHASSIS)
PANASONIC TC21M1R (Z4 CHASSIS)
PANASONIC TC2205 (U2 CHASSIS)
PANASONIC TX2 (ALPHA 1 CHASSIS)
PANASONIC TX21M1T (Z4 CHASSIS)
PANASONIC TX21T1 (ALPHA 2)
PANASONIC TX21V1 (ALPHA 3 CHASSIS)
PANASONIC TX24A1 (ALPHA 2 W)
PANASONIC TX25T2 (ALPHA 2 CHASSIS)
PANASONIC TX28A1 (ALPHA 2 CHASSIS)
PANASONIC TX28G1 (ALPHA 2 CHASSIS)
PANASONIC TX28W2 (ALPHA 3 CHASSIS)
PANASONIC TX28W3 (EURO 1 CHASSIS)
PANASONIC TXC74 (ALPHA 1 CHASSIS)
PANASONIC U4 CHASSIS
PANASONIC Z3 CHASSIS
PANASONIC Z4 CHASSIS
PANASONIC Z5 CHASSIS

Panasonic alpha 1 chassis

Lack of colour at top of screen: The fault occurred when playing back pre-recorded video tapes. The cure is to change C604 from $0.047\,\mu F$ to $0.33\,\mu F$.

Power supply fault: The mains input circuit and rectifiers were OK as there was $320\,V$ across the reservoir capacitor, but the power supply produced no output voltages. Checks on the outputs revealed that the h.t. rectifier D851 was short-circuit.

Tripping: Also, the SR2KN avalanche diode D854 was short-circuit. The usual cause of this situation is the STR54041M chopper chip IC801. This time, however, the cause was C808 (10 μF, 50 V) which had gone low in value.

Panasonic alpha 2 chassis

Beating bars would appear over text: While soak testing some of these sets prior to delivering them we noticed that if they were left running on teletext white beating bars would appear over the text. Then a customer complained about this, so we investigated. We found that the trouble was prone to arise when advertisements appeared on ITV and Channel 4. It could also be provoked by interrupting the signal and then restoring it. Our theory is that the text artificial sync generator switches on when signal interruptions occur and that when sync pulses return the ASG doesn't lock to them. On speaking to Panasonic we found that there's a modification kit to overcome a design fault in the MPU chip. The part number is TZS803001.

Eratic colour: When we switched the set on there was no colour at all, although the colour was perfect once the back had been removed. The set was put on the soak bench and minutes later the picture flashed orange, green then blue. The colour-killer finally put a stop to the colourful display. Checks on the 8.86 MHz crystal X601 and its associated trimmer capacitor were fruitless. The chroma chip IC601, the delay line DL601 and the matching coil L603 were also innocent. Voltage checks were then carried out around IC601. A fluctuating voltage at pin 5 led to a check on C612 (0.01 μF). It read about 5 kΩ on my meter. A replacement cured the problem.

Intermittent colour: Sure enough we found that after a few hours the colour would sometimes flicker off then back again. Where to start? Close inspection of the picture showed that there was barely enough colour saturation even with the bar graph giving a maximum reading. With the colour set at maximum it became evident that line pairing was present in red areas. It was as if the delay line circuit was giving problems. This proved to be the case: the chroma delay line DL601 was defective, a replacement restoring normal saturation without the intermittency.

Picture bowed in at the sides: Checks in the EW diode modulator circuit failed to reveal anything amiss. We had to go a little further back, to the control circuit, where we found that C754 (180 pF) had developed a slight leak. Replacing this capacitor restored the picture's straight edges.

Dead, whine from power supply: This set was dead apart from a whine that came from the power supply. The 2SD1441RL line output transistor Q551 was short-circuit and the fusible resistor R567 in the feed to the stage was open-circuit. When these items had been replaced the set was still dead, the SR2KL overvoltage diode D854 having gone short-circuit. Removing this diode and disconnecting the line output stage enabled me to confirm that the h.t. voltage was much too high. The cause was IC801 (STR54041M).

Failure of line output transistor: Failure of the 2SD1441RL line output transistor Q551 is not uncommon in these sets. When you get this problem, always resolder the line driver transformer T531 – otherwise the replacement transistor may fail after a short time.

Video disappears intermittently: It would leave only the sound. As we've had similar problems before we carried out a quick check on the waveforms around the M51326P scart switching chip IC2601. Video should enter at pin 5 and reappear at pin 12. In the fault condition it didn't. Temporarily linking the two pins proved the point.

No picture, no tube heater supply: Inspection around the line output transformer revealed a crack in the PCB, although this didn't affect the heater supply – what did was the fact that the winding between pins 7 and 10 was open-circuit, presumably caused by the impact that had damaged the board. As a test we wound a small coil of wire around the transformer's core to provide a heater supply.

Intermittent black lines on the right-hand side of the picture: On one occasion both sides of the picture were affected. This should be regarded as an early warning sign with all Alpha 1 and 2 sets. It means that the line output transistor Q551 is in imminent danger of going short-circuit because of dry-joints on the secondary side of the line driver transformer T531.

No line or field sync from cold: But when the back was removed everything returned to normal. Eventually, after many days of cold starts, we declared C3550 on the text panel to be faulty. A new 1 μF, non-polarized capacitor cleared the fault.

Tuner faulty for 5 seconds at switch-on: For the first few seconds of each day this set behaved as though its tuner unit was faulty. The picture would come on all snowy, but very soon there would be a quick flash and the selected programme would appear. During the first few days that we had this set on test the tuner unit proved to be OK, as did IC171 which generates the tuning voltage. Over the course of a further few days we tried all the capacitors in this area, one by one. The culprit turned out to be C15. A replacement 2.2 μF capacitor restored normal operation.

No picture: The cause was that the brightness control voltage was missing at pin 20 of the TDA3505 video processor chip IC602. The decoupling capacitor C309 (10 μF, 50 V) was leaky.

Panasonic alpha 2 W chassis

No sound or vision, power supply ticking: This set was almost dead: there was no sound or vision, but the power supply was making a ticking noise. Checks showed that the 155 V supply's protection diode D854 was short-circuit, indicating an overvoltage problem, while R567, the fusible resistor that's in series with the supply to the line output stage, was open-circuit. We replaced the protection diode, removed R567, connected a dummy load and powered the set via a variac. A check on the voltage at TPE1 as the input was increased showed that there was no regulation – the voltage would have passed 160 V if we'd let it. A new STR54041 chip restored regulation, and the h.t. voltage could now be stabilized at 155 V. So a replacement was fitted in the R567 position, the dummy load was removed and the set was tried with the full mains input. R567 went open-circuit and the line output transistor Q551 went short- circuit. A new line output transformer, transistor and R567 completed the repair.

Dead set: This beast had come from another dealer, who'd given up. It had come to him as a dead set, so he'd replaced the regulator chip IC801. This made no difference. He then found that the overvoltage protection diode D854 was short-circuit. He'd removed it and tried again. This is where his problems started. Without the protection, the voltage on the 160 V rail shot up to about 300 V. The reservoir and smoothing capacitors started to smoke. Then the line output transistor winced and went short-circuit, blowing the fusible link R567. This is where we came in. We fitted another STR45051M chip in position IC801 and checked the circuit around pin 5, which is where the regulation control takes place. R815 (5.6 Ω) was open-circuit, and we also replaced C808 – as a precaution. After replacing D854 then Q551, C851 and R567 in the line output stage we gingerly powered up the set. Fortunately the friendly rustle of e.h.t. preceded a very good picture.

Picture distortion: When this set had been on for a few hours the right-hand side of the picture would creep in by a couple of inches and there would be slight EW distortion. Although there were no signs of dry-joints at its pins, experience suggested that the first step should be to resolder the line driver transformer. This is all that was required.

High pitched squeal from the power supply: There was no line output stage operation. We found that the 2SD1441 line output transistor Q551 was short-circuit and the thermal link R567 in the h.t. feed to the stage open-circuit. As no other shorts or defects could be found we fitted new components. After a 3-hour soak test the set shut down and the squeal returned. Why does this always happen after you've told the customer that his set has been fixed? Anyway this time we fitted replacements then carried out a thorough check on the soldered joints in the line timebase. As one of the line driver transformer's legs looked suspect we removed the transformer, cleaned and tinned its legs, then refitted and resoldered it. A long soak test proved that all was now OK.

Panasonic alpha 3 chassis

Set was dead but making a noise: It was not an overload whine but an unloaded power supply buzz. R555, the fusible link in the h.t. feed to the line output stage, was open-circuit. No shorts could be measured, however. Overriding R555 for the purpose of testing showed that initially, at switch-on, there was a burst of e.h.t. that was clearly too high – the focus spark gap also fired over. The cause of the problem was the line output transformer T551.

Would become very dead intermittently: In the fault condition the standby LED glowed but nothing else happened. It took many weeks of soak testing before we discovered that C1251, which decouples the reset line to the main microcontroller chip IC1213, was leaky. As a meter check on the capacitor produced a reading of 700 Ω, it was understandable why the set was sometimes inoperative. This capacitor was one of those infamous 10 nF, 50 V ceramic ones that earned a deserved reputation as prime suspects a few years ago.

Very intermittent colour fault: When the symptoms were present they lasted for only a short period and were quite difficult to see. We eventually found that it was easier to try to establish the fault area by soak testing the set in the teletext mode. In this mode you could see a very slight variation in the blue level. With a scope connected to the collector of the blue output transistor Q352 we found that the d.c. voltage was varying by a volt or so. The culprit turned out to be a 330 pF ceramic capacitor, C352. It provides frequency-selective decoupling at the emitter of Q366 (Q352 and Q366 are connected in series).

No tuning memory: This was because the chip select signal to the memory chip from pin 53 of the microcontroller chip was missing. We found that L1209 had gone open-circuit. Its part number is ELEXT100KA.

Panasonic euro 1 chassis

Picture would flash at random: A white raster was alternating with a normal picture. In moments of stable operation we also noticed that there were no on-screen graphics. As these are added in RGB form we changed the VDU2416 video processor chip. This of course made no difference. An overdue check on the teletext operation produced a white raster with random graphics characters spread around the screen. After fitting a new TPU2763 teletext chip (IC1761) normal operation was restored.

Weak monochrome picture, poor sync: When checks were made around the SAD2410 video AD converter chip IC1601 we found that while good video was present at the input some of the data outputs appeared to be of low amplitude. Replacing this chip cured the fault.

On-screen menu failure: All that would be displayed over the picture were random numbers and characters. As the information for the on-screen menus is held in an EPROM on the digital board, this was the first thing we checked. Another reason for homing in on it was the fact that it's a plug-in chip. But a replacement made no difference at all. We eventually traced the cause of the trouble to the TPU2735 teletext processor chip, despite the fact that teletext worked perfectly.

Dark picture with weak sync: The first thing we did was to check whether the AV inputs were similarly affected. They were, which suggested that the cause of the fault was in the digital circuitry. A textbook video waveform entered the SAD2140 analogue-to-digital converter chip IC1601, but data checks after this were less informative.

Our usual approach to these sets is to revert to guesswork. So we replaced IC1601. Fortunately this cured the fault.

Intermittent light raster, no sync: It would never happen with the back off! After some inspired guesswork I decided to replace the SAD2140 video AD converter chip IC1601. Fortunately this solved the problem.

Intermittent loss of colour: It came as no surprise when, after about an hour, the colour disappeared completely. By going into Service Mode One we were able to discover that the colour VCO adjustment couldn't be pulled in. As a first guess as to the cause, we decided to replace the PAL master oscillator crystal X1656. Fortunately this cured the problem.

Would shut down after 10 seconds: The set would restart briefly when a channel button was pressed, and whilst running would change channels. We carried out a series of these brief running tests in order to make some measurements and after a few we noticed that the c.r.t.'s heaters were glowing. So the line output stage was in operation, although no picture appeared on the screen. The cause of the trouble was field collapse. We found that the 27 V supply to the TDA8175 field output chip IC561 was missing because R561 (1.5 Ω) was open-circuit. The chip and the fusible resistor had to be replaced. Part number for the resistor is ERQ12HJ1R5.

Dark, blank raster only: The set's owner claimed that the picture was poor before this symptom occurred. Scope readings in the digital circuitry failed to produce any clues, but a burnt finger told me that the ACVP2205 adaptive comb filter chip was faulty. When this had been replaced the picture, although restored, was decidedly solarized. This second fault was eventually cured by replacing the SAD2140 S-VHS analogue-to-digital converter chip IC1601.

Nicam sound problem: The stereo sound was low and distant, and sometimes came from one channel only. The FM sound was good, however, from both channels. Unfortunately the Nicam chip proved to be innocent. Our next stab in the dark was successful. We replaced the AMU2481 multi-sound processor chip IC1431.

Teletext and graphics problems: After a few minutes black lines would develop across the text screen or the on-screen graphics boxes, then increase until black eventually predominated. Obvious guesses like the text processor i.c. and text memory chip proved to be incorrect. The culprit turned out to be the DPU2553 deflection processor chip IC1501.

Picture geometry errors: Very intermittently this digital TV receiver would develop large picture geometry errors, leading to failure of the line output transistor. When it had been on soak test for several hours we noticed that the height and width had increased dramatically. Fearing for the safety of the line output transistor, we switched off and resorted to guesswork. We suspected the DPU2553 deflection processor chip IC1501, and replaced it. But later that day the fault returned. Eventually, after many hours of uninspired guesswork, we discovered that the cause of the fault was the MCU2600 master oscillator chip IC651.

Off-air channel fault: This set worked happily enough in the AV mode, but when an attempt was made to select an off-air channel it would shut down then come back on again with a display of snow. The MSP2410–08 multisound processor chip IC1401 turned out to be the culprit, although the voltages around it and the data lines all seemed to be OK.

Panasonic euro 2 chassis

Field fold-over: The problem with this second-generation digital receiver was field fold-over at the top of the screen, bringing the c.r.t. current sampling lines down into the display area. The cause of the fault was traced to D456, a 16 V zener diode (type MA2160) in the field output stage.

Goes into standby after some time: The fault could be instigated sooner by covering the ventilation slots in the back cover. Unfortunately any of the processor chips in a digital TV chassis can cause this sort of problem. So we set out on a course of elimination. The digital video processor IC was replaced first. Later that day the heavily insulated set cut out. The multi-sound processor chip was tried next, with the same result. The teletext processor chip (TPU3040–18) was eventually found to be the culprit.

Stuck in standby: Checks around the main microcontroller chip showed that there were some signs of life here, but although most conditions around the digital chips were OK the set couldn't be lured into life. The set was eventually made to work by disconnecting the serial data and clock lines to the EAROM chip IC1203, although the picture geometry and the customer set-ups were poor. A new EAROM chip (X24C016P-P1) cured the fault.

Set would cut out when hot: Fortunately on this occasion there were some clues. Just before the set lapsed into standby, its picture would break up into lines – as if the AD converter's frequency was varying. The cause of this can be crystal X601 or the digital video processor chip IC601. Although we opted for X601 first, the culprit was IC601 (type VDP3108–30).

Panasonic TC1485 (Z3 chassis)

Dead set, h.t. rail OK: There was not much else. The l.t. supply from the chopper transformer T801 was quite low. This usually indicates that the power supply is running at the wrong speed because of no line drive. A check at pin 15 (X-ray protection) of the M51407SP i.f./colour decoder/timebase generator chip IC101 confirmed that the protection circuit was in operation, as there was a voltage there. After disconnecting various feeds to the sensing circuit transistor Q502 was pronounced guilty. But why? After much voltage checking in the associated circuitry R560 (270 kΩ) was found to be open-circuit.

Tries to start then shuts down: Check whether the value of R560 (270 kΩ) or R552 (27 kΩ) has gone high.

Intermittently dead: To start with the set worked perfectly. Subsequently it lapsed into an inoperative state. A check showed that there was no power-on signal from the system microcontroller chip IC1101, but further checks in this area revived the set which continued to work faultlessly all day. Next day the set couldn't be prodded into life. Everything appeared to be fine except that IC1101's reset line seemed to be slightly lower than usual. Suspicion then fell on a small yellow component, C1122 (0.01 μF), which proved to be leaky. A replacement restored reliable operation.

Dead set with D816 (R2G) short: Suspect that the S550103 chopper i.c. is producing a high output voltage intermittently and needs replacement. The h.t. voltage at TPE1 should be 103 V ± 1.5 V. A faulty STR50103 chip can result in the voltage rising to 150 V or more.

Set stuck in standby: This set was stuck in standby. We found that the microcontroller chip IC1101 didn't produce the power-on command (pin 6 high) because the chip was not being reset. IC1104 produces the reset action when its supply reaches 5 V, but the latter was low at 4 V. On checking back to the 5 V regulator transistor Q804 we found that the reference zener diode D810 was leaky.

Brightness increase causes cut-out: Tests showed that the contrast didn't decrease as the picture became overbright, although the beam limiter operated to shut down the line drive. We found that R316 (47 kΩ), which applies the beam limiting control voltage to the video section of the receiver, was open-circuit

Panasonic TC1785 (Z3 chassis)

Dead set, no power supply voltages: In the circumstances it's a good idea with these sets to remove Q806, the standby switch transistor, then wind the set up via the variac. This time the set started up, with all the voltages correct. Q806 was then checked and found to be open-circuit. A new 2SA683 transistor put matters right.

The colour could not be controlled: The colour was permanently at maximum, although the on-screen indicator moved. A voltage check at pin 31 of the microcontroller chip IC1101 produced a reading of 3.25 V which didn't vary when adjustment was attempted. Disconnecting pin 31 proved that the chip was OK. Control is carried out at pin 21 of IC101 on the i.f./decoder panel. When this pin was connected to chassis a monochrome picture was obtained. So this chip also worked correctly and the fault was between the two i.c.s. It was traced to D1149 in the pull-up and biasing circuit. This MA165 diode was leaky.

Tuning unreliable: The set drifted from station to station at will. You could retune and store the stations, but you still had to retune from cold. As expected a new tuner unit made no difference at all. Checks in the tuning voltage circuit soon brought us to a likely culprit: the 10 nF ceramic capacitor C17 was leaky.

Field scan fault: The top of the raster was expanded while the bottom was compressed, but with no fold-over. The cause was C458, which is listed in the manual as being 4.7 μF, 50 V but turned out to be 10 μF, 50 V.

Picture size fluctuates: This set suffered from a nasty looking and sounding intermittent fault. The picture size would fluctuate quite dramatically, with arcing in the neck of the c.r.t., then the set would shut down to standby. When the fault occurred, the h.t. voltage rose to about 155 V. The cause of the trouble was the STR50103A chip in the power supply. Unfortunately the activity in the tube's neck had led to an intermittent blue gun short. We were able to sort that out by judicious use of the B&K 467 tube tester.

Panasonic TC21M1R (Z4 chassis)

No video with faint unsynced lines in the background: Good video came out of IC601 but it was attenuated going through board H, where the video switching and sync separator are to be found. Checks around the sync chip IC521 showed that the sync pulses leaving pin 14 were poor. The 180 pF ceramic capacitor connected to this pin turned out to be leaky, a replacement restoring the excellent picture.

Set comes on with field collapse: The field scanning would then open out but the set would shut down at irregular intervals. During these irregularities the field drive from board C remained constant, so attention was turned to the LA7837 field output chip IC451. Suspicious as always of the small yellow capacitors in these sets we checked the ones connected to this i.c. and soon found that C462 was slightly leaky. A 470 pF replacement restored normal operation.

Dead set, no operation: This set came in dead, i.e. there was no operation. When the optocoupler in the power supply was overridden a blank raster appeared, with no sound. The main microcontroller chip was totally inoperative, so our next line of investigation was to check around the oscillator, resets, feeds etc. This eventually led to removal of the ST24C02A EEPROM chip IC1202, which restored normal operation. A replacement was put on order.

Slow increase in brightness: We've had two of these sets in recently with the same fault, a slow but sure increase in brightness, culminating in uncontrollable brightness wilh flyback lines. When the first one came in we carried out checks on the RGB output stage and c.r.t. first anode supplies but found no variations here. A check on the grid network on the c.r.t. base panel, however, showed that R380 (680 kΩ) was open-circuit. In both cases removal of this resistor showed that one end had not been properly inserted through the PCB.

Intermittent field collapse: The fault appeared at random and didn't seem to be affected by time or temperature. Must be a capacitor, we thought! When the fault was present the trigger input at pin 2 of the field output chip IC451 was missing. In addition the voltage at this pin rose to 5.8 V. Attention was turned to the field oscillator, which is part of the video processor chip IC601 on board C. Although the voltages here were correct there was no ramp waveform at pin 34, to which the oscillator's timing components are connected. The culprit was the 3.3 µF tantalum capacitor C402. Remember – beware of blue tants!

Tuning drift: A fairly common fault, for which, as usual, the cause was the tuner unit. When using the portable appliance tester to carry out the insulation resistance test for the final safety check, however, the reading we obtained was infinity – you would normally expect to get a reading of around 9.5 MΩ. The resistor between the live and isolated sections of the chassis was open-circuit. It's R814, an 8.2 MΩ safety component.

Panasonic TC2205 (U2 chassis)

Mysterious faint light bands on the screen and volume variations: The two h.t. reservoir capacitors C854 (195 V supply) and C852 (160 V supply) were found to have only one leg each! They are 10 µF and 100 µF respectively, 250 V.

Stuck in standby: We dived for the chopper power supply rectifier diodes D852/3 etc., the usual cause, but they were all intact. After much time spent checking around we decided to replace the line driver transistor Q501. This restored a bit of life but the trouble was by no means cured. In the end we came to the conclusion that the line output transformer was probably faulty, something that's virtually unknown in Panasonic sets. Fitting a replacement cured the trouble.

Tube cut off: Panasonic manuals can cause almost as much trouble as a faulty set. In this case the c.r.t. was cut off and we eventually traced the cause to the diodes in the 12 V regulator transistor's base circuit – they were both leaky. This is in fact a not uncommon problem with these sets. But the circuit diagram in the manual shows both these diodes as ordinary types, the parts list says they are both zener diodes (no voltage given) while the board layout diagram shows one as being a zener diode and the other not. In fact they are both 6 V zener diodes and the blanking circuit is fussy about this – the correct types should be fitted. We proved the point by using two 6.2 V zener diodes.

Dark area would appear at bottom: After several hours' use a dark area would appear across the bottom of the screen. The size of this area increased with time, moving upwards to obliterate more of the picture. If the contrast control was set to maximum the darkness would retreat back to the bottom again. Use of a hair dryer and freezer enabled us to prove that the field output chip IC401 (TDA1104SP) was the cause of the problem.

Severe picture disturbance, sync: This took the form of wildly incorrect line and field sync was traced to one of those purple capacitors. C854 was open-circuit.

Panasonic TX2 (alpha 1 chassis)

Dead set: The 2SD965R standby switching transistor Q802 in the power supply was short-circuit.

Dead set, mains fuse intact: It may save time to go straight for the STR54041M chopper control chip IC801 – it's the usual culprit.

Defective line output transformer: After replacing the LOPT we found that the set worked but the line output stage made a shrieking noise and the picture verticals were ragged. This was our fault: the single-turn white wire around the transformer's limb had been connected to chassis in reverse phase. There's a mark on one of the wires to indicate that it should be connected to a similarly marked solder point on the chassis. But because of component density in the area this is difficult to see.

Dark defocused bars moving slowly down the screen: The fault would occur when this set was switched on from cold. Increasing the brightness made the effect worse, but the symptom cleared after a few minutes to leave a good picture. The cause of the trouble was an ageing picture tube. We were able to minimize the effect by careful adjustment of the tube's drive and first anode voltages.

Panasonic TX21M1T (Z4 chassis)

No sound and only half a picture: Sure enough the line scanning was locked but the start was shifted about half way across the screen. Checks around IC601, where the video and line outputs are obtained, failed to reveal anything amiss. Eventually the culprit turned out to be the

.001 μF capacitor C503 across the line shift control, it was leaky. Replacement restored the sound and the complete picture.

Tuning and sound problems: This set displayed four symptoms: the tuning would appear to drift off very slightly, the sound was muted, there was no volume on-screen display (the picture ones were all OK) and if the sweep tuning was started it didn't stop when a channel was found. The cause of all these symptoms was that the 'signal/noise' switching circuit, which consists of Q1272/3/4 and the associated components, was giving a low (no signal) output all the time. All the components in this circuit were OK, however. It seemed that the cause of the trouble was insufficient sync drive at the base of Q1272. The sync separator that provides this drive is in IC521, which is on the teletext PCB. Comparison with another set showed that the output at pin 14 of this chip should consist of 8.3 V pulses sitting on a d.c. level of 1.7 V. In the faulty set the amplitude of the pulses was only 5.5 V. We found that the cause of the trouble was C524 (180 pF) which is connected between pin 14 and chassis; it was leaky, measuring just over 4 kΩ. As a result pin 14 couldn't rise to the full supply level, as it should. The reason for the symptoms was that the set failed to detect that it was on a signal. So the a.f.c. was switched off and the sound muted, which also removes the on-screen display.

Intermittent loss of chroma: The fault in this set was very intermittent – we had to run the set for several days before it started to occur. The symptom was that the picture's chroma content would intermittently drop to a low level or disappear altogether. As soon as the back cover was removed the fault cleared, but we found that we could instigate it by some careful flexing of board C. When the fault was present the input to the chroma processing chip IC601 (pin 5) disappeared, although it was still present at the output from the chroma source changeover switching chip IC2651 (switches between tuner-derived chroma and S-video input chroma). The cause of all this trouble was an intermittently open-circuit 0.01 μF surface-mounted coupling capacitor, C2651.

Tuning memory problem: There was a strange fault with this set: if you stored a channel in a tuning memory position that already stored information the same information would be stored in all other locations in use. The cause of the fault lay on the text PCB, which had been subjected to previous damage. Link J10 was broken. In other circumstances it would be worth looking for a dry-joint in this area.

Panasonic TX21T1 (alpha 2)

Very poor picture: The contrast was very low irrespective of the customer control setting. The reason for this is that the beam current limiter (or ABL as Panasonic prefer to call it) is working overtime, limiting the contrast control voltage applied to the colour decoder chip. The culprit is R562 which goes open-circuit. It's a very small 0.25 W, 130 kΩ resistor: the replacement supplied is a fair amount larger than the original.

Picture goes dark intermittently: The trouble with this set was that the picture would go off intermittently and the screen would go dark. With faults like this, one of the first things to do is to check the voltages supplied to the video processing chip by the contrast and brightness controls. In this set the contrast control voltage should be 1.5 V for minimum contrast and 4.3 V for maximum. When the fault developed the voltage was low at 1 V. Now the chip could have been pulling the line low but we decided to eliminate other possibilities first. The electrolytic C311 seemed a likely suspect and on a recent Panasonic course we were told that it had given trouble. Not this time, however: the fault was still present when it had been replaced. Only when C626 was disconnected did the fault disappear. A new 10 nF, 50 V ceramic capacitor provided a complete cure.

No sound: The voltages around IC253 seemed to be within reason but there was no life and the situation didn't feel like a faulty chip. As the voltage at pin 2 seemed to be slightly low we disconnected C269. Up came the sound. C269 (10 nF, 50 V) read about 3 kΩ when checked with a meter.

Sync lost after a few minutes: This was a brand new set, straight out of the box. After working for a couple of minutes picture sync was lost then the screen went blank, the power supply shut down and the set squealed. We turned the set off and allowed it to cool. The initial symptom was the giveaway: when the fault developed we squirted the TDA2579A line generator/sync chip IC501 with freezer. This restored normal operation for a couple of minutes. A new chip provided a permanent cure.

Field collapse symptom: This set displayed the classic field collapse symptom – a white line across the screen. As checks around the TDA2579 timebase generator chip IC501 showed that it was not producing a field drive output it was replaced. This made no difference. The guilty component was sulking quietly nearby. Need you ask? – a leaky 0.01 μF capacitor, C403. It's in the field feedback circuit. A replacement restored full field scanning.

Sound would mute after 1 hour: This set operated perfectly for an hour or so. Then the sound would mute and no controls would work. There didn't seem to be any problems around the main microcomputer chip IC1203 and the DAC chip IC171. The data and clock signals fluctuated normally – usually the data freezes if one of these chips is faulty. Eventually, after much hair tearing, it transpired that the PCD8582P memory chip IC1202 was faulty.

Intermittent whistling at switch-on: Sure enough a high pitched whistle came from the set when we switched it on, stopping as soon as we touched it and then not to return until next day. In fact any attempt to touch the set cured the fault until next day. Many days later the cause of the fault was traced to a dry-joint on the line output transformer's overwinding – the point that provides sync between the line output stage and the chopper power supply.

No text although channel flags OK: This indicated that part of the teletext circuitry was working. Approaching this fault with the knowledge that the many-legged devices (chips) are more reliable than the small, brown, two-legged ones (ceramic capacitors), we soon discovered that the culprit was C3517 (10 nF) at pin 15 of IC3501 – it feeds teletext data to IC3502. A replacement restored all those nice little characters on the screen.

Faulty tuning: The picture looked as if the tuning was slightly off: there was a mainly dark screen with zigzag lines, although some signs of chroma would flash up intermittently, and the sound was muted. As the tuned channel readout was correct we took a look at the i.f. signal at the input pins (8 and 9) of the vision and sound i.f. chip IC101. It looked fine – a stable field-rate signal could be seen. But the demodulated output at pin 3 wasn't recognizable as a video signal. We scoped the 39.5 MHz waveform at the pins (22 and 23) for the synchronous demodulator's tuned circuit and found that it was missing. As the coil wasn't open-circuit and the capacitor wasn't short-circuit we replaced the chip. This produced no improvement. The capacitor was the next item to be replaced, but still no luck. When a new coil was fitted we had a healthy 200 mV peak-to-peak waveform at both pins and normal operation.

Picture went red after short while: We checked this and that and froze everything on the c.r.t. base panel, none of which helped. It turned out that a disc ceramic capacitor, C353 (220 pF), was going leaky. It decouples the emitter of the red output transistor.

Blue disappeared after half an hour: The grey scale was OK. There was a B – Y waveform at pin 2 of the TDA4510 colour decoder chip IC601, but it was of low amplitude. The relevant d.c. voltages were also slightly wrong. We tried replacing the chip and the chroma delay line, to no avail. Many small components were then tried before the cause of the fault was finally found. The culprit turned out to be C612 (0.01 µF) which is connected between pin 5 of the TDA4510 chip and the 12 V supply. A replacement restored correct colour.

Text information flashes on screen: After about an hour's use and whenever a different channel was selected the channel information derived from the text would flash unlocked across the screen. There was also loss of teletext synchronization. The cause of all this was C3511, which of course is a 10 nF ceramic capacitor.

Panasonic TX21V1 (alpha 3 chassis)

No microcontroller activity: Although this set powered up there was a distinct lack of any microcomputer chip activity, with no display on the front panel, no control response, no sound and only a dark, blank raster. Checks around IC1203 and IC171 revealed no obvious problems, however. In such cases it's often a good idea to disconnect the serial data and clock lines to the text panel, where other microchips connected directly to IC1203 and IC171 usually lurk. With these disconnected the set sprang to life, producing sound and a perfect picture. When the text micros were isolated one by one the MAB8461PW13S chip IC3507 proved to be the cause of the problem. Replacing it restored normal operation on all functions.

Dead, no line drive: We found that R851 (2.2 Ω) in the supply to the line driver transistor was open-circuit.

No Nicam sound: Most of the Nicam faults we get are caused by incorrect adjustment or a faulty i.f. unit. This set failed to produce Nicam sound and the display showed neither the Nicam symbol nor the mono one. On investigation we found that the 6.55 MHz oscillator wasn't running. The cause of this was eventually traced to the 0.01 μF capacitor C2549 which is connected to pin 9 of IC2502, the carrier APC detect pin. After replacing this capacitor and carrying out 6.55 MHz adjustment the set worked correctly.

Panasonic TX24A1 (alpha 2 W)

On-screen display not line locked: If the volume or any other function was adjusted the relevant bar came up on the screen but it was as though the line hold control was misadjusted: the picture in the background was fine, however. We were all of the opinion that the cause of the fault was on the teletext panel. Sure enough when text was selected with the handset the decoder had real trouble processing the information. The page header kept corrupting, the selected page would only very rarely update and most of the display was garbage. Curiously if mix was selected the text line hold also went out, with the background picture very much line locked. We replaced the SAA5231 VIP chip IC3501 and checked the supplies and the 6 MHz clock, all to no avail. A check on the video input to the chip showed that this was perfect. We were sure that the cause of the fault was somewhere around this chip, probably something to do with the data slicing or sync separator sections. So the capacitors connected to the relevant pins were replaced. This eventually paid off: the culprit was C3511 (220 pF) which is connected to pin 24, the pulse generator pin. Since then we've heard from Panasonic that this is a known fault with Alpha 1 and 2 series receivers.

Intermittent 'flashing' and going off: Naturally it worked perfectly until it was returned to the customer's house, whereupon it displayed 'speckled' white bands like mains interference, the colour flashed on and off and it finally lapsed into standby. Back at the workshop we removed Q802 to prevent the set switching off. The fault was clearly evident and because of the apparent arcing we changed the line output transformer. This failed to improve matters. Scopes were then hooked up to h.t. lines, l.t. lines and data lines but nothing untoward was detected. Despair was fast setting in when a workshop colleague, intrigued by the fault, thoughtfully tapped the screen. This cleared the fault. Could the tube be faulty? When the fault next returned I crept up carefully, tapped the screen – and again it cleared! I then had a bit of inspiration. Could one of the tube's fixing bolts be loose? Sure enough the one bolt that held the earthing springs for the tube's Aquadag coating was loose. There were no further problems once it had been tightened.

Overloaded power supply: It was squealing and the h.t. line was very low at only 25 V. A fault in the line output stage was suspected, and indeed disconnecting the supply to it and connecting a dummy load in its place produced the correct h.t. voltage. No obvious shorts could be found, and the line output transistor seemed to be fine. So the line output transformer was suspected – we've had it fail on a few occasions in this chassis. Before replacing it, however, we thought it a good idea to fit a new output transistor, just in case. When checked out of circuit the old transistor gave the same readings as the new one. But they do give odd readings – about $100\,\Omega$ across the base-emitter junction, presumably because of an internal resistor. Anyway after fitting the new 2SD1441 the set sprang to life, with the correct h.t. voltage etc. This didn't provide a complete cure, however. According to the set's owner it went off again an hour after being returned. It came back with the same symptoms – low h.t. and squealing. When another 2SD1441 line output transistor had been fitted the set worked all right, and once again the old transistor compared perfectly with a known good one. We assumed that the cause of the fault was base-emitter junction breakdown, but why? – with the internal resistor you can't check the junction in the normal manner. Tracing the base print connection back to the line driver transformer solved the problem: the chassis side of the winding was dry-jointed. Resoldering the transformer connections provided a complete cure.

Sound muted at switch-on for 10 minutes: The fault wouldn't return until the next day. On the first day we established that the fault was on the mute line. On the next day we reached another board, and on the following day yet another. We finally arrived at the front M board, but after painstaking tests this board proved to be faultless. The culprit was C1206 on the main board. It's the standby 5 V reservoir capacitor, in the supply to the front control panel.

Dead set, a difficult power supply fault: We were asked to look at this set as a colleague had run out of ideas with the fault he now had. The set had originally been dead. After replacing R815 ($5.6\,\Omega$ fusible), Q801, Q802 and the STR54041 chopper chip it would start up and then immediately turn off again. This was because the wire attached to plug V2 had been fitted to socket H16 (factory preset) instead of socket M2 on the front PCB – a case of more haste, less speed!

Safety resistor blows after 4 hours: After about 4 hours' operation the safety resistor (R567) in the h.t. feed to the line output stage would blow. This would be accompanied by various shrieks and whistles, and at this point the line output transistor Q551 would be very hot indeed!

Extensive checks were carried out in both the power supply and the line output stage, all to no avail. After many hours of frustration and heartache we eventually found that C502 in the line oscillator circuit had been fitted incorrectly, with one leg loosely attached to the adjacent solder.

Field collapse: We quickly discovered that the TDA2579A timebase generator chip IC501 was not producing any field drive output. A new TDA2579A chip appeared to put that right, but two weeks later the set was back again with the same fault. This time C403 (0.01 μF) which is connected to pin 2 (field feedback) of IC501 was found to be leaky. We've had no further trouble since this item was replaced.

Power supply squealed, nothing else: In this case you usually find that the line output transistor is leaky and its fusible feed resistor open-circuit, the basic cause of all this being dry-joints at the pins of the line driver transformer. All four pins should be resoldered, otherwise the set will bounce. This is becoming a common fault.

Intermittent loss of line hold: The fault took almost three days to put in an appearance. Strangely, when the fault was present the sharpness control varied the amount of line slip. Inevitably, removal of the back cover cleared the fault for another few days and restored the sharpness control to its usual function. Many days later we found that resoldering the connections to the line output transformer (T551) seemed to cure the fault.

Went to standby after switch-on: There was a brief burst of life when this set was switched on. It then went to standby, leaving three coloured blobs at the centre of the screen. To assist with fault-finding we disconnected the standby switching to the power supply, then found that the 1.2 V supply in some parts of the decoder etc. was missing. The cause of this was L303 (10 μH) in part of the 12 V feed being open-circuit. With the standby switching disconnected the set displayed its true symptom: a raster but no sound or vision.

Shut down after one second: The LED indicator still displayed the programme number. The power supply incorporates an optocoupler which is used for standby switching. When the primary side of this was shorted out the set came to life, with no signs of overloading, funny smells etc. We decided to start by checking the various supplies to control board M. The 12 V supply at pin 4 of plug M2 was absent. The cause was simply an open-circuit coil, L303 (10 μH). A replacement restored normal operation.

Panasonic TX25T2 (alpha 2 chassis)

Dead set, no power supply operation: We've had a batch of these sets in recently with the same complaint – dead with no power supply operation. The cause of the fault was that in each case D851 (C2408M) was either short-circuit or leaky. It provides the 160 V supply for the line output stage. Incidentally does anyone know of a cure when these sets lose the channel 0 or 1 tuning memory, in all cases the tuning memory being reset to 00?

Field collapse: Check at pins 3 and 7 of the AN5521 field output chip, where the voltage should be about 25 V. If pin 3 is at zero, R848 will probably be open-circuit. It's a small component that looks like a wirewound resistor but is actually an 800 mA ceramic fuse. It is also hard to find. Look behind the two large capacitors at the back of the field output chip's heatsink.

Nasty field fault: A particularly nasty field fault occurs when R469 goes high in value. The symptoms are bad distortion, line doubling and horizontal striations present all at once. This resistor is part of a damping network that's connected across the field scan coils. Its purpose is to prevent line-frequency waveforms reaching the field scan circuitry, induced by the close proximity of the line and field scan coils. This line-frequency signal can show up as black lines across the upper part of the field scan. The value of R469 can vary from that given in the service manual. It's usually 150 Ω or 180 Ω. Check by reading the bands. Fitting a wrong value resistor will produce a similar fault symptom.

Bright flyback lines top of screen: Something was upsetting the field output chip (IC451) in this receiver. About twice a year bright flyback lines would appear across the top of the screen. A new field output chip would put matters right. Fortunately there's now a permanent cure. Kit TZS5EN001 consists of a replacement scan coil and output chip. Apparently some earlier scan coils used in this model proved to be a bit too much of a burden on IC451, leading to its premature failure.

Panasonic TX28A1 (alpha 2 chassis)

Shut down after 20 minutes: This set would not come to life again unless pins 3 and 4 of the ON3105 optocoupler D811 were linked together. The set would then work perfectly. D811 is used as an isolator to couple the power-on signal to the live side of the power supply. In the fault condition this signal didn't get there. A new ON3105 put matters right.

Audio trouble: There was no sound for the first half hour after switching on. We'd not seen one of these sets before and didn't have the manual, so we had to play this one by ear! A slight buzz from the speaker confirmed that the amplifier was working, and no signal at the scart socket suggested that the cause of the fault lay near the intercarrier sound chip. Some likely chips were heated and cooled to no avail. Then quite by accident some freezer caught IC1204, a 3-pin 5 V regulator. Its output remained the same when it was heated and cooled, but a scope revealed that an enormous amount of 100 Hz ripple was present at its output when freezer was sprayed in its direction. IC1204 is fed from a small mains transformer and bridge rectifier whose 1000 μF, 16 V reservoir capacitor C1206, which is behind IC1204, had dried up. A replacement cured the fault. When we obtained a circuit diagram we discovered that a muting circuit is linked to the output from IC1204.

Chrominance, no luminance: With a modern set the luminance signal path can be quite involved, what with S connectors and YC processing for colour transient improvement (CTI). The latter area was where our luminance signal went missing – within the TDA4565 CTI chip on its subpanel. A replacement restored the full display.

Panasonic TX28G1 (alpha 2 chassis)

Picture would bow in after half a day's soak test: Fortunately the fault became a permanent one after a couple of days, enabling us to discover that the culprit was a small, yellow capacitor connected to the base of transistor Q751 in the EW circuit: C754 (180 pF) was leaky.

Left-hand sound output distorted: We traced back to PCB H, where the output at pin 1 of IC2401 was distorted though its input at pin 22 was without distortion. The d.c. level of the input was low, however, because the 100 pF ceramic capacitor C2442 had a 5 kΩ leak. The circuitry in the right-hand channel is identical. So C2441 could cause a similar problem.

Panasonic TX28W2 (alpha 3 chassis)

Intermittent diagonal black and white stripes: The type of intermittent fault we least like produces some obscure complaint and lasts for about 2 minutes every other day. So how about this one: the symptom consisted of diagonal black and white stripes that appeared maybe once a day! We left the set to its own devices on the soak test bench. Hours later a colleague drew my attention to diagonal bold black and white stripes across the screen. I quickly checked the text display, which was faultless, but by now my 2 minutes were up as the symptom receded. Many days later the cause of the fault was traced to an intermittently leaky zener diode, D329, which stabilizes the supply to the TDA4670 transient improvement chip IC302: as the supply fell the lines appeared. The part number is MA4068.

Surge limiter resistor R822 blown: This set's fault proved to be a bit of a problem. R822, the 4.7 Ω, 10 W surge limiting resistor associated with the mains bridge rectifier, had blown. As no obvious shorts could be measured we fitted a replacement. Then, filled with apprehension, we switched on. Instead of the friendly rustle of e.h.t. as the set came on it squeaked and blew R822 again. This time the safety resistor R555 in the feed to the line output stage had also expired. So R822 was again replaced but R555 was left open-circuit. Up came the 150 V h.t. supply, but when the line output stage was reconnected both resistors blew again. The growing pile of 10 W resistors convinced me that I had to be brief with my next checks. R822 was replaced but the line output stage was left disconnected. The line drive waveform was then checked. It was bizarre, consisting of just high-frequency spikes. A new TDA2579A timebase generator chip (IC501) was fitted and another quick check was made: the waveform was as before. Eventually we found that C501, a friendly 0.1 μF brown Mylar capacitor in the line oscillator circuit, was leaky. A replacement, along with a new line output stage feed resistor, restored normal operation and an excellent picture.

No colour: Checks around the decoder showed that the chroma signal was present and correct, so attention was turned to the TDA3505 video control chip TC303 The voltage at the d.c. colour control pin (16) was low at 0.035 V instead of about 3 V. What could be the cause? Our first check was naturally on the 0.01 μF ceramic decoupler C644, which was virtually short-circuit (80.5 Ω). With any other make the parallel zener diode would have been a far more likely suspect!

Intermittent loss of picture/sound: Dry-joints around the prescaler chip inside the tuner were the cause, a good solder-up putting an end to the trouble – proved by a lengthy soak test.

Panasonic TX28W3 (euro 1 chassis)

Yellow picture: Sure enough the blue was missing. We gingerly checked to see if the RGB drives were emerging from the digital pack. Fears of chasing around many-legged flatpack chips receded when we found that the drives were present and correct. One of the blue drive biasing resistors, R3374 (100 kΩ), turned out to be open-circuit, a replacement restoring the normal colour range.

Set lost signals as it was watched: This set intermittently lost signals, on any preset channel, as you watched it. It was not a memory problem – the data for each preset remained correct. Calling up the self-test display during the few seconds when the fault was present showed that there were no errors, so no help here. Despite the fact that the tuner address was OK the fault was in the tuner, although obviously nothing to do with the bus. A new tuner restored correct operation.

Set comes on with no sound or picture, and returns to standby: Although this is a digital TV chassis we feel that the picture it produces is inferior to that provided by its analogue predecessor, Model TX28W2 (Alpha 3 chassis). Interesting that it produces the same faults! If the set comes on with no sound or picture, returning to standby a few seconds later, R561 (ERQ12HJ1R5) is open-circuit: it's the fusible resistor in the supply to the TDA8175 field output chip IC561, which goes short-circuit.

Right-hand channel distorted: IC1401 was first accused, as it's the Nicam and normal sound processor chip. A replacement proved that it was innocent, however. The culprit was eventually found to be C1423 as it had never been soldered in at one end. It's part of IC1401's oscillator circuit.

Panasonic TXC74 (alpha 1 chassis)

After 20 minutes top of picture would brighten and drop down 2 inches: The field scan became badly distorted. We could find nothing amiss in the field drive and output stages despite extensive tests. So attention turned to the scan coils which are unfortunately bonded to the A59EAK00X01 tube and are not available separately. Fortunately a local dealer was able to supply a tube from a scrapped set. Carefully removing this and fitting it to the faulty set solved the problem.

Line output transistor kept blowing: This set was brought to us because the line output transistor Q551 kept blowing. The one fitted had indeed been damaged, with quite a heavy leak between its collector and emitter. We fitted a replacement then left the set running on the soak-test bench. After about an hour the width came in slightly, but when the set was approached the fault cleared for another hour! This behaviour repeated itself for most of the rest of the day. Fearing for the life of Q551, we checked the line drives and output stages carefully for dry-joints. Sure enough there were some rather obvious dry-joints at the secondary winding connections of the line driver transformer T531. Resoldering these restored Q551's life expectancy.

Colour goes funny, worse when cold: As the set had been running for several hours, it was not possible to make a diagnosis at the customer's house. We took the set back to the workshop and, when it was nice and cold, the fault was certainly there. A whistling noise came from the set, and the picture was breaking up into horizontal lines. Scope checks at the secondary side of the power supply failed to reveal anything amiss, so we treated the set to some freezer squirts on the primary side. When C808 (10 μF, 63 V) was cooled the set went berserk. A new capacitor restored normal operation.

Panasonic U4 chassis

Intermittent memory loss: These sets suffer from intermittent memory loss. We've received various modification kits in the past but none of them have really solved the problem, although I have found that resoldering the crystals associated with the MAB8440 and SAB3035 chips on the print side of the PCB has provided a cure – the legs are much smaller than the holes, and the joints become dry. More recently Panasonic have released a kit that modifies the appropriate M board to the U5 design. This seems to cure the problems, but unfortunately one of the first we received and fitted didn't work at all as the MAB8441 chip was faulty.

Intermittent tuning problems: Loss of the tuning voltage and/or no search were traced to the SAB3035 chip. We've had dry-joints cause this problem with both these and certain Zanussi colour sets.

Dead set, no line output power: The 8.2 Ω resistor that feeds the line output stage was open-circuit. As with many Panasonic sets, a quick look round for faulty electrolytics can be fruitful. In this case C555 was found to have a leg missing due to corrosion. This wasn't the end of the story, however. When we switched on there was a normal picture but no sound. The intercarrier sound chip, an SN76622AN in this case, was hot enough to fry eggs on. A replacement restored the sound.

Defocused and overbright picture: This suggested that there was an internal short in the tube – we've had this before with Panasonic sets fitted with Mullard tubes. This time, however, the fault was caused by a faulty line output transformer. There was excessive leakage between the focus and first anode supplies because of spillage which had entered via the rear cover ventilation slots. A new transformer was required.

Burst of life at switch-on, then off: There was a burst of life at switch-on, then the set went off as the protection system came into operation. A check showed that the h.t. rose to far too high a level. The feedback to the error detector transistor is developed across C808 (47 µF, 16 V) which had dried up and was virtually open-circuit.

Panasonic Z3 chassis

Reduced height and poor linearity: In both cases the faulty component was C404 (4.7 µF, 50 V). It's used in the feedback circuit between the field output and ramp generator stages.

Dead, but initial rush of e.h.t.: This was an odd fault. The set was dead but there was an initial rush of e.h.t. Then almost immediately the set would become inert, with just a buzzing from the power supply. The latter was up and running, with correct h.t., but there was no line drive because the protection circuit was operating. As there were no obvious problems in the deflection circuits and the components in the protection circuit all read correctly, we removed the protection transistor Q451 which was biased on. This revealed the fact that there was field collapse. We soon traced the cause to the MA2270 zener diode D4539 in the field output stage. Once full scan had been restored the set was left on soak test. After about 10 minutes we heard it fire over and saw that the point of distress was in the tube's neck. The h.t. and e.h.t. were fine, the cause of the trouble being a faulty tube. It seems that it had killed off the original zener diode.

No colour: We turned our attention to the circuitry around IC101, where amongst other things the colour decoding is carried out. Voltage checks drew our attention to a small ceramic capacitor connected to pin 29. C616 was leaky of course, being one of the dreaded 10 nF, 50 V ceramic-type capacitors.

Dead set, power supply buzz: It was an unloaded buzz as opposed to an excess load whine. Checks showed that the h.t. was present but there was no line drive. It was being killed by the X-ray protection system in IC101 (M51407SP). As cold checks around the nasties failed to reveal any faults we shorted pin 15 of IC101 to chassis and tried again. The set now worked normally. So the thing to do was to find out what was triggering the protection system. When the base of transistor Q502 was disconnected the protection stopped. This suggested a beam current fault, but the picture was clearly OK. In fact Q502 was being incorrectly biased because R560 (270 kΩ) was open-circuit.

Noisy, occasional fizzing sound: The h.t. was going high, and the STR50103A chopper chip was sensitive to freezer. A replacement restored normal operation.

Panasonic Z4 chassis

Intermittent loss of sound: When the fault occurs a sharp tap on the cabinet will cure it. But with the back removed no amount of banging, twisting and pulling will instigate it – which is not surprising when you realize that the dome speaker system is part of the back cabinet assembly, and that if the speaker wires are lightly pulled they both come off! Retightening the connecting plugs to the speaker cures the fault.

No sound, picture shifted: The H-centre control R509 had no effect. The cause of these problems was traced to C503 which decouples the horizontal centring voltage. It's a 10 nF disc ceramic capacitor and had developed a 1.2 kΩ leak.

Bright green screen at switch-on: After a few seconds a normal picture would appear, but once or twice a day the picture would cut out entirely. The cause of the trouble turned out to be capacitor C350 on the tube base panel. Our thanks to Panasonic Technical for assistance with this one.

Panasonic Z5 chassis

Lack of height plus some fold-over: We soon discovered that the 30 V supply, which is derived from the line output transformer, was low at 20 V. The feed resistor R521 had risen in value from 8.2 Ω to about 20 Ω.

Low h.t., constant tripping: When the feed to the line output stage was disconnected the symptoms remained exactly the same. C823 (10 nF, 500 V) in the snubber circuit was eventually found to be the culprit. It was leaky.

No signals: When the contrast was turned up, just a little noise occasionally appeared on the screen. A dead i.f. strip was suspected, but a meter prod applied to the tuner's output produced a little noise, while applying the prod to the output from the SAW filter produced a lot of noise. So attention was turned to the tuner unit. The pin functions are printed on the PCB, which is handy but unfortunately misleading. The pin marked 30 V is actually the tuning voltage, which should vary. It did not. A quick scope check on the series data and clock lines showed a change in activity when the tuner's buttons were pressed. When the tuner was taken apart it was soon apparent why it didn't work: the 4 MHz crystal wasn't soldered to the board. Note also that the only supply to the tuner is MB (11 V). BT is not connected.

Red LED flashing quickly and a whistle from the power supply: We checked the line output stage for short-circuits. As there were no obvious ones we connected a 60 W bulb as a dummy load between pin 9 of the line output transformer and the live earth section – note that in this chassis the line output transistor is on the primary side of the power supply, with the line output transformer providing mains isolation, so care is required over selection of the correct chassis point when carrying out tests. As we'd forgotten to disconnect the line output transistor, when we switched the set on it started up with the bulb alight, all functions working and correct voltages. This suggested a power supply rather than a loading fault. We eventually found that R821 (330 kΩ) was open-circuit, upsetting the trip arrangement.

Philips

PHILIPS 10CX1120
PHILIPS 2A CHASSIS
PHILIPS 2B CHASSIS
PHILIPS 3A CHASSIS
PHILIPS ANUBIS A CHASSIS
PHILIPS ANUBIS A-AC CHASSIS
PHILIPS ANUBIS B-AA CHASSIS
PHILIPS CF1 CHASSIS
PHILIPS CP110 CHASSIS
PHILIPS CP90 CHASSIS
PHILIPS CTX CHASSIS
PHILIPS CTX-E CHASSIS
PHILIPS CTX-S CHASSIS
PHILIPS FL1.1 CHASSIS
PHILIPS G110 CHASSIS
PHILIPS G8 CHASSIS
PHILIPS G90 CHASSIS
PHILIPS G90AE CHASSIS
PHILIPS GR1-AX CHASSIS
PHILIPS GR2.1 CHASSIS
PHILIPS NC3 CHASSIS

Philips 10CX1120

Dry-joints: T9629 (BU271) in the battery converter unit went short-circuit due to dry-joints around the line output stage.

No results: On investigation the line oscillator and driver stages were found to be working but the drive was insufficient to turn on the line output transistor. T534 (BC637) had low gain.

Tuning problem: We've had the same very intermittent and obscure fault on a number of these 10 inch colour portables. Fortunately the first one we had to work on had been taken in part exchange, so repair wasn't urgent – in fact it took about 6 months! The list of symptoms is as follows: the channel display shows 88; the picture drifts slightly off tune and you can't change channels; the picture goes dark and the sound level drops; some channel display segments light up brighter than others; the set won't switch on from cold; the line output transformer screeches and the output transistor gets red hot, eventually burning out. On one occasion this first set worked for days without even blinking . . . We eventually got a clue when we found that with the fault present there was excessive ripple on all the supply lines. Naturally the mains bridge rectifier's 150 µF reservoir capacitor C621 was the first suspect, but a replacement made no difference. Neither did replacing an endless number of smoothing capacitors. At one point we even tried a new line output transformer. The breakthrough came when the ripple was seen to be at line frequency, something we'd not previously noticed. A check around the line output transformer showed that the voltage at the cathode of the 16 V supply rectifier D551 had risen to 20 V. Now the supply from the chopper circuit hadn't changed, neither had the flyback pulse. So the only way in which the 16 V supply could rise would be if pin 12 of the transformer wasn't connected to chassis. Under the fault condition we were able to measure about 1.2 V between pin 12 and the tuner. Linking these two together cured the fault. This set uses double-sided print and the top and bottom earth planes weren't properly connected – there was a resistance of a few ohms between them. This was probably due to a poor soldered-through joint, but we weren't able to locate it. The solution was to make another connection. There's a convenient large hole at the rear edge of the PCB, between the scart socket and the line output transformer. Scrape away the varnish on both sides and link through with some thick desoldering braid. Don't use the semicircular hole or the back won't go on, and don't try linking up earths anywhere else because this will create earth loops, starting other faults like buzz on sound. This modification cured all the faults and I've since had several other sets that have been repaired in the same way. We wonder how many of these sets there are lying around in service departments because no one has the time to track down the cause of this obscure condition?

Very erratic field scanning: Sometimes there was cramping across the centre of the screen, occasionally complete field collapse. The culprit turned out to be C582, a ceramic capacitor that's connected to pin 20 of the TDA1770 field timebase chip – another habitual 10 nF offender.

Philips 2A chassis

Noise but no tuner oscillator: What looked like a simple case of noise but no tuner oscillator operation was eventually traced to a hairline crack in the print between pin 4 of the tuner and pin 7 of socket M1. As a result, the tuner was deprived of its tuning voltage.

Intermittent failure to start: As the fault was very erratic we decided to disconnect the supply to the line output stage and provide a bulb as a dummy load. The set then started every time. When the line output stage was reconnected we were back to the intermittent tripping. This state of affairs continued for several days until we noticed a spitting noise that came from the set while it was in the corner of the workshop. Investigation showed that flashes were coming from inside C2609 (9.1 nF) in the diode modulator circuit, and when it was removed we found that one leg was badly charred. Fitting a replacement cured the problem, but while we were working on the set we noticed that the components in the 18 V regulator circuit had been running very warm. We replaced these, spacing the resistors off the panel slightly. We've since had several of these sets that showed signs of overheating in the same area.

Dead set or intermittently dead: Check C2609 (9.1nF, 1.5 kV) by substitution. This component is in the EW diode modulator circuit. Note that it's not present in sets fitted with FS tubes while with conventional 90° tubes its value is 5.6 nF.

Tips when changing a tube: Watch out if you have to change the tube in recent Philips TV sets. The one we encountered, fitted with the 2 A chassis, had fixing nuts with left-hand threads!

Line output transistor shorted: This is becoming a stock fault with these sets. The reported fault is a dead set and you will normally find that the BU508 line output transistor is short-circuit. If so examine C2609 carefully for bulges and arcing around its legs – this will destroy the BU508. At the same time check whether R3601 (5.6 Ω safety resistor) is open-circuit. It supplies the EW modulator. If it is you will usually find that C2616 has gone open-circuit.

Dead, protection thyristor fired: This set was dead. On investigation we found that the protection thyristor 6727 had fired because the voltage at L5601 in the EW modulator circuit was too high. We noticed that C2608 (4.7 μF) was missing, but a check on a few stock sets showed that this is not fitted in models that have 45AX tubes. All the flyback tuning components and supplies were OK. Attention was then turned to the EW modulator, where the driver transistor Tr7599 (BD234) was found to be non-conductive. D.c. checks revealed that R3598 hadn't been pushed through the print at the factory.

Would switch off after 20 minutes: This set would switch off after exactly 20 minutes and no amount of tapping would bring it back to life. When the fault was present there appeared to be a short between pin 17 of the line output transformer and chassis. A freezer test eventually showed that C2609 (9.1nF) was short-circuit when warm. Normally this capacitor cracks and goes black, giving a visible clue.

Channel selector fault: We had an unusual channel selector fault with this set, as though the memory was useless. A replacement selector, part number 459 61047, provided a complete cure.

Shorted line output transistor: These sets can give you a bit of trouble if the problem is a short-circuit line output transistor. C2609 (9.1 nF) in the diode modulator circuit is a known offender. Sometimes we have also replaced the EW modulator diodes D6609 (BY228) and D6610 (BYW95C), the protection capacitor C2618 (1.5 nF) and the 140 V h.t. reservoir/smoothing capacitors C2697 and C2701 (both 47 μF – we use higher voltage ratings in these two positions). It has not been possible to pinpoint the primary cause of the trouble but we find that replacing these components avoids any comebacks.

Intermittent loss of raster: The problem with this set was intermittent loss of raster. When the fault occurred the voltage at pin 7 of the TDA3561 colour decoder chip (contrast control voltage) fell to 0 V and a 60 V peak-to-peak signal appeared at R3493. C2496 at the earthy end of the e.h.t. supply was going open-circuit intermittently.

Intermittent tripping: We resoldered the usual joints around the line output transformer but the problem persisted. It wasn't until we saw the width flutter in and out that we were able to pinpoint the cause. This was the line output stage tuning capacitor C2609, which is shown as 9.1 nF in the circuit but was 7.5 nF in the set.

Tip on changing field output chip: When changing the TDA3654 field output chip in this chassis don't forget to refit C2565 (390 pF) which is on the print side of the board. If you leave it off the output stage tries to become a long wave transmitter and you get moiré patterning on the screen.

Crackling on sound: The set had been a long-term inhabitant of the soak-test bench. The problem occurred even with the focus and e.h.t. leads disconnected, suggesting a fault in the line scan circuit. We changed the components one by one and finally proved that the flyback tuning capacitor C2618 (1.5 nF) was the culprit.

2 AT fuse blown: Do check beyond the bridge rectifier where you'll find that the 2 AT fuse 1651 has blown. The usual cause is diode 6664 (BYD335). C2664 (1.5 nF) should also be replaced.

Dead set, no h.t. from power supply: There appeared to be no starting voltage at the base of the BUT11AF chopper transistor. All the components in the feed checked out OK, however, and nothing seemed to be holding the voltage down. No shorts could be found across the power supply outputs. Luckily we checked the standby thyristor 6727 which was short-circuit. A new BT151 brought the set to life but we refitted the faulty one in order to be sure that we weren't being fooled – it shouldn't have shut down the power supply completely. It does though! Mark your circuit diagram accordingly.

Mains fuse blasted, chopper shorted: We had three of these sets in one week with the 2 A mains fuse blasted and the BUT11A chopper transistor T7686 short-circuit. D664 (BYD33J) was also faulty. The cause of all this was that C2664 in the snubber network had split in half. It's 1.5 nF, rated at 1 kV. Replacing these components restored the sets to normal operation. In sets that have come in for other faults we've seen this capacitor to be split or bulging. Obviously we change it before the chopper transistor fails.

Crackling noises at switch-on from cold: This caused loss of memory settings for the brightness etc. The set was now dead and we were expecting a battle with the power supply. There was a.c. from the switch, but no 300 V across the main reservoir capacitor C2659. In fact the a.c. supply wasn't reaching the bridge rectifier. This foiled us for a while until we realized that one half of the degaussing thermistor is in series with the bridge rectifier. It had cracked after arcing for some time. Hence the intermittent memory loss.

Severe patterning: We've had a number of these sets that suffer from severe patterning because of a missing capacitor. It's C2565 (390 pF), which should be mounted between pins 1 and 5 of the field output chip IC7570. In each case it has been clear that the chip has been changed recently. This device (TDA3562/4) overheats and can quickly fail if C2565 is missing. Presumably engineers are failing to refit this capacitor following chip replacement. A possible reason is that it's not shown on the provisional circuit although it does appear in later ones.

Line timebase intermittent squeal: When the set was first switched on there was intermittent squealing from the line timebase. It ceased once the set had been running for a few minutes. Resoldering the line output transformer pins provided a complete cure.

H.t. and e.h.t being pulled down: The h.t. and e.h.t. came up for an instant at switch-on, then the main h.t. decayed to a steady 30 V with the other secondary power supply voltages low. It seemed certain that the cause of the trouble was in the line output stage because disconnecting the feed to the line output transformer brought all the supply lines back up. Checks showed that the output transistor, the EW modulator diodes, the secondary rectifiers fed from the transformer and the tuning capacitors were all OK. The only thing left was the transformer. A new one restored normal operation.

Dead set, black mains fuse: The obvious things to check were the chopper transistor and the mains bridge rectifier diodes. They were all OK. In a situation like this, where no obvious short can be found, we usually remove the degaussing thermistor and try again – this device can be responsible for violent fuse blowing. With the thermistor out the set remained dead, but this time the mains fuse remained intact. We then spotted a small blue capacitor, C2664 (1.5 nF), with a split down the side. It turned out to be open-circuit. So did the associated BYD33J diode D6664. With replacements fitted the set remained dead. Checks showed that there was no output from the mains bridge rectifier – and no a.c. input either! Study of the circuit diagram soon showed what had happened – Philips dealers will be aware of this. We'd left the degaussing thermistor out, intending to replace it after finding the cause of the fault. In this chassis the thermistor is in series with the incoming mains feed to the bridge rectifier, acting as a surge limiter as well. Replacing this item restored the set to full working order.

Erratic tuning system: This set was brought in to have the back-up battery replaced, which is now a very common requirement. After doing this the tuning system behaved erratically. The set would search all right, but all was not well when a station was found. The tuning system stopped its search as it should, but would then alternate on either side of the correct tuning point, drifting in and out of tune continuously. Tuning is carried out by the SAB3037 CITAC chip. Its supplies and the 4 MHz clock signal at pin 21 were fine and as the tuning system stopped its search on finding a station we concluded that the channel identification input signal was OK. On finding a signal the SAB3037 chip is supposed to carry out fine tuning by doing an a.f.c. test. This can be monitored at pin 7. The fault lay here. It seemed that the test was being carried out but the chip ignored the outcome. Fitting a new chip and tuning in restored normal operation.

Random squawks on sound: The sound was modulated at random by squawks and squeaks that could be reduced by critical adjustment of the volume control. We found that C2151, a 0.1 μF capacitor in the damping network across the output of the TDA1013A audio amplifier chip, had only one lead soldered.

Picture too wide: Here's a cautionary tale – don't always believe the circuit diagram! This set had been to another dealer because it was dead. He had solved that problem but then found that the picture was too wide and couldn't be adjusted. He'd tried everything! The line output transformer, coils 5601 and 5611 and all the capacitors in the relevant sections of circuit had been replaced. The resistors had all been removed and checked out of circuit. The transistors in the EW correction circuit had also been replaced. To check that the correct parts had been fitted I compared the faulty set with a stock one. The stock set didn't have a capacitor in the 2608 position. When this capacitor was removed from the faulty set the picture size was brought back to normal. Just because a capacitor is shown in the circuit diagram doesn't mean that it was fitted in later production!

Dead set with no indicators alight: This is becoming a common fault. You find that the mains fuse and the 1 Ω surge limiter resistor 3654 are open-circuit while two of the bridge rectifier diodes, also the BUT11A chopper transistor 7687 and the CNX62 optocoupler 7668, are short-circuit. After replacing these components check for a short-circuit reading across the 280 V reservoir capacitor 2659. If all seems to be well the set can be switched on with confidence. On one occasion we found that the 1N4148 diode 6689 was also short-circuit.

2 AT main fuse blown: If you come across one of these sets with the 2 AT mains fuse 1651 blown don't immediately go for the chopper transistor. It's quite common to find that the chopper transformer's tuning capacitor C1664 (15 nF, 1 kV) has gone short-circuit. A totally blank raster with no sound but with the e.h.t., focus, first anode and l.t. supplies present may have you fooled, but not for long. Replace the back-up battery, reset the analogue controls and the panic will be over.

Half-inch gap on left side of picture: There was a half-inch gap at the left-hand side of the screen and just the slightest hint of bent verticals. Adjustment of the width control didn't improve matters and we eventually discovered that the cause of the trouble was the 'lower' of the two diodes in the EW diode modulator circuit, D6610 (BYW95C).

Occasional shutdown and power supply whine: A slight tap on the PCB in the line output stage area would bring the fault on. Close examination showed that the tuning capacitor C2609 was dry-jointed. Resoldering this and several suspect joints in the same area cured the fault.

Power supply dead: The power supply was dead. Checks showed that there was 0.6 V at the base of the BUT11AF chopper transistor and over 300 V at its collector, but the circuit wouldn't oscillate. As there were no shorts across the secondary windings of the chopper transformer attention was turned to the snubber network connected to the primary winding. D6663 (1N5062) was found to be leaky – 150 Ω both ways.

Set ticked in standby: Everything else was pefect, but the ticking wouldn't go. We were convinced that the cause of the trouble was in the power supply, and after a long and finally rewarding search the culprit turned out to be C2690. It's a 1 µF, 100 V non-polarized capacitor that's connected between the earthy side of the chopper driver transistors and the non-isolated chassis.

Overbright raster with flyback lines: The first anode control had no effect. We found that the feed resistor R3473 (910 kΩ) had gone open-circuit.

Black lines on the picture: It looked as though every other line of picture content was missing. After a long time had been spent carrying out scope checks etc. we finally found that C2050 (47 µF, 50 V) was open-circuit. It's off the 12 V rail, in the i.f. can.

Power supply slow to start: Check whether R3670 still reads 33 Ω. When its resistance value goes high the power supply takes a few seconds to start.

Chopper transistor short-circuit collector-to-base: If you encounter one of these sets where the BUT11 chopper transistor has gone short-circuit collector-to-base don't omit to check whether L5687, which is in series with its base, is open-circuit. The power supply will run when this coil is open-circuit but there will be no regulation. So with a dummy load the 140 V line will rise to an excessive level.

Excessive width: After soldering some dry-joints to cure an intermittently dead set we were faced with excessive width, the width control having no effect. We found that the −26 V supply to the width control was missing as the 15 Ω safety resistor R3602 was open-circuit. This was in turn because of a solder splash that shorted C2602 to chassis.

Tripping power supply: One of these sets with remote control came to us with a tripping power supply. We found that the cause was a short-circuit in the MAB8441 microcontroller chip on the front control panel.

Dead set, overvoltage thyristor firing: Check whether C2698 (4.7 μF) is open-circuit. For field cramping at the bottom of the picture, replace C2575 (4.7 μF). If the problem is ragged verticals and the 140 V h.t. supply is low, replace the h.t. reservoir and smoothing capacitors C2697 and C2701. They are both 47 μF types.

'Bang, dead!': There was a short-circuit across the mains bridge rectifier's reservoir capacitor C2659 but the chopper transistor Tr7687 was all right. The short-circuit was caused by diode D6664 (BYD33J) and pulse capacitor C2664 (1.5 nF, 1 kV) – the capacitor had split in half.

Short-circuit chopper transistor: The usual repair job put that right but there was a standby problem. When the set was cold it would go into the standby mode but the LED flickered. When the set was warm it would still go into standby but wouldn't come out: the power supply would buzz loudly and the LED's flicker rate was faster. We eventually found that R3689 (39 Ω) was open-circuit. It's in series with D6689 in the chopper transistor's drive circuit. Both components were replaced, although the diode measured OK on test.

Low h.t. 25 V instead of 140 V: When the power supply was unloaded the h.t. rose to 140 V but any load, for example a lamp or leaving the line output stage connected, reduced the reading to 25 V. We changed all the transistors and replaced the optocoupler, but this made no difference. Eventually we found that the 82 Ω safety resistor R3690, which links the chopper transistor's switch-off circuit to chassis, was open-circuit.

Raster cramped bottom, stretched top: The cure was to replace C2575 (4.7 μF, 63 V) in the field linearity feedback loop.

Dead set; ticking from power supply: With the line output stage disconnected and a dummy load connected in its place the power supply produced the correct 140 V h.t. output. When the line output stage was reconnected the protection thyristor Thy6698 fired. The 6.8 V zener diode D6700 was leaky.

Ticked in standby: Replacing C2690 (1 μF, 100 V) cured the fault (see earlier fault).

Normal sound, blank screen: When the first anode control was turned up there was a full raster with flyback lines. Scope checks around the TDA3561A colour decoder chip IC7300 showed that there was a video input at pin 10 but no RGB outputs at pins 12, 14 and 16. A new chip restored the picture.

No sound or vision: If the 140 V h.t. supply line is at 40–50 V and a screaming noise comes from the power supply, check whether the h.t. reservoir and smoothing capacitors C2697 and C2701 are open-circuit. They are both 47 μF, 250 V types.

Intermittent loss of sound and vision: If there is also a buzzing noise coming from the power supply, check whether the protection circuit is being activated – by monitoring the voltage at the gate of thyristor Thy6698. If the thyristor is being triggered it's possible that one or other of the series connected zener diodes D6699 or D6700 is leaky. They are both type RD6.8 V.

Set was 'eating' line output transistors: The dealer who brought it to me had fitted several – the correct BU508 V type and had also replaced the line output stage tuning capacitor. The usual dry-joints had been attended to. We decided to check the line drive waveform. It didn't look too bad, but every so often there was an odd shake. Although the line driver transformer connections appeared to be all right, we decided to resolder them. When the soldering iron touched one of the legs the solder fell away, leaving a rather blackened tag poking through the board. We removed the transformer, cleaned the legs then refitted it. For good measure we also replaced the damping components R3633 (6.8 kΩ) and C2633 (1.2 nF). The set then ran with no further problems.

Dead set, shattered mains fuse: This set also had an open-circuit surge-limiter resistor (R3654). The BUT11AF chopper transistor, D6664 (BYD33J) and two of the bridge rectifier diodes were short-circuit. In addition to these items we replaced the CNX62 optocoupler and the

2.2 nF, 2 kV pulse capacitor C2664, just in case. A check for dry-joints then revealed a beauty at one leg of the 9.1 nF flyback tuning capacitor C2609. After resoldering this we switched on and found that the set worked normally. It surprised us that the line output transistor had survived all this.

Dead set: The BU508V line output transistor was leaky. A replacement quickly failed, accompanied by squealing. By unloading the line output transformer's secondary windings in turn, we found that the set operated with a blank raster when D6644 was lifted. We eventually discovered that the TDA2579/N5 timebase generator chip IC7535 was faulty.

No results with low h.t. output: The h.t. recovered to 140 V when the H-scan drive plug M17 was removed. This fault can be caused by a faulty line output transformer.

Dead set, faint squeal from the power supply: When the supply to the line output transformer was disconnected (pin 7), the h.t. voltage returned to the correct level. As no faults could be found in the secondary supplies obtained from the transformer we suspected the transformer itself. The primary winding (pins 5 and 7) was OK, but our check indicated that there was a fault between this winding and chassis, at pin 18. Resistance readings were normal, but a new transformer restored the set to life, proving the tester's worth.

Picture looked like a mosaic: As there's nothing digital here, we suspected and replaced the TDA3561 colour decoder chip. Fortunately this cured the fault. We've since had the same problem with an Hitachi set that uses the same chip.

Intermittent tripping: As checks on the h.t. voltage were inconclusive we replaced the optocoupler in the power supply. This cured the fault – proved by a long soak test.

Field cramp: For field cramp at the bottom of the raster, replace C2575 (4.7 μF, 25 V). It usually goes open-circuit.

Tripping: TR7598 was short-circuit collector-to-emitter and R3601 (5.6 Ω) was open-circuit.

Standby, mute LEDs flash: Very intermittently the standby and mute LEDs would flash and the channel display would go off. After a lot of work we found that the 7 V supply to the search tune/control PCB, at pin 7 of M3, was fluctuating. The symptom was so infrequent that the cause couldn't be traced to a particular component. Replacing C2716 (1500 μF), D6726 (1N4148) and D6642 (BYD33G) seems to have stopped the flashing lights, however.

Lack of height: Lack of height with cramping and centre crossover distortion was the problem with one of these sets. After rather a lot of checks we traced the cause to R3573 in the field feedback circuit. With 110° sets its value should be 51 kΩ – in our faulty set the resistor measured 120 kΩ. With 90° sets the value should be 2.7 kΩ.

Poor field sync and occasional line jitter: The cause was traced to C2500 (1 μF, 50 V). It's connected to pin 6 of the TDA2579/N5 timebase generator chip IC7535.

Contrast couldn't be turned down: Although the contrast control voltage at pin 2 of plug M1 (the output from the CITAC module) varied, there was no variation at pin 7 of the TDA3561A colour decoder chip. When diode D6492 (OF449) was lifted to isolate the beam limiter circuit from the contrast control normal operation was restored. D6492 was leaky.

Unstable picture: There was rolling and occasional line pairing on scene changes. It seemed logical to check the electrolytic capacitors associated with the TDA2579 timebase generator chip. Replacing C2550 (1 μF), C2551 and C2552 (both 22 μF) produced a stable picture, although the originals checked out correctly when tested with a capacitance meter.

Smeared colours: This set produced a good monochrome picture. But the colours were smeared and broke up at the left-hand side of the screen, and there was too much saturation. The culprit was C2263 (2.2 μF), which is the reservoir capacitor for the ACC circuit. It's connected between pins 4 and 5 of the TDA3561A colour decoder chip IC7300.

Philips 2B chassis

Chopper transistor short-circuit: We have a large number of the version two of this chassis out and are beginning to get a steady stream of them back in the workshop with the BUT12 chopper transistor short-circuit. In all the cases we've seen there have been dry-joints on the chopper transformer. We haven't had a set fail after resoldering these joints.

Philips recommend, however, that if you experience further BUT12 failures for no apparent reason the device should be replaced with a 2SC3973B and the value of R3671 should be changed to 22 Ω, 5 W. When the BUT12 fails it sometimes goes short-circuit collector-to-base, which rather upsets the rest of the power supply. Usually it damages T7686, T7685, D6686, D6672, D6671, D6670, D6690, R3687, R3670 and C2690.

Dead set, 140 V supply was low at 122 V: There was also no channel display. We made the usual checks in the line output stage but as everything here seemed to be OK we checked the other supplies for shorts. This showed that the 20 V and −20 V supplies each read 10 Ω to chassis. The TDA1521 audio output chip was short-circuit.

Set refused to start: Due to insufficient bias at the base of the chopper transistor this set refused to start. The start-up bias is provided by R3656/7, C2658 and R3686/7. We checked all these components and although they seemed to be OK we replaced them. Still the set refused to start. Many hours were spent on the power supply, checking and replacing the various semiconductor devices, all to no avail. Many wonderful theories were put forward and tested, but the one thing about which we were relatively certain was that the bias was being damped by something in the chopper transistor's quite complex base circuit. Eventually, when it seemed that the best thing to do was to hide the set, pretend it no longer existed or replace every single component in the power supply, we got around to checking C2670 (68 nF). It read correctly and no leakage could be detected using our component tester. We nevertheless replaced it. When the on/off switch was operated a strange noise came from within the set – it was the rustle of e.h.t. The set was working and has continued to do so.

No sound: EEPROM X2402 is probably faulty.

Protection crowbar firing: Check whether the print by pin 1 of the line output transformer is open-circuit.

140 V supply normal with a 60 W dummy load but low with a bright picture: Transistor 7685 (BC547) is leaky.

Set stuck in standby: Here are two faults with the same symptoms but different causes. The first set could be made to function only when the gate of thyristor 6727 was disconnected. While checking around the standby control circuit we eventually discovered that D6729 had reverse leakage. The set worked normally when this diode was disconnected at one end. It actually feeds the mute circuit, to prevent plops in the standby mode. Sufficient voltage was being passed back via the diode to trigger the thyristor and put the supply into the standby mode.

In the second set the power supply was tripping. If the programme button on the front panel was held in the set would try to start but the LED display would do strange things. As the power supply proved to be OK we disconnected D6734 to disable the standby command. The set then started up but the display was haywire and none of the front controls did what it was supposed to do. Scope checks showed that there was a lot of noise on the microcontroller's data lines. This disappeared when the EEPROM X2402 was removed. Fortunate that – it was the only one of the three chips in the control system I had in stock! Fitting the replacement cured all the problems. These sets require the correct option code to be programmed in: 26 for a Nicam set, 18 for a non-Nicam version. When the set had been retuned and the correct option had been programmed in everything was back to normal.

Picture blanked by incorrect pulses from the field output stage: All the power supplies were present and correct, including the e.h.t., but the picture was being blanked by incorrect pulses from the field output stage. If the set was left on for a long time the picture would begin to appear, about 2 inches from the bottom of the screen. It would reveal itself one line a second, until the screen was full. The cause of the fault was C2571 ($100\,\mu F$, $25\,V$) in the field flyback boost circuit (note that the value is $68\,\mu F$ with $90°$ sets).

At switch-on from cold the picture would appear with the top half black: Or the picture would have a 3 inch. black band across the top half of the screen. As the set warmed up, the black area would slowly shrink until it disappeared completely. After this the fault wouldn't return until the following day. At first a teletext fault was suspected. But a scope check at the blanking pin (28) of the TDA4580 video control chip IC7300 produced a steady zero voltage d.c. display, so we could rule out the teletext area. Over to the contrast control pin 19. Once again nothing amiss. When we checked at the sandcastle pulse input pin 10, however, the field blanking section was seen to be about ten times its normal width, thus blanking off much of the picture. As the fault cleared, the width of the field blanking pulse shrank until it returned to the correct mark–space ratio. The field blanking pulse comes from the field output chip via diode D6564. We'd already tried using a can of freezer, but not in the field output stage area. Freezer checks here brought us to D6570 and C2571, which provide the flyback boost. We replaced the capacitor first (100 μF, 25 V) and that put an end to an intermittent, head-scratching fault.

'Fizzing and smoke': The cause of the complaint was the 1.5 nF, 2 kV (110° version) capacitor across the line output transistor – it had split. A replacement restored life to the set, but there was field collapse. As we could find nothing amiss initially we replaced the field output chip. This made no difference. We then did what we should have done in the first place and carried out some meter checks around the TDA8370 timebase generator chip IC7550. It soon became apparent that there was no 12 V supply at pin 22, which meant that the chip was running on its start-up supply. R3535 (10 Ω) was open-circuit because C2536 (100 μF, 25 V) was short-circuit. Replacing these two items brought everything back to normal.

Philips 3A chassis

Power supply shut down (but was OK): This was a rather annoying fault as time was wasted due to an error in the manual. The power supply had shut down but was OK as it worked with a dummy load (60 W bulb) connected across the 140 V h.t. line in place of the line output stage. We found that when the line output stage was connected the protection circuit operated, fusing thyristor Ty6698. Checks in the line output stage eventually revealed that the 315 mA fuse F1601 was open-circuit. It's in the feed to the EW correction cucuit. So we removed and checked the BD678 EW driver transistor Tr7599 which was leaky. When we looked in the equivalents book for a suitable replacement we discovered that it's a Darlington device – the manual shows it as being an ordinary npn transistor. Thus the transistor wasn't faulty. A new fuse cured the problem.

Very bad geometry: With these sets the geometry and various options are controlled digitally: adjustment is via a service remote control unit with the set in the service mode. It's unusual to have to carry out any adjustments unless major repairs have been carried out in certain areas. This set had very bad geometry and when it had been put in the service mode it seemed that all the options had been changed. But the set hadn't been anywhere for repairs. When the geometry had been readjusted and the parameters had been stored in memory all seemed to be well. But the same thing happened a few weeks later. The parameters are stored in an X2404 EEPROM, IC7900, which we decided to replace. After doing this you have to reprogramme the entire set: all the system options – Nicam, teletext, channel and programme numbers etc. have to be stored in the memory. Since doing this the set has not been back.

No picture: We found that the sandcastle pulses were missing, though they returned when the PAL decoder board was removed. The cause of the trouble was a short from pin 10 of the TDA4580 chip to chassis – a new chip was required.

Power supply failure: Power supply failure in these sets is far less common than with other chassis in the series. However, if you do get problems with shorted BUT12A chopper transistors and resoldering the chopper transformer doesn't provide a cure, check for cracked print around pins 2 and 11 of the transformer. You'll find that the power supply works at up to about 180 V from a variac but destroys the transistor when the input voltage is slightly higher. A recent set with this problem had a hairline crack in the print about half an inch from pin 11. The transistor fails because its switch-off circuit doesn't work correctly. It's thus driven hard on with a high mains input.

Power supply fault: If you have a power supply fault and find that the BC369 transistor Tr7686 is faulty, take extra care in checking that the pin connections of your replacement match the circuit requirement. Check with the circuit diagram: take the writing on the PCB with a pinch of salt.

Dead, power supply squealing: Voltage checks at the gate of THY6698 showed that this protection thyristor was firing. Various diodes etc. that feed the thyristor were next checked. This brought me to transistor Tr7499, which senses the 26 V, 200 V, e.h.t. and d.c. protection. We then found that the 200 V supply was missing at D6638 (which is fed from the LOPT), because fusible resistors R3638/9 were open-circuit. Some of these sets differ in the way the 200 V supply is protected. This one had two 27 Ω resistors in parallel (part number 4822 052 10279). Others have different value resistors or a Wickman fuse. A common cause of the fusible devices blowing is dry-joints at the LOPT.

Philips anubis a chassis

'Dead – blows the mains fuse': This very new-looking set appeared on the bench recently, equipped with yet another even smaller chassis. The reported fault was 'dead – blows the mains fuse'. Attention was immediately turned to the now obligatory BUT11AF chopper transistor which was short-circuit. We suspected that there might be something more sinister and a quick call to the friendly man at Philips revealed that a service kit has been issued to deal with the problem. The reference number is SBC7021, part number 4822 310 20491. Fitting this restored normal operation.

Squeaking from line output stage: The h.t. was low at about 40 V. As the power supply worked correctly when the line output stage was disconnected and a dummy load was substituted we decided to carry out some checks in the line output stage. For want of something better to do we changed the transformer. This made no difference at all. We drew a blank with various other components, then hit on the idea of disconnecting the scan coils. This produced the line scan collapse symptom. It couldn't be the scan coils, could it? It was, and the c.r.t. had to be replaced as well – they come as an assembly.

No power: It was soon apparent that the h.t. voltage was being dragged down by a fault in the line output stage. A d.c. resistance check between the collector of the line output transistor and chassis produced a reading of 20 Ω. Isolating the pins of the line output transformer proved that this was the culprit, a replacement then curing the problem.

Dead set, power supply OK: Although the power supply delivered the correct h.t. voltage to the line output stage one of these sets remained dead. On investigation we found that the 5.6 kΩ feed resistor R3444 in the supply to the line driver stage was burning up because the driver transistor (Tr7440) had a steady 9 V at its base instead of line drive pulses. At switch-on the TDA4504 i.f./timebase generator chip IC7015 should produce short-generation pulses at pin 29 to get the line output stage going. Once this happens, the line output transformer provides the supply to the chip. As the transformer was faulty, this didn't happen. Tr7440's base voltage was being held high because pin 29 of IC7015 was virtually open-circuit and the 2.7 kΩ pull-up resistor here (R3359) is fed with 9 V from the chopper circuit. Presumably the symptoms would be the same if IC7015 developed a fault that stopped it producing line drive pulses.

Philips anubis A-AC chassis

Blank raster and no sound: The on-screen display worked OK. An external video signal fed in at the front sockets proved that the set was in the external mode with the u.h.f. programme positions. A check on the status voltage at pin 18 of IC7015 produced a reading of 3 V instead of 0 V. This voltage comes from transistor 7877, where we found that the earth line was floating because the jumper by the scart socket had never been soldered.

Thumping noise from loudspeaker in standby: This new set worked all right until it was put into the standby mode. The standby LED then pulsed and a thumping noise came from the loudspeaker. When standby is selected thyristor Thy6570 conducts and the +5 B supply rises. This voltage increase is sensed by the 6.2 V zener diode V6568, which switches transistor Tr7553 on. The cause of the fault was that D6568 was leaky: a new LLZ F6V2 diode was needed.

Apparently dead, with only 20 V instead of 95 V from the power supply: With the feed to the line output stage removed, by disconnecting plug M5, the voltage on the 95 V line rose to the correct figure. With M5 reconnected, the voltage at the base of transistor Tr7555 could be seen to pulse at switch-on. This showed that the overvoltage protection circuit was being activated. As this happened only when the line output stage was operative, the problem was clearly associated with a line output stage derived supply. We found that zener diode D6555, which monitors the +5 V B supply, was leaky.

Would flicker in bright scenes: This set worked fine with dark scenes. When the picture was bright, however, it would flicker and the power supply output voltages would fall. Was the line output stage taking too much current, or was the power supply unable to provide sufficient current to meet the line output stage's normal requirements? Fortunately we were able to make some comparisons with a stock set of the same type. By loading the good set's power supply with bulbs, we found that its 95 V output could provide 700 mA before the supply began 'chirruping' and the voltages began to fall. The faulty set could provide only 500 mA. At least we knew that the fault was in the power supply. Fitting repair kit SBC7021 (part number 4822 310 20491) made no difference, neither did bridging the various electrolytics. We finally checked the gain of all the power supply transistors that hadn't already been replaced. Tr7554 (BC337A) was leaky.

Philips anubis B-AA chassis

Set worked but alarm LED kept flashing: This set worked but the alarm LED was flashing and ER4 was displayed on the screen. The ER4 display indicated a teletext fault, and when text was selected only lines came up. After checking the supplies to the text board and finding that everything was OK we checked the oscillators. 1701 (27 MHz) wasn't working: a new crystal put this right.

Horizontal grey line superimposed on picture: Scope checks in the field output stage showed that the field scan waveform had an oscillation on it. C2413 (10 nF) was found to be open-circuit.

Programming the ST24C02P EEPROM: If you replace the ST24C02P EEPROM IC7685 you'll need to programme the new device via the on-screen menus. The only snag is that with an empty EEPROM the language is set to Norwegian and the child lock will be active. When this is the situation the word 'Barnlas' is displayed. To clear the child lock using Norwegian select special functions option C on the main menu. The next menu will show Barnlas (child lock) as option A: set this to Fran (off). Once the child lock has been cleared you can change the language to English and set up the other parameters.

Philips CF1 chassis

Only worked for a few minutes: This set worked for only a few minutes from initial switch-on when new. Checks around the chopper power supply didn't reveal anything obvious so we disconnected the 95 V output and provided it with a dummy load bulb. A scope showed that there was an initial switch-on kick at the base of the BUT11F chopper transistor. All the semiconductor devices in the power supply were checked and found to be perfect. We disconnected pin 4 of the feedback optocoupler and still the power supply didn't run. This at any rate proved that the fault lay on the primary side of the transformer. We then started to check each component in turn and when C2317 (47 nF) was bridged the power supply came to life. When it was unsoldered we found that one leg was loose in the body. A replacement put matters right. A second of these sets failed very quickly from new. The problem was lack of line sync due to a faulty TDA2577A sync/timebase generator chip.

Dead, power supply would not start: We set about measuring all the usual things but couldn't find anything amiss. A short across one of the supply lines was suspected, but there was nothing that was measurable. As a last resort we unhooked the secondary side of the chopper transformer and connected a dummy load in the form of a diode and bulb. The power supply then came to life. When the 185 V supply rectifier D6310 was reconnected the power supply shut down, but replacing it didn't alter the situation. What was happening was that the protection capacitor C2310 (1.5 nF) in parallel with D6310 was going short-circuit under load.

Blank unmodulated raster, no sound: We found that there was no output from the TDA2541 i.f. chip, so this was replaced. There were still no signals. A check on the components around this chip then revealed that C2147, a disc capacitor, was leaky. Replacing this restored normal operation.

Field collapse: This was accompanied by sound from a foreign radio station. This indicated that there was an i.f. fault as well. A look at the circuit diagram showed that the field timebase and i.f. sections of the receiver are both supplied from the same source. Tracing back revealed that R3583 (1 Ω) was open-circuit.

Dead set, C2310 short circuit: We found that the 1.5 nF protection capacitor C2310 across the BYD33M 190 V rectifier D6310 was short-circuit. When this happens the chopper power supply shuts down. To check voltages in this chassis you have to remove the two tin covers. Be sure to link the solder points where the covers are fitted, otherwise you can have the situation where the BUT11F chopper transistor has no emitter earth connection. Beware – the covers are not at earth potential.

No sync, no sound, flyback lines present across the entire screen: The culprit was the TDA2577A sync/timebase generator chip. We had a TDA2577 in stock but when this was fitted there was normal sound but field collapse, so the A suffix is obviously important.

Failure of line output transistor: The problem with this set was failure of the BUT11AF line output transistor after a few hours despite resoldering the line output and line driver transformers. It was becoming an expensive one! After much examination of the board we found that R3503 (220 Ω) was dry-jointed. It's connected to the collector of the line driver transistor.

Rolling: There was in fact lack of field sync, as the picture could be made to roll down or up. Most of the sync processing is carried out in the TDA2577A chip but, since this chip is becoming hard to obtain, we decided to check any relevant peripheral components. R3379 and C2377 form the field sync pulse integrator: when checked C2377 (10 μF) was found to be almost open-circuit.

No colour: There was no colour, with the voltage at the wiper of the customer colour control very low. When the control was set to maximum the picture darkened slightly. This suggested a short across the control line. Disconnecting the feed to the TDA3560/N3 colour decoder chip (pin 6) restored the control voltage, which could now be varied from about 1 V to 3 V. With a new chip fitted the colour was back, although we were then told that the symptom will occur if the reference oscillator is off frequency. The old chip had hit the bin before we heard this – why do people always tell you these things after you've thrown the bit away?!

No raster, e.h.t. OK: R3585 (1 Ω) was open-circuit.

Intermittent picture faults: When we tested the set we found that there was poor line sync from cold and that following this the field scanning would be unstable, with varying periods of jitter or perhaps a field roll. C2368 (4.7 μF) was found to have gone very low in value. It's connected to pin 6 of the TDA2577A timebase generator chip IC7375.

Philips CP110 chassis

Chopper transistor failure problems: We've had a few cases where, when the chopper transistor has failed, two of the diodes in the bridge rectifier and the chopper control chip have failed as well. In this particular case, however, the 210 V supply rectifier had gone short-circuit. When we switched on again we were rewarded with an overbright monochrome picture: the transistors in the green output stage had gone short-circuit and the resulting high voltage had taken R3445, R3410, D6406, T7413 and R3416. Had there been a thunderstorm perhaps? The problem with another of these sets was field fold-over that varied as the chassis was flexed. No dry-joints could be seen, so attention was directed to the top of the board. The area around C2574 was most sensitive to probing and when this electrolytic was removed the cause of the problem could be seen – it had lost its electrolyte all over the board.

Dead but fuse hadn't blown: There was 300 V at the collector of the BUT11 chopper transistor. A scope check at the base of this transistor showed that the control chip was providing drive for a split second, but of too large amplitude. The BUT11 was found to be open-circuit base-to-emitter.

EW fault: This was traced to R3599 (47 Ω) being open-circuit. While investigating this fault we noticed that two of the transistor types have been transposed in the circuit diagram, i.e. T7600 should be shown as type BC558 instead of type BF819 while T7601 should be shown as type BF819 instead of type BC558.

No sound, only a blank raster with the channel display showing E: This display indicates that the microcomputer thinks an external input is present at the scart socket, though there wasn't. An external input is sensed on the scart status line, which is normally at 4 V in the TV mode but in this case was at 0 V because the microcomputer chip was short-circuit to chassis internally. Some versions use the TMP47C432AP-8188, part number 209 72038, while others use the version suffixed −8189, part number 209 87305 (the manual lists only one of these). If you fit an −8188 type in place of an −8189 no teletext functions will be available. Guess how we found out!

No sound or vision, displayed F1: We disconnected the scan coil plug, connected a dummy load and found that the 140 V h.t. supply was low at 50 V. To check whether the power supply was at fault we disconnected the collector of transistor 7726. The 140 V returned, so the power supply was OK. We've had quite a few faulty microcomputer chips so the next stop was at pin 14 of IC7840. But this was at 3.7 V which is normal for TV operation. In between these two items there's an inverter transistor, 7739 (BC548). When checked it was found to be open-circuit base-to-emitter.

Intermittent power supply shutdown: The picture would go off for a split second, the display flickering in sympathy. A light tap on the panel would instigate the fault, which would occur even with the line output stage disconnected. We've had a similar fault caused by a faulty chopper (SOPS) transformer with loose foil windings, but the panel wasn't sensitive in that area. In fact the problem was nearer to the mains switch − the posistor in the degaussing circuit was arcing internally.

Repeated failure of the BUT11AF chopper transistor: This is a common problem with these sets. It can usually be cured by taking the following steps. Replace the BUT11AF and the TEA1039 chopper control chip, increase the value of C2661 to 2200 μF, remove C2657 if fitted, replace the CNX62 optocoupler and fit a 39 Ω resistor (part number 4822 050 23909) in parallel with coil L5656. If the mains bridge rectifier has failed, fit 1N5061 diodes. Finally check that resistors R3658 (120 Ω) and R3659 (100 Ω) in the chopper transistor's base circuit haven't changed value.

'Flashing' picture: For a 'flashing' picture check whether R3415 (470 kΩ) or R3412 (180 Ω) on the tube base panel is open-circuit.

Line output transformer failure: We've had several of these sets in with line output transformer failure. After fitting the replacement we were left with no EW correction. The cause is R3599 (47 Ω, 1/8 W metal oxide) which surprisingly, is on the tube base panel.

Picture 'pulsed' on and off: This one was unusual. Sure enough the picture blanked out and then returned at roughly one second intervals. The h.t. supply was rock steady at 140 V. Scope checks around the TDA3562A colour decoder chip showed that it was blanking its RGB outputs. Heating the chip cured the fault so a replacement was fitted, but this made no difference. Now this set uses auto grey-scale correction and the circuit for detecting the black level is on the c.r.t. base panel. Checks took us to transistor Tr7413 on this panel where we found that R3415 (470 kΩ) in its base bias network was open-circuit.

Picture would slowly appear: This fault applies only to later versions of the chassis that have transistor Tr7672 in the power supply. The symptoms are as follows. No picture when the set is first switched on but after a few minutes a low-contrast picture begins to appear, gradually improving until, after half an hour, the contrast is back to normal. Tr7672 was conducting when the fault was present. It was being turned on because of excessive ripple on the 140 V line. C2670 and C2621 had dried up.

Mains rectifier capacitor failure: We're experiencing, and have done for some time, a high failure rate with the mains rectifier's reservoir capacitor C2656 (150 µF, 385 V). It tends to go low in value, with the result that there's no start-up supply to the TEA1039 chopper drive chip IC7669. Pin 9 is the place to check. Around 14 V should be found here. If this supply is low or absent, replace C2656 despite being able to measure 300 V or more across it.

Set refused to start: We went through all the usual things without success. Scope checks then showed that there was a very damped waveform at the collector of the BUT11AF chopper transistor Tr7665. We eventually found that the BYD33D rectifier diode D6672 in the 32 V supply on the secondary side of the circuit was faulty, reading about 1.5 kΩ each way. A replacement restored normal operation.

Set tripped: This set came from another dealer, the complaint being that it tripped. Once the chassis had been withdrawn it was obvious that someone had had a long and meaningful relationship with the set. The power supply had been rebuilt, with a new i.c., chopper transistor, driver transistor and optocoupler. Most of the rectifier diodes and small electrolytics were also new. But the power supply wouldn't run cleanly: it hiccupped continuously. As the conditions in the primary side of the power supply seemed to be OK, attention was turned to the secondary side. Fuse F1653 had blown, but a replacement lasted only a couple of minutes. This fuse protects the 15 V and 12 V supplies. The voltages were correct, but the current increased the longer the set was left in operation. Many components had been replaced in this area, including C2671 (1500 μF) which had been fitted the wrong way round. It was fitted as the printing shows, but these are positive supplies and the positive side of the capacitor was shown connected to chassis! Fitting a new capacitor solved the problem, but when we tried it the set wouldn't go into standby. This time C2735 was found to be the wrong way round, although the printing here is correct.

No vision for 10 minutes, sound OK: In the fault condition e.h.t. was present and the tube's heaters were alight, but a check at the c.r.t. cathodes showed that the tube was cut off. A fruitless search in the luminance-chroma circuits and the microcomputer section got us nowhere. We eventually found that there was excessive ripple on the l.t. and h.t. supplies when the fault was present. Further investigation in the power supply led us to the CNX62A optocoupler Tr7670 which turned out to be the culprit.

Dead power supply: This set had a dead power supply. There was 300 V at the collector of the BUT11AF chopper transistor but the start-up voltage at pin 9 of the TEA1039 chopper control chip was low at barely 1.2 V. This voltage, which should be about 11 V, can be checked only at switch-on. It comes from the junction of C2656 (150 μF) and C2661 (2200 μF) which are connected in series across the 300 V supply. The chip was checked by substitution and proved to be OK. We noticed that the 300 V supply decayed very quickly at switch-off, which provided the clue. C2656 was virtually open-circuit, a replacement restoring normal operation.

Low h.t. supply: If the 140 V h.t. supply is low at 60–70 V, check whether C2661 (2200 μF) is open-circuit. This is the reservoir capacitor for the chopper control circuit. Fit a Philips replacement, part number 4822 124 21511 – standard capacitors won't work in this position.

No sound or vision, F1 displayed in the channel display: We found that the line oscillator had stopped as there was no supply at pin 12 of the sync/i.f. module. The cause of the trouble was a crack in the print beside the line output transformer.

No picture initially, then poor: This set had a very misleading fault: there was no picture for a few minutes, then when it did appear it looked as though the tube's emission was low. We had the feeling that the tube was probably OK, however, and so it was. The cause of the fault turned out to be the 22 µF, 250 V h.t. reservoir capacitor C2670. For good measure we replaced the associated 22 µF smoothing capacitor C2621 as well.

Power supply destroyed in standby: There's a modification to overcome the problem. If it hasn't already been done, change C2661 from 1500 µF to 2200 µF and remove C2657 (if fitted). Then replace the faulty parts – usually the bridge rectifier diodes, the BUT11AF and BC337 transistors in the chopper circuit, the TEA1039 chopper control chip and the fuse. After doing all this we put one of these sets on soak test. Horror of horrors, it went dead again after a few minutes. It turned out that the two resistors in the chopper transistor's base circuit, R3658 (120 Ω) and R3659 (100 Ω), had gone high in value. Replacing them, also the BUT11AF transistor and the TEA1039 chip, got the set working again – for what we hope will be a long time.

Dead set with F1 displayed: This indicates that the 12 V supply is missing. In fact, however, the power supply was in the standby mode. The microcontroller chip had told it to start, but because transistor 7739 in the standby circuit had a base-emitter leak the power supply didn't start up.

Set would flash on and off: This set would start up at switch-on but just as the picture began to appear the set would flash on and off, quite violently but irregularly. The power supply and optocoupler chips were tried without success. We noticed that the fault became less frequent as the set warmed up. Tapping the power supply in the area of the degaussing posistor produced the fault, so a replacement was fitted. The set was then left to soak test. At switch-on from cold next day it worked perfectly.

Blank screen, sound OK: We turned up the first anode control expecting to see a white line across the middle of the screen but instead were greeted by a full blue raster with flyback lines. Checks on the tube base panel proved that Tr7405 (BF423) in the blue output stage was short-circuit. A replacement and a good resoldering of all the transistors on the panel produced a nice picture – after readjusting the first anode control of course.

Mains fuse shattered: Also, two of the diodes in the bridge rectifier circuit were short-circuit. After replacing all four bridge rectifier diodes and fitting a new fuse the set was still dead. A check showed that the output produced by the bridge rectifier was 180 V instead of 290 V. Assuming that we'd put the diodes in the right way round, there was only one possible cause, the reservoir capacitor. A new 150 μF, 385 V capacitor restored the power.

Would change channels up but not down: The front pushbutton control panel would change channels upwards but not downwards. All the other functions, i.e. tuning, volume, brightness etc., worked normally. We've had front control panel problems before with these sets, only some functions working and others not etc. The cause of the trouble is the keyboard foil, where it slots into the socket on the main board. This set was no exception. Reseating into the socket doesn't always work: renewing the foil provides a permanent cure. The replacement foil assembly is much better than the original, because the foil solders directly on to the main board. This could become a 'stock problem'.

Would become dead intermittently: We'd had this set in the workshop on a couple of occasions before without the fault showing up. This time, however, the set failed after being soak tested for four and a half hours. We removed the back and checked the h.t. voltage, which was high at 156 V. Next we checked for line drive. It was missing at the collector of the BC337 line driver transistor TR7630 because this nice little device was now open-circuit. A new transistor restored normal operation, with the h.t. correct at 140 V.

No raster, e.h.t. and tube heaters OK: Although there was e.h.t. and the tube's heaters were alight there was no raster. The cause of the fault was in the field output stage, the tube blanking system being in operation. R3570 (8.2 kΩ), the field output stage biasing resistor, was open-circuit.

Bottom half of raster missing: The top was cramped. A check on the voltages in the field output stage revealed that R3572 (1.5 Ω) had failed. It's connected between the emitters of the field output transistors.

Set would go off after a while: The period would vary from 10 minutes to 5 hours. The set would then stay off, coming back on again after a similarly indeterminate period. The cause was a dry-joint at the collector of the chopper transistor Tr7665. Dry-joints were also developing at the pins of the chopper and line output transformers.

Would pulse in and out of standby: At switch-on this set would pulse in and out of standby for up to 10 minutes. During this period the audio would also pulse in and out, remaining for progressively longer until the picture appeared and all was well. Voltage checks showed that the outputs from the chopper circuit were all on the low side, although the rectified mains voltage was OK. Favourite culprits for this type of problem are the TEA1039 chopper control chip and the CNX62 optocoupler, but replacements made no difference. Attention was next turned to the electrolytics. Be warned! Avoid the practice of bridging any capacitors directly connected to the TEA1039 chip while the set is operating. This will quite likely result in the destruction of the BUT11AF chopper transistor, the TEA1039 chip and other components. After fruitlessly checking all the electrolytics on both the primary and the secondary side of the circuit we noticed a small subpanel that's not shown in our service data. This board, located adjacent to the chopper transformer, is a modification that provides overvoltage protection associated with the 15 V supply. As this is the supply that's monitored for regulation purposes, it seemed logical to suspect that a component on this panel could be the cause of the trouble. A small 100 µF capacitor looked a likely suspect, being mounted just a few millimetres away from a large wirewound resistor. Bingo! Replacing it provided a complete cure. To avoid a repeat performance it's a good idea to fit the replacement capacitor on the print side of the subpanel.

Failed power supply: We found that the fuse was intact and the BUT11AF transistor was OK. So a check was made at pin 9 of the TEA1039 chopper control chip. The start-up voltage here should be 9 V but was only 2 V. A new chip brought the set back to life.

Short range remote control: All the remote control functions operated correctly – but only if you stood within two feet of the set. The cause of this was C2967 (100 µF, 10 V) inside the remote control receiver can. It decouples the supply to the receiver chip.

Faulty after new back-up battery: This set exhibited fault symptoms we've not come across before with the CP110 chassis after replacing the battery. When we depressed the mains on/off switch the power supply could be heard firing up, but this was not accompanied by the usual healthy e.h.t. rustle. The display segments were all illuminated, showing 88, and a faint raster with reduced height and width appeared. The lower half of the raster was noticeably darker, which seemed to indicate severe power supply hum. Neither sound nor picture were evident. Normal operation was restored when the battery was removed. So we came to the conclusion that there must be a faulty component that connected it to the 6 V supply and the system microcontroller-memory chip. Isolating diode D6901 proved to be the culprit.

Ragged verticals: Check C2633 (100 μF) by replacement. It decouples the supply to the line driver stage.

Loud whistling on warm-up: This set is used, with a VCR, in the lecture theatre of our local D.G. hospital. The lecturers had complained that there was a loud whistling once the set had warmed up. Hot-melt glue on the line linearity coil ensured silence.

Would flash cyan every few seconds: As the set warmed up the picture would flash cyan every few seconds. Heating the TDA3566 colour decoder chip IC7260 would clear the fault, but a replacement failed to cure the problem. When we fitted the tube PCB into another set we got a picture that flashed green, suggesting a fault in the auto grey-scale circuitry. Voltage checks showed that the base of transistor Tr7413 was at 4 V instead of 5 V. The culprit turned out to be its bias resistor R3415 (470 kΩ) which is connected to the 200 V rail.

Poor start-up: But this set worked perfectly in the workshop. It turned out that the mains voltage made the difference – it was 235 V in the workshop and 244 V at the customer's premises. We connected the set via a variac and tried various input voltages. This proved that it wouldn't work above 240 V. The culprit turned out to be C2690 (100 μF, 50 V) on the mains overvoltage subpanel. It read only a few microfarads. After fitting a replacement the set worked fine, starting perfectly at all inputs between 200 and 265 V from our variac.

'Teletext fault': When we selected text we found that most of the page was correct but the letter at the far right-hand side of each row was wrong. After checking the supplies to ensure that the voltages were correct and free of ripple, we checked the RAM's data and address lines. No problems could be seen here, so we decided to try a new ROM chip (μPD4364). This cured the spelling errors.

Early version. Poor tube?: The owner described all the symptoms of a very poor tube: the picture took several minutes to appear, and when it did it was very dull – even with teletext. Our tube tester disagreed, however. We soon found that the contrast control had very little effect on the voltage at pin 6 of the TDA3562A colour decoder chip, but disconnecting the beam limiter made no difference. A transistor (T7672) that's not shown in the manual is connected to pin 6. It's fed from the 143 V rail via a pair of resistors that are also not shown. Ripple on the h.t. line was switching this transistor on, and hence the beam limiter. Replacing the 22 μF h.t. reservoir and smoothing capacitors C2670 and C2621 cured the trouble. We've since had the fault with another of these sets. Later versions appear to have this circuit deleted.

Dead except for 'F1' in the display: A quick check showed that fuse F1653 had failed. A replacement restored some life, but the h.t. was pulsating. So I disconnected the supply to the line output stage, connected a light bulb as a dummy load, switched on and attempted to put the set into standby. Although the display showed standby, the bulb still glowed brightly and pulsated. A replacement CNX62 optocoupler (IC7670) provided a complete cure, proved by reconnecting the line output stage and switching on. This must be the only Philips chassis in which the power supply doesn't self-destruct when the optocoupler fails.

Dead set, shorted line output transistor: When we removed scan plug R13 and tested the power supply with a dummy load we were pleased to find that it ran normally. But when a new line output transistor had been fitted the results were awful. The line output transformer emitted a raucous scream, and the display consisted of a dark grey raster with a vertical band of lines near the left-hand edge. This slowly changed to a very low-contrast picture, with no colour and false line lock. A new line output transformer made no difference. Concentrating on the false line lock, we checked the waveforms around the i.f./sync module and discovered that the sandcastle pulses at pin 5 were incorrect – the short, high-amplitude pulse was delayed with respect to the longer blanking pulse. Adjustment of the line hold and phase presets simply made matters worse, and a new TDA2579 (plain or A version) sync/timebase generator chip didn't provide a cure. Philips don't show the circuit of this module in the CP110 manual, but the same circuits appear in the CP90 manual, where the sync and i.f. parts are separate. We finally traced the cause of the problem to the 3.9 V zener diode D6067, which had developed a leak.

Chopper transistor short-circuit: After replacing this, the TEA1039 chopper control chip and a few other items, we disconnected the feed to the line output stage, connected a bulb as a dummy load and switched on. The lamp glowed for a fraction of a second then went out – the power supply had tripped. Further checks showed that the cause of the fault lay in the optocoupler circuit. This was narrowed down to the standby switching. In the standby mode the BD438 transistor Tr7727 switches on, supplying more current to the optocoupler. The outputs from the power supply then fall. In fact in standby the transistor was supplying too much current. When its collector was disconnected the voltage at this point was found to be 32 V instead of 8–10 V. Although there was no measurable leak, a new BD438 transistor cured the problem.

'Takes 5 minutes to come on from cold': The well-known cure is to replace the 22 μF h.t. reservoir and smoothing capacitors C2670 and C2621. We changed them as one measured 18 μF on the meter, but this only improved the start-up time to 2 minutes. The cure this time was to replace the 100 μF, 50 V electrolytic capacitor mounted on a small PCB between the chopper transformer and C2656.

Ragged verticals initially: When this set was first switched on the verticals were ragged, the fault clearing after about 15 minutes. The cause was traced to C2633 (100 μF, 25 V), which decouples the supply to the line driver transistor Tr7630. D.F.

Fuse shattered after mains switch replacement: This set had originally come in for a new mains switch, as the original one wouldn't latch. After replacing it we switched on confidently but the mains fuse shattered. As we couldn't find any sign of a short-circuit we next tried running the set up via a variac. At about 160 V the fuse went spectacularly and smoke came from a small subpanel in the power supply. As this panel is not mentioned in our service manual, we were on our own. Tr7691 (BC368) and D6691 (18 V zener diode) had both failed, and C2656 (150 μF, 385 V) was short-circuit. Someone had been playing here, because C2656 was fitted back-to-front! When these items had been replaced the set still wouldn't work. A new TEA1039 chopper control chip finally restored normal operation.

Short-circuit chopper transistor: If the BUT11AF chopper transistor Tr7665 in the power supply is short-circuit, take the following action. Replace Tr7665 (BUT11AF), IC7669 (TEA1039), and D6657–6660 (four 1N5061 diodes). Fit a 39 Ω resistor (part number 4822 050 23909) in parallel with L5656. Remove C2657 (2.2 nF) if fitted. If C2661 is 1500 μF, change it to 2200 μF (part number 4822 124 21511). Check the degaussing posistor and replace it if in any doubt. Check that R3658 (120 Ω) and R3659 (100 Ω) haven't changed value. If the power supply won't start up, C2656 (150 μF, 385 V) is probably low in value.

Took a while for picture to appear: The picture was faint then brightened as the set warmed up. A check on the h.t. gave us the clue. Replacing the 22 μF reservoir and smoothing capacitors C2670 and C2621 cured the fault.

Philips CP90 chassis

No picture, just traces of sound: The voltages in the SOPS power supply were all OK and after a great deal of time had been spent we found that there was a tiny hairline print crack between pin 7 of the line output transformer and the junction of R3495/C2594. As a result there was no 15 V at the earthy end of the e.h.t. circuit, with obvious effects in the beam limiter circuit.

No sound or picture: We have had several cases of this. R3623 (8.2 Ω) goes high in value, fuse 1640 stays OK but diode D6667 (1N4148) simultaneously goes open-circuit. This gives roughly half of 95 V and 163 V and the set is in the standby mode – when D6667 is open-circuit the voltage across winding 1–2 on the SOPS transformer is not rectified. This is why TS7677 (BUT11) doesn't conduct and the set remains in standby.

Field collapse: Check whether R3623 (8.2 Ω) is open-circuit. This removes the 163 V supply to the base circuit of Tr7571 in the field output stage. R3623 is by the line output transformer.

Earphone socket problems: This has proved to be a very reliable chassis to date. We have, however, had a couple with low, garbled sound due to the earphone socket plug and you find that everything is OK. In both cases the owner preferred bypassing the socket to replacing it.

Low h.t. with a noise coming from the power supply: The line output transformer had shorted turns.

Intermittent full-white raster: A dry-joint at the earth connection to the combined focus/first anode control module was the problem.

Dead power supply: The start-up voltage was being lost at the junction of R3661 and R3673 because transistor T7673 was conducting due to a leak between pins 4 and 5 of the CNX62 optocoupler. On another of these sets there was a power supply fault that confused the micro-computer chip. The symptoms were no results with a pulsing power supply. If a dummy load was connected in place of the line output stage the h.t. produced was low at 60 V instead of 95 V. The cause turned out to be the BZX79-F6V2 zener diode D6702 in the voltage comparator circuit – it was leaky. Out of curiosity we refitted the faulty diode to find out what was causing the pulsing effect. Thyristor 6726, which conducts in the standby mode, was being intermittently triggered by the microcomputer which was rather confused as it was getting the power fail signal and no reset pulse.

Loud humming noise in standby: This set was brought into the workshop during a cold spell – back in February. Its owner was in the habit of putting it into standby before switching off at the mains. The result with this chassis is that at power-on the set comes on in standby. If the room was cold the set gave a loud humming noise in standby. The hum disappeared when the set was in use. With the set on the bench, in the standby mode, we found that cooling transistor 7727 and thyristor 6726 increased the h.t. voltage and produced the noise. Replacing 2703, 6726, 3726, 2726, 6727 and 7727 in turn had no effect. The control voltage from the microcomputer chip was OK. Time for a change of plan! The shop still had a new set in stock, so comparisons were carried out. Cooling had the same effect! Then we noticed a supplement to the manual. This showed that the circuit around T7727 had been altered after serial number PM03 and BA03. Updating the circuit to the later specification cured the problem. Change R3727 from 180 to 120 Ω, and 3729 to a BZX79F4V3 zener diode with an 820 Ω resistor in series, cathode of the zener diode to T7727.

Failure of first anode supply: We've had several of these sets with no picture due to failure of the first anode supply. In every case the cause has been an internal defect in the line output transformer.

Set apparently struck by lightning: The symptoms we had were no output from the power supply and a corrupt channel indicator display. The 5 V supply to IC7840 was present, but we judged the device to be a little too hot for comfort. Replacement brought the set back to life with good colour pictures and sound. On test we found that the 1.5 V back-up battery 1901 didn't hold its charge, as a result of which there was memory loss of the analogue functions. IC7840, type TMP47C432AP8188, is available from Willow Vale. Be sure to order the 8188 numbered device as this denotes programming specifically for this set.

Dead set, overvoltage protection circuit operating: When we used a variac to reduce the mains input voltage we discovered that the 95 V line was uncontrolled. The optocoupler was OK and transistor 7637 was conducting hard as it should have been to turn on transistor 7653. But this latter transistor didn't conduct because chip resistor 3668 (15 Ω) was open-circuit.

Dead set, no power supply output: There was a short-circuit across the 95 V h.t. line but it didn't go when the scan coil plug was disconnected. C2695 (2.2 nF), the protection capacitor connected in parallel with rectifier diode D6695, was short-circuit.

A blank raster and no sound: We found that the 11.5 V supply was missing at the cathode of thyristor 6726 in the standby circuit. It wasn't switching on therefore. The thyristor checked out OK but we found that diode D6733 (BAV19) was open-circuit. Replacing this restored normal operation.

Field collapse: There was a small amount of scan below the field collapse. A quick check showed that there was no 95 V supply to the field output stage. The cause was a crack in the print by R3275.

Intermittent switch-on when hot: This was a tricky fault. If the set was left alone it would work faultlessly all day and every day. If it was switched off and then on while hot, however, it would occasionally stay in the off mode. After many, many tests and component substitutions we found that D6670 (1N4148) was the cause. No fault could be found with it after removal, either when cold or hot, but a replacement cured the fault.

Very dark picture: Checks showed that the voltage at the contrast control pin of the colour decoder chip was very low. If D6490 was lifted the voltage came back up and a good picture was obtained. This diode provides a link to the beam current sensing circuit connected to pin 7 of the line output transformer. Obviously something was amiss here. The culprit was the 33 nF capacitor C2495.

Dead set, power supply whistling: The overvoltage crowbar thyristor was firing. By running the set with reduced mains input (via a variac) we discovered that the 95 V from the power supply was rising to over 100 V, pointing to a fault in the error amplifier stage. Checks here showed that transistor 7701 (BC548) was short-circuit base-to-emitter.

Keeps blowing the line output transistor: A new BUT11AF transistor would last only a couple of days. The e.h.t. was normal, no dry-joints were visible and the transistor's base drive waveform was OK. We then noticed that the set was a late model fitted with the A51AEL30X05 tube. This called for a few component changes, including the use of a BUT12A line output transistor. The correct transistor lasted long enough for line tearing to be seen before we managed to switch the set off. The line output transformer was faulty.

Dead set, no power supply activity: Initial checks showed that there were no short-circuits across the outputs, also that there was no start-up voltage (normally 2.1 V) at the collector of transistor 7673 in the chopper control circuit. This transistor was being turned on because the optocoupler was leaky. A new CNX62 restored normal operation.

Replace the line output tuning capacitor: When servicing these sets we'd recommend replacing the 1 kV 1.5 nF line output stage tuning capacitor C2619 with the beefier 2 kV version, Philips part number 4822 122 32501.

Only half an inch of field scan: There was only half an inch of field scan, in the centre of the screen. The voltages around the field output transistors were all OK and the drive from the TDA2579 chip was correct. Chassis return resistor R3582 (3.3 Ω) was open-circuit.

Dry-joints: When checking this chassis for dry-joints one place where you might not think to look is in the i.f./sync can. Dry-joints can occur in this can, especially around the TDA2579 chip.

Set would go dead after 3 hours: If the set was then switched off and on again it would work for another few hours. We removed the main panel and heated it with the trusty hairdryer. When the set failed the main h.t. rail voltage had fallen to about 20 V, suggesting that the set was in the standby condition although the standby LED was not alight. Checks showed that pin 14 of the microcontroller chip had indeed gone low and that if the voltage at the reset pin 33 was momentarily reduced the set would start up again. A replacement microcontroller chip restored normal operation. What had fooled me initially was the fact that the standby LED did not light up. This was because the standby command hadn't been given by the remote control unit. Thus the latch within the microcontroller hadn't turned the LED on.

Field collapse: This was because R3623 (8.2 Ω) was open-circuit. This surge-limiter resistor is in the 163 V line output stage-derived supply. It's used by the RGB output stages and also provides base bias for the upper field output transistor.

Lack of contrast: The set displayed a picture that lacked contrast because preset R3944 (5 kΩ) was broken.

Dead set, no standby LED or channel indicator display: Check for oscillation at pins 31 and 32 of the TMP47C432 microcontroller chip IC7840. If there's no oscillation check C2934 and C2935 (both 27 pF) which can become leaky.

Bright line across screen: Unusual these days, as field collapse usually causes blanking by upsetting the sandcastle pulses etc. In this case, however, the 8.2 Ω safety resistor R3623 in the 163 V supply was open-circuit, removing the feed to the RGB output stages and the field output stage bias. It seems odd that the same resistor should be involved with both feeds, since its failure will produce uncontrolled maximum beam current and field collapse, with tube damage unless the set is switched off pretty quickly.

Dead set, line output transformer shorted: In this case the cause was a poor connection at the scan coil plug/socket. As the pin had blackened, the lead/plug and socket were all replaced.

No picture or sound, the main power supply was working: The tube's heaters were aglow, but there was no picture or sound. A check showed that the microcontroller chip's 6 V supply was missing. The cause of this was D6733 (BAV19) which was open-circuit.

Went off intermittently: We put the set on test and after 10 minutes it did exactly as reported. A visual check on the PCB showed that there was a perfect dry-joint at pin 9 of the line output transformer. After resoldering all the pins we gave the set a soak test for a couple of days then declared it cured.

Intermittent failure to switch-on from cold: We disconnected the scan coil plug and connected a 60 W bulb as a dummy load across C2696. In this condition the h.t. was correct at 95 V every time, suggesting a fault in the line output stage. Experience led us to check the overvoltage protection circuit, however. It's based around thyristor Thy6696. As everything seemed to be OK on those occasions when the fault didn't appear, we decided that it would be safe to remove zener diodes D6699 and D6700 in turn. When D6700 was removed there was correct start-up every time, although a replacement 15 V zener diode failed to cure the fault. We suspect that there was a certain amount of line frequency ripple at the cathode of D6733, because replacing C2703 (330 μF, 25 V) cleared the fault. As a check we measured the original component's capacitance, which measured 270 μF, then put it back in which restored the original fault.

Dead set, number 3 shown: After a few seconds the display changed to F1. Checks in the power supply revealed that the 95 V and 22 V outputs were both missing, the former because the BYW95B rectifier D6695 was open-circuit and the latter because the 400 mA Wickman fuse F1690 had failed. Wonder which went first?

Stuck in standby: The microcomputer chip wasn't happy. As diode D6934 was leaky the chip's clock didn't run correctly.

Field collapse: The cause was traced to R3623 (8.2 Ω) being open-circuit. It's in the rectifier circuit that produces a 163 V bias supply for the field output stage.

Dot pattern visible in grey areas: Check whether C2691 (330 μF) has fallen in value. It's the reservoir capacitor for the +22 (19.5 V) supply.

No sound or picture: We usually tap around in case of dry-joints, but this didn't reveal anything. Visual inspection of the PCB then showed that the line output transistor T7677 was dry-jointed all round. We resoldered the connections and did the line output transformer as well. When the set was switched on after this there was perfect sound and vision. What amazes us is how some sets ever work with such defects, but we're always grateful for a simple job.

Intermittent tripping and patterning on the screen: It was difficult to narrow down the cause of the patterning to a particular section of the set, because there was both a dot pattern of the sort caused by arcing and the more usual herringbone effect. The latter also varied wildly, being reminiscent of sound-on-vision. These effects were all caused by C2691 in the power supply. This 330 μF, 25 V electrolytic had gone low in value.

Dead set symptom: This seems to be something of a stock fault with these sets. D6665 (1N4148) becomes leaky, producing the dead set symptom. It's connected to pin 5 of the CNX62 optocoupler in the power supply, via R6665 (4.7 Ω).

No line oscillator operation: If the 95 V supply is present but there's no e.h.t. etc. check the 19 V supply. You'll find it marked 19 V by C2691. If the voltage is low, check C2691 (330 μF) by replacement.

Herringbone pattern from cold: Scope checks showed that there was ripple on the 19 V supply. Replacing C2691 (330 μF, 25 V) cured the fault.

Dead sets, h.t. supply present: We've had two remote control sets with the same fault symptom – the set appeared to be totally dead but the 95 V h.t. supply was present. The 5 V supply was missing, however, as the standby thyristor was switched off. The cause of the trouble was in both cases lack of line drive, in one set because the reservoir capacitor (C2691, 330 μF) in the 19 V supply had dried out and in the other set because the TDA2579 timebase generator chip in the i.f. can was faulty.

It's easy to be fooled into thinking that the cause of this trouble is a fault in the power supply or microcontroller circuit. A quick test is to shunt a diode across the standby thyristor Thy6726 – the display should work and show an error code.

Dead set, 95 V h.t. present: An unusual fault with this remote control model. The start-up 22 V supply was found to be very low because C2691 (330 μF, 25 V) was open-circuit.

Occasional failure to turn on: In the fault condition a whistle was just about audible from the chopper circuit and the voltage on the 95 V h.t. line was only 1.5 V. These symptoms arose because the crowbar thyristor Thy6696 was conducting. A check on the three zener diodes in its gate circuit showed that they were all OK, and a new thyristor made no difference. Replacing the electrolytic capacitor C2700 (4.7 μF), which decouples its gate, put matters right – even although the capacitor's value measured correctly when checked.

No picture: This was because the tube's first anode voltage was missing. The line output transformer was faulty. It's the first time we've had one fail in this way.

Width changed as brightness raised: When the brightness was turned down the overvoltage trip came into operation. The cause of the fault was found in the chopper control circuit, where R3665 (4.7 Ω) was open-circuit and D6665 (1N4148) was short-circuit. These two components are connected in series. We also replaced the CNX62 optocoupler.

Strange interference dot pattern on the screen: Don't examine the i.f. department as we did. Replace C2703 and C2691 in the power supply. They are both 330 μF electrolytics.

Field bounce: Checks showed that this originated in the i.f./sync panel. A new TDA2579 timebase generator chip cured the problem.

Top of picture flickers and wavers to one side: Replace the following capacitors in the i.f. can: C2044 1 μF, 63 V; C2045 22 μF, 35 V; C2073 33 μF, 16 V; C2087 1 μF, 63 V; C2093 22 μF, 35 V; and C2098 1 μF, 63 V.

Field collapse with an uncontrollably bright line: This was because the 163 V video h.t. supply was missing. Safety resistor R3623 (8.2 Ω) was open-circuit. The 163 V supply also provides bias for the field output transistors.

White picture with flyback lines: This set produced a nice picture but intermittently, maybe once every few hours, the picture would turn to white with flyback lines, as if the first anode control had been turned up too far. When this occurred the picture was just visible in the background. We envisaged a long soak test, but fortunately spotted a tell-tale dark ring around the chassis pin of the line output transformer, indicating the start of a dry-joint forming. Slight movement of the pin while watching the screen, using insulated pliers, proved that the symptoms could be made to come and go at will. The set worked all right after resoldering all the transformer's pins. A word of warning: when you've replaced the back cover after a repair and find that the picture is dark and dull, you've broken the rear-mounted 5 kΩ contrast preset R3944 – its shaft protrudes through a small hole in the back cover. Guess how we found out!

Philips CTX chassis

Arcing at the e.h.t. cap: We've found that a lot of faults in these sets are caused by arcing at the e.h.t. cap – this happens even when the set is used in a dry environment. The arcing occurs at the point where the e.h.t. lead enters the cap, not around the edge of the cap. Figure 4(a) illustrates the problem: even when the plastic nut is fully tightened there's a slight gap because the cable is so thin. The remedy we use is shown in Figure 4(b): adding the outer insulation from an old e.h.t. lead gives a perfectly sealed connection, eliminating the cause of the arcing. The tape is used only to keep the lead tidy and has no insulating role. I find that this works very well, with no further problems.

Power supply pumping: A portable fitted with this chassis made several visits to the workshop before we got to the bottom of the trouble. Each time the complaint was that the set would 'tick', with momentary loss of picture and sound. Although we saw it once we weren't fast enough to make a diagnosis! The problem finally got so bad that the fault was more on than off. The power supply was pumping, and on each pump cycle the h.t. came up to about 80 V. In a darkened room we could see some sparking inside the stalk of the e.h.t. 'flower' connector to the tube bowl. Fitting a new lead and connector solved the problem for good.

Intermittent no start or would go off intermittently: We eventually found that the slider of the set-h.t. potentiometer R3325 was dirty.

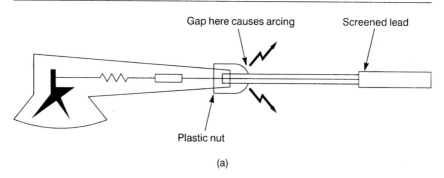

Gap here causes arcing   Screened lead

Plastic nut

(a)

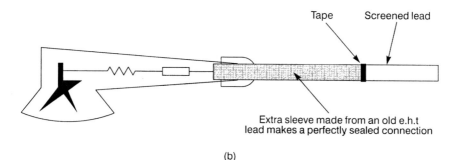

Tape   Screened lead

Extra sleeve made from an old e.h.t
lead makes a perfectly sealed connection

(b)

Figure 4 *Method of curing e.h.t. arcing in the Philips CTX chassis. (a) Cause of the problem. (b) Recommended cure*

Picture just fades away, sound remains OK: The cure is to resolder all the line output transformer pins – they'll usually need it. The ones that cause this particular fault are 2 and 8 which provide the tube's heater supply.

Tripping: We found that the BYV95A 26 V rectifier D6590 in the line output stage was short-circuit.

Excessive height: This chassis normally uses a TDA2577 as the sync/ field generator chip. Don't try a TDA2577A in this position you'll end up with excessive height.

No picture: There was just a blank raster when the brightness control was turned up. The voltage at the TDA3560 colour decoder chip's contrast control pin 7 was low at 1 V because C2565 (39 nF) in the beam-limiter circuit was open-circuit.

Sound OK, raster was blank: There was 0.5 V at pin 7 of the TDA3560 colour decoder chip. A scope check at pin 7 of the line output transformer (bottom end of the e.h.t. overwinding) showed that very high amplitude line pulses were present, suggesting that C2565 (39 nF) was open-circuit. Bridging it with the workshop 47 nF test capacitor made no difference, however. Close examination revealed that C2565 had a fractured connection at one end and a replacement restored the set to normal operation.

Low h.t.: The h.t. in this set measured only about 60 V. We eventually found that one of the 6.8 V zener diodes 6325/6323 had a forward resistance of around 500 Ω instead of the customary 1000 Ω.

Kept losing its memory: The 2.4 V cell 1777 had been fixed in position but never soldered. All was OK after soldering.

Flat 2.4 V remote control battery: This set came in with the now very common flat 2.4 V battery on the remote control board. When this battery is flat the set becomes very forgetful of the various analogue settings, i.e. tuning, volume, colour etc. After replacing the battery we found that there was no sound. The TDA2611A audio output chip's supply was missing because R3170 was open-circuit. When a replacement was fitted it quickly burnt up and went open. The cause of this was the audio output coupling capacitor C2182 (470 μF, 16 V) which had gone short-circuit. We then remembered having had to replace this electrolytic on many occasions before, so we fitted a 25 V type as a precaution.

Bad patterning on the screen: There was a very odd fault with this set, bad patterning on the screen. It was like CB radio interference. It went away when the colour was turned down, so something was affecting the chroma circuits. There was no ripple on the 12 V supply to the TDA3560 colour decoder chip, and a replacement chip made no difference. We noticed that the patterning seemed to change in sympathy as people in the picture talked. Something was amiss with the sound rejector trap. Shorting it out made no difference, proving that 6 MHz sound was entering the colour decoder chip. A replacement obtained from a scrap set provided a complete cure. When the old one was examined the coil's continuity was found to be OK so we came to the conclusion that the internal series tuning capacitor was open-circuit.

Line tearing: It was much worse in the VCR position. The dealer had already changed the TDA2577 chip and the VCR switching components 3374, D6374 etc. As the tearing varied with the picture content we homed in on the sync separator section. C2378 (1 μF) had dried up.

Coloured patterning: The set with which we had the fault was a Pye 42KT2142/05T: it had very severe colour patterning. The faulty coil is shown as 5153 on the circuit diagram but was 5150 in our set. We were able to rob a coil from a scrap KT3 two-chip decoder panel – Toko type 91792.

Dead set: There was 320 V at the collector of the BUX84 chopper transistor but the voltage at the collector of the BF422 driver transistor Tr7353 was only 7 V instead of 100 V. As this transistor was clearly without drive we moved back to the TDA2577 timebase generator chip IC7375 which also controls the chopper. Its supply pin 16 was at only 0.6 V instead of 9 V. A resistance check from this point to chassis produced a reading of about 2 Ω. C2395 (100 μF) was short-circuit.

No colour at switch-on from cold: As the set warmed up, the colour level increased gradually. It reached the point where it was barely adequate even with the colour control at maximum, then varied between this level and zero. The cause of the trouble was high-resistance connections to the TDA3560 colour decoder chip's holder.

Dead set: Check whether R3394 or R3395 is open-circuit. These 27 kΩ resistors are rated at 2.5 W.

Loss of memory (tuning, brightness): This set had remote control. It came to us twice for much the same fault. On the first occasion there was no tuning memory. Simple, the memory battery was flat. A week later it was back again, this time with no brightness or colour memory. It's the only time we've had to replace the M5840A chip.

Off-tune picture, sound buzz: There was an off-tune, slightly distorted picture with buzz on sound, higher channels being worse. Teletext consisted of gobbledegook. The scan tuning wouldn't lock, although you could use the fine tuning. The cause of the trouble on this occasion was a leaky capacitor within i.f. transformer 5158. It's best to replace the whole can. Don't be tempted to tweak cores: this may appear to help, but the teletext doesn't like the result.

Difficulty getting chopper circuit to work: When you've changed just about everything in the power supply and it still won't start up, try replacing C2323 (220 μF, 10 V). It worked for us.

Dry-joints around line output transformer have arced: We've had a number of these portable sets in which dry-joints around the line output transformer have been allowed to arc and the set has become quite dead. Resoldering still leaves you with no 12 V supply because you'll find that the 2.2 Ω fusible resistor R3585 has gone open-circuit.

Philips CTX-E chassis

No results: Voltage checks threw suspicion on the line output transformer. This was proved to be faulty by disconnecting the anode cap and removing the solder from pin 7 (the earthy end of the e.h.t. section). On doing this the set came to life and the c.r.t. heaters lit up. When a new transformer was fitted the set again went dead, only this time when the solder was removed from pin 7 there was arcing between this pin and the print. This led us to assume that the new transformer was all right, so attention was turned to the power supply. We found that D6317 (BA317) which removes the drive from the TDA2577 chip when the trip operates had a reverse leak of 80 kΩ.

Dry-joint at pin 12 of the line output transformer: A recurring fault with these popular sets. The result is no timebase operation. As a matter of course now when one of these sets passes through the workshop we resolder all the line output and chopper transformer connections using high melting point solder. One of these sets came in with no picture but with the e.h.t. and focus supplies present. We found that R3585 (2.2 Ω) was open-circuit, though there was no sign of any discolouration.

No sound or vision, e.h.t. present: A flat memory battery was the first suspect but it wasn't the cause this time. When the first anode control on the tube base panel was turned up a small raster appeared. A check showed that the supply to the line output stage was low. When the set-h.t. control was turned past the half-way point the voltage fell! R3327 (120 kΩ) was open-circuit.

Power supply and line output fault: This set was brought to us in pieces from another dealer. We had to rebuild the power supply because the transistors were short-circuit, having been fitted back-to-front. We sorted all this out and got the 125 V h.t. right with a bulb as the power supply load. When the line output stage was connected, however, the power supply wouldn't start. After replacing the line output transistor the h.t. pulsed up to 80 V, with a discharge apparent in the line output transformer. We were about to order a new one when we decided to disconnect the focus and e.h.t. leads in case of shorts. When the e.h.t. lead was disconnected the h.t. rose to 125 V and there was e.h.t. from the transformer. A check on the e.h.t. lead showed that it was dead short between the screen and inner conductor. A new lead wasn't all that was required, however. We had to replace the field output chip and several safety resistors in the 12 V supply. The set then worked well and we were relieved to see that the tube still had life in it.

No sound: This was caused by the absence of the l.t. supply to the audio output chip. R3170 (3.3 Ω) was open-circuit for no apparent reason.

No raster: When the set was switched on the tube's heaters glowed and the e.h.t. came up. There was no raster, however, while a loud hum came from the speaker. Safety resistor R3585 (2.2 Ω) was open-circuit, removing the 12 V supply to the signal circuits.

Failure to come on, bad memory: This set would sometimes fail to come on. We also noticed that the stations were not being memorized. As a result we replaced the back-up battery. There was no further trouble after we'd done that.

'Went bang then dead': Sure enough the mains fuse had blown apart – only the end caps remained in the fuseholder. The bridge rectifier had gone short-circuit, but when this had been replaced and a new fuse had been fitted the set remained lifeless. There was h.t. at the collector of the chopper transistor 7355 but it wasn't being switched on. By chance we noticed that resistor 3317 (100 kΩ) was discoloured and when tested it turned out to be open-circuit. It provides the BC548 transistor 7322 in the chopper control circuit with base bias. We decided to check this transistor which also proved to be open-circuit. When these two items had been replaced the set at last came to life. Sound blurted out, the e.h.t. rustled up and the tube's heaters glowed, but there was no picture. Turning up the setting of the first anode control on the tube's base panel showed that the cause of this was field collapse. As the TDA3651A field output chip's 26 V supply was present a replacement i.c. was fitted. At last all was well.

Insufficient width with bowing at the right-hand side of the screen: We found that the fusible resistor R3483 (6.2 Ω) was open-circuit. As a result there was no drive to the EW diode modulator.

On/off switch arced, set dead: The on/off switch arced and spluttered but the set remained dead after fitting a new switch. We found that the BU508A line output transistor had failed. When we switched on again after fitting a replacement the fault was field collapse. A new TDA2577 chip put that right.

No colour, unresolvable grey-scale tracking errors: Once the B&K tube tester had showed that all three guns were firing well our quest for the missing chroma began at pin 8 of the TDA3560 colour decoder chip, where we found that there was a nicely built sandcastle. The voltages around the chip were low, however, because the 9 V supply at pin 1 was low. The 10 Ω feed resistor R3222 had risen in value to 410! A replacement restored the colour and correct tracking.

Persistent line tearing/pulling when showing videos: This can occur even on the VCR compatible channel. Check the setting of the a.g.c. preset R3144. It might save you a wasted morning!

No sound or vision, screaming noise coming from the line output transformer: The BU508 line output transistor goes short-circuit if the set is left on for more than a few seconds. Try replacing C2351 (4.7 μF). It's in the chopper transistor's base drive circuit.

Dead set, no 125 V output from PSU: The TDA2577 chopper control chip was producing a square-wave drive output at pin 11 but this didn't reach the chopper transistor. The BF422 driver transistor Tr7353 was open-circuit between its base and its collector.

No results: Since the line output transistor is driven by a secondary winding on the chopper transformer, the TDA2577 sync/timebase generator chip produces the drive for the chopper circuit. A check at pin 16 showed that the 9 V supply was missing. The two 27 kΩ, 2.5 W resistors in series with the feed, R3394 and R3395, were both open-circuit. As everything else checked out all right (no shorts) a couple of replacements were fitted. The set then worked a treat.

Dead set: Checks showed that the BU508AF line output transistor Tr7562 was short-circuit. A replacement was fitted and some dry-joints around the line output transformer were attended to. After that the set was working again. But a week later it was back, with Tr7562 once more short-circuit. Very unusual for these sets. As there were no obvious faults we replaced the BU508AF transistor, also the flyback tuning capacitor (C2567), and resoldered the chopper and chopper driver transformers. When we tentatively switched on all was well. The waveforms were all correct, and Tr7562 ran only slightly warm. But we were not convinced. When we switched the set on from cold next morning there was a scream from the line output stage and all went dead. We disconnected the line output transistor, which was again short-circuit, and connected a 60 W bulb across C2330b, along with a digital meter set to the peak storage mode. At switch-on the lamp lit quite brightly for a few seconds then settled down with a soft glow. The digital meter showed that the peak voltage had been around 220 V, so excess voltage was doing the damage. There's a soft-start network, consisting of C2351 (4.7 μF) and R3351 (220 Ω), in the chopper transistor's base circuit. When checked C2351 produced a low reading of 2 μF. A replacement allowed the power supply to start up normally. Then a new BU508AF and a long soak test proved that the fault had been cleared.

Tripping: H.t. rectifier D6583 was short-circuit.

Jagged verticals: This was caused by a faulty e.h.t. lead.

'No picture, sounds like a radio': There was e.h.t., and when the first anode control was turned up we found that there was field collapse. It was obvious that the sync/drive generator chip was running from its start-up supply. The cause of the trouble was loss of the 12 V supply because the 2.2 Ω fusible resistor R3585 was open-circuit. A replacement restored normal operation.

Arcing from first anode, now dead: A check on the 125 V h.t. supply showed that it was low at 70 V. The voltage at the collector of transistor T7330 in the trip circuit was correct at 6.8 V, so this eliminated possible problems in the excess current protection system. We removed for testing transistors T7322 and T7323 in the chopper pre-driver circuit. T7322 (BC548) was open-circuit. A new e.h.t. lead and cap were required of course.

Tripping: This was cured by replacing D6564 (BY448) and D6482 (BYV9SB).

Intermittent loss of stations and LED display: One of these sets suffered from intermittent loss of stations and the LED display. The cause turned out to be nothing more than our old friend dry-joints at the line output transformer.

Philips CTX-S chassis

High brightness, off-line frequency: Strange symptoms here: high brightness, off-line frequency and when the set was switched off you got off-tune vision and sound momentarily. As a start we checked the brightness circuit which was found to be working correctly, then the colour decoder chip. After wasting much time we decided to scope the input to the TDA2540 i.f. chip. It seemed clear that the problem was in this area. After carrying out d.c. checks we decided to resolder the chip and the i.f. coils. This did the trick and we kept the set on soak test for a few days to be sure.

Faulty chroma delay line: In recent months we've had a number of these sets in with faulty chroma delay lines (type DL701). Symptoms range from intermittent loss of colour to permanent no colour or alternatively severe Hanover blinds. The latter fault can be tracked down to the delay line by tapping its case. As in the later KT3 chassis it pays to clean the pins of the chrominance chip. These can also be responsible for the no colour symptom when dirty.

No sound or picture: It's easy to get caught out sometimes by missing the obvious. A customer brought in one of these sets with the complaint that there was no sound or picture. We'd just dealt with another one with field collapse, so we quickly advanced the first anode control and saw that there was a line across the centre of the screen. New field output and timebase generator chips were fitted before we did what we should have done first, check the supply lines. The 12 V supply was missing due to an open-circuit safety resistor.

Set would work when main board was flexed upwards: Otherwise the set was dead because there was no h.t. The chopper transistor's collector leg had sheared off the body.

Philips FL1.1 Chassis

Shut down at switch-on: This set gave a chirp at switch-on then shut down with the standby, on and stereo lights lit. There was 0.7 V at TP56, which indicated that the hardware protection circuit was operating via the PROT (protection) line – this line monitors the sound, line output, beam current and EW modulator circuits. Scope checks showed that the fault was in the EW correction circuit, where transistors 7542 and 7540 were short-circuit. There's a mistake on the circuit diagram here: transistor 7542 is drawn as a pnp device but the type number shown is BC848C. The correct type is given in the parts list – BC858C.

No functions; flash at switch-off: This nearly new set had been got at by another dealer. When it was switched on all the LEDs on the front lit up. None of the functions operated but there was a flash from the screen at switch-off. The set couldn't be put into the service mode. Checks around the control panel then showed that the data lines were low and inactive. This is available only as a complete replacement panel, so a new one was ordered. Fitting it restored normal operation. The correct options for UK use have to be programmed in. When this was done as per the manual the set could be tuned. The on-screen diagnostics then revealed that there was a fault in the Nicam decoder which had to be replaced. Further faults were found in the audio stages, and after replacement of three more chips normal sound was obtained. All this because someone had had a go!

Dead set, went into protection mode at switch-on: This set came from another dealer with the complaint that it was dead. When switched on it went into the protection mode. You override this by shorting pins S24 and S25 together momentarily so that fault-finding can start. When we did this the h.t. came up and there was line drive. Scope checks showed that the line protection circuit was operating and after a few minutes we observed line collapse on the screen. A check across the line scan coil plug produced an open-circuit reading. Oh dear! If the scan coils were faulty a new tube would be required as well – the two come as an assembly. There's a PCB with a socket on the scan coils: investigation here with a meter proved that the coils were OK, and when the PCB was unclipped a hairline fracture was seen. Bridging this with tinned copper wire cured the problem.

No results: Causes can be many and varied, but if the set is in standby and when you switch it on via the remote control handset the mute, stereo and standby LEDs light up try checking the resistance between the 141 V line and chassis. If you are lucky you will get a low-resistance reading because C2504 is leaky (my circuit says it's 2404, but the PCB is marked 2504). It's a blue, disc-ceramic capacitor similar to the type used in the 2A and CP90 chassis: the value varies between models but is around 1 nF. Handle the PCBs carefully when they are in the slides: if they are left hanging out too far without support cracks can occur at the edges of the large signal panel. Faults we've had so far include a 'dead' set with all the front LEDs on, caused by lack of the POR signal because of an open-circuit in the track from the collector of Tr7272 to the small-signal panel jumper, and intermittent low width with no EW correction because of a crack in the track between R3608 and R3607.

Went to standby at switch-on: If you pressed the channel buttons on the remote control handset the e.h.t. came up then died. After this the LEDs at the front started to flash. The cause was a dry-joint on L5521. Resoldering it put matters right.

Philips G110 chassis

Picture shifted over: This set would intermittently come on with the picture shifted over and a black vertical bar in the centre of the screen. We finally found that there was a leak from pin 12 of the TDA2579 timebase generator chip to chassis due to a solder bridge under chip capacitor C2460.

No sound: As it came from another branch we first tuned it to our transmitter. This is where the cause of the fault lay: the set would scan the band but wouldn't lock to a channel as it scanned past. A bit of jiggery-pokery enabled us to stop the tuning at the required point, but there was no sound. There was also no on-screen display and the set went to standby after 10 minutes or so. We assumed that the latter symptom occurred because the set thought it was in the 'sleep' mode, and a problem on the video identification line was suspected. This would fit in nicely with the other faults. We carried out a meter check on the signal identification voltage at pin 13 of the TDA2579A timebase generator chip IC7470. The voltage should be very low, a few mV above zero, with no aerial input, rising to 9V when a signal is present. In this set the voltage remained low, signal or no signal. Thus the microcontroller chip thought that the set wasn't tuned in, muted the sound and operated its '10 minutes to power off mode'. The set wouldn't lock because the microcontroller depends on this line going high to halt the tuning. As we've had a number of defective TDA2579A chips, causing this particular fault amongst others in sets from several manufacturers, we fitted a replacement. Everything then seemed to be OK, so the set was returned. A couple of days later it bounced, with exactly the same fault present. We went through the same procedure and replaced the chip. The only other item we thought might cause the trouble was the 0.1 μF chip capacitor C2451, which is connected to pin 18 of the i.c. After removing it we fitted a conventional capacitor in the same position. This definitely cured the fault.

Power supply breakdown: Philips supplies a complete repair kit – you must replace all the parts supplied. We recently had one of these sets come in from another dealer who said that although he'd fitted the power supply kit the set would shut down after an hour or so, just as though it had been switched off. Sure enough the set did exactly as he said. When we checked the 140V supply we found that there was virtually no voltage here while the supply from the mains bridge rectifier was down to about 20V (instead of 280V). Two of the bridge rectifier diodes were going open-circuit when warm. We replaced all four and had no further problems after that. When we spoke to the dealer he said that he hadn't bothered to change the diodes although they are part of the kit, because they had measured all right. A lot of frustration could have been avoided if he had heeded the manufacturer's instruction to change all the parts in the kit.

Severe crossover distortion left-hand channel: Sometimes you jump to conclusions that give you extra work. As this Nicam set had what sounded like severe crossover distortion in the left-hand channel we went straight for the TDA1521 output amplifier IC7270. A replacement failed to cure the fault and we then discovered that the distortion was present only with Nicam reception: mono f.m. from a signal generator produced perfect results. The obvious thing to do was to check the audio stages in the Nicam section. There are two LM833 operational amplifiers, IC7350 and IC7351, one for each channel. They receive their inputs from the TDA1543 DA converter chip. Checks with a signal tracer showed that the distortion was present at the first stage amplifier in the left-hand channel. As luck would have it we didn't have an LM833 so a new TDA1543 was tried. This cured the fault.

Dead set: The mains fuse (1600) was shattered, the BUT 18AF chopper transistor (7625) was short-circuit, the CNX83A optocoupler (7614) had failed and the four bridge rectifier diodes (6602–5) were short-circuit. It's the fourth time we've had this problem. Replacing the defective components has so far restored normal operation without any bouncers, but we can't help wondering whether there's something else lurking there!

Power supply would not start: This set had come from another dealer who had given up. Basically it was dead. The power supply had been repaired, using the official service kit, but stubbornly refused to start so we ran the set with a reduced mains supply – about 90 V a.c. from the variac – and connected a dummy load in place of the line output stage. We also shorted across pins 1 and 2 of the optocoupler to disable the control circuit. In this condition the power supply produced a normal 140 V h.t. output. This suggested that in normal operation the optocoupler was being turned hard on, shutting down the power supply. There's a protection circuit that consists of Tr7655 and Tr7656, with several feeds – from the line output, EW, beam current and audio circuits. When this circuit operates the optocoupler is turned hard on, shutting down the pulsewidth modulator. It was operating, as about 0.8 V was measured at the collector of Tr7656 – there would normally be zero or a slightly negative voltage here. As the line output stage had been disconnected, the first three feeds could be ignored. Only the sound protection system was still in operation, via the LLZ-C20 diode D6657. A check on this chip diode showed that it measured 500 Ω both ways. The set worked normally when a replacement was fitted, the short was removed from the optocoupler and the line output stage was reconnected.

Chopper transistor failure: We've had a couple of cases where the BUT18AF chopper transistor fails again a few hours, days or even weeks after fitting the recommended power supply repair kit. There's a further recommended modification that can be carried out. It consists of fitting two BYD73B diodes (Philips part number 4822 130 60778) in series between the base and emitter of the BUT18AF transistor, with the anodes on the base side. You must also ensure that Tr7654 is a BC817, not a BC847. This modification certainly seems to work: I've had no further problems after doing it.

Set wouldn't power up: It's not an uncommon complaint with this chassis. Unfortunately if the chopper transistor fails, a power supply rebuild is necessary. A further problem is that the majority of the components are surface mounted. This wasn't going to be necessary here however. The h.t. was low at 20 V, but obviously the power supply was working to some extent. We suspected a fault in the line output stage, and by disconnecting the h.t. feed to the transformer and fitting a dummy load the full h.t. was obtained. But there was nothing amiss in this area. When pin 3 of the transformer was disconnected the set started up and we had full h.t. and e.h.t. There was no 14 V supply, however, as pin 3 provides the feed, hence no sound or raster. What was happening was that the timebase generator chip was working in the start-up mode. It receives an initial supply from the chopper circuit, the 14 V supply taking over once the line output stage gets going. Thus with the 14 V supply disconnected the chip wasn't working at its full capacity. We had, however, established that the fault was connected with the 14 V rail. The problem is that it feeds the 12 V regulator, whose output goes all over the place. To isolate the cause of the fault we had to disconnect various feeds. In fact the faulty component was the BC858 transistor Q7284 in the audio protection stage – it's another surface mounted device. A tip for this chassis: before you start, get your SMD transistor kit ready.

Shorted chopper and LOP transistors: If you get one of these sets with short-circuit chopper and line output transistors, remove and inspect C2546 (8.2 nF, 2 kV). Replace it if there is any evidence of heat stress.

Dead set: The owner of this Nicam set came to me with a tale of woe: it had apparently been elsewhere for six months and been returned as not worth repair because of major failure in the power supply. The set was very dead indeed. Anyone who has worked on these sets will know that they use surface mounted components in the power supply, and that the print doesn't take kindly to rough handling. Fortunately the state of the print wasn't too bad, but much work had been done in this area. I decided to start at the beginning and fit the recommended power supply kit. When this had been done a dummy load was fitted in place of the feed to the line output stage and the set was connected to a variac. At around 110 V the power supply started up then faded away again. We soon discovered that the voltage provided by the mains bridge rectifier was low. This was because the wrong degaussing thermistor had been fitted – one section is used as a surge limiter. When the correct type had been fitted the power supply started up but wouldn't regulate as the mains input was increased. Checks around the optocoupler revealed that Tr7652 (BC857C) was open-circuit. After fitting a replacement the power supply stabilized but the set-h.t. control was at one end of its travel. We checked the comparator circuit but everything here was OK. Eventually we found that R3652 (220 Ω) had almost doubled in value, affecting the operation of Tr7654 (BC817) and thus the current flowing through the optocoupler. A new chip resistor put this right and the set was returned to its delighted owner.

Intermittent start and no standby: The on/off switch sometimes had to be operated half a dozen times before the set would come on. Everything then worked normally until the set was put into the standby mode, when it shut down without even the standby LED being alight. In the fault condition the h.t. voltage fell to almost zero and a slight whistle came from the chopper transformer. This suggested that the power supply was in the protection state, which was confirmed by measuring the voltage at the base-collector connection between transistors T7656 and T7655. The voltage here should be zero or slightly negative but was 0.7 V, enough to turn on the transistors and thus the optocoupler, shutting down the power supply. Disconnecting the line output stage made no difference to the fault condition, so we checked the 20 V zener diode D6657 which monitors the 16 V supply. It read perfectly when checked with the diode test facility of a digital meter but produced a 1 MΩ reverse leakage reading when checked with the good old AVO. A new diode cured both problems with the set. We've had this diode fail before, but on previous occasions the set shut down completely. All this confirms our belief that digital meters are no substitute for a good analogue meter when it comes to testing diodes and transistors.

Would die in standby mode: This set had been fitted with the recommended power supply kit and worked well – unless you put it in standby. You would then, after 2 to 3 seconds, be rewarded with a pop and a short-circuit BUT18AF chopper transistor. There was obviously a drive problem in the standby mode, but how would one check this? We switched to normal operation and carried out some scope checks in the power supply. These showed that the BUT18AF's base drive waveform was incorrect – the switch-on delay was missing. Replacing D6612 and D6614 again cured that (they are part of the kit, so maybe we had a dud one). But the set still failed in standby. We eventually replaced D6646, D6649, Tr7655, Tr7656 and Tr7654 together. This cured the fault. There remained the problem of the bill . . .

Chopper transistor blows at switch-on: If you encounter a G110 that blows the chopper transistor at switch-on, even when all the bits in the repair kit have been fitted, but the power supply works all right if brought on slowly by use of a variac, change the on/off switch and the posistor in the degaussing circuit.

Chopper transistor short-circuit: If you find that the chopper transistor is short-circuit it's best to fit the SBC7020 kit (part number 4822 310 20489) rather than try to repair the power supply using hit-and-miss methods. I've had quite a few G110s in which dealers have tried fitting the kit but hit problems. These are usually of their own making. The leaded diodes 6602, 6603, 6604, 6605 and 6617 have their wires tightly crimped to the print: the usual method of removal, by heating and levering with a soldering iron, will result in damaged print. The best way to remove them is to cut their wires close to the holes on the top side of the board. The diodes will then fall away. Use desoldering braid to remove the solder and the remainder of the wire. Take care when replacing surface mounted diode 6637. Each diode has its anode and cathode marked with an 'a' and a 'k', but it's easy to confuse diode 6637's cathode marking with the collector marking of transistor 7656 and end up with the diode fitted back to front. Be economical with the solder when fitting surface mounted devices (SMDs), and don't have your iron too hot. SMDs are easily damaged by heat, and it's easy to make solder bridges that cause short-circuits. If the power supply is dead after you've fitted the kit, make sure that you haven't introduced a solder bridge between the anode of 6670 and one end of 6669. Check the values of resistors 3615, 3616, 3619, 3652, 3653 and 3654. Disconnect the line scan coil plug and fit a dummy load. Start up the power supply gradually, using a variac: check that the 140 V line doesn't rise excessively as the input voltage is increased. A problem in the line output stage can blow up the power supply – this will be proved by the line output transistor being short-circuit. Any arcing that occurs in the mains plug, the on/off switch or the degaussing posistor can also upset the power supply. Either way, before you return the set give it a soak test in standby and with the picture – for as long as you dare!

Faults caused by lightning damage:

Dead set: The BUT11AF chopper transistor and its TEA1039 control chip short-circuit, with R3659 (100 Ω) and R3658 (120 Ω) burnt.

Channel display keeps going to 'E': Microcontroller chip 7840 faulty.

No display, with the 140 V line low at 60–80 V: The 6 V regulator transistor 7738 (BC558) open-circuit.

Display goes to F1 a few seconds after switching the set on: Transistor 7739 (BC548) short-circuit.

Dead, power supply seemed OK: When we powered the set it whined. The power supply ran perfectly with the feed to the line output stage disconnected and a 60 W bulb connected across the h.t. line as a dummy load. It turned out that the line output transformer was the cause of the trouble: it's the first one I've had to replace. The picture was a bit strange when the replacement had been fitted. Part of the on-screen display was shifted to the left, with the first letter missing. It soon became apparent that the width was at maximum and the EW correction circuit wasn't working. Further checks brought me to the 10 kΩ width control R3525 which was open-circuit. A replacement and setting up corrected the display.

Dead apart from buzzing sound from power supply: We disconnected the h.t. feed to the line output stage – by pulling out the scan coil plug – and connected a dummy load in its place. This proved that the cause of the trouble was in the line output stage rather than the power supply. No shorts could be detected with the scan coil plug removed: when the plug was refitted, a short appeared across the line output transistor. The scan coil coupling/S correction capacitor C2550 was short-circuit.

Dead set: The power supply was shut down and we found that there was a low-resistance reading across the line output transistor. We checked various components before coming to the 390 nF, 250 V line scan coupling/S correction capacitor C2550 – the large blue one. It read 10 Ω both ways and on removing it we found that there was a small, black pinhole burnt into the case, near one of the leadouts. A replacement brought the set back to life.

Would go dead if switched to standby by remote control: This set would start up and run perfectly all day. But if it was switched to standby by remote control it would go dead, with no standby LED illumination. Switching it on and off at the mains would then restore normal operation. The culprit turned out to be the 20 V zener diode D6657.

Totally dead, 148 V h.t. low at 20 V: A replacement line output transformer restored the set to life, but the picture was far too wide. In addition the width varied with the brightness of the scene – this turned out to be a vital clue. Disconnecting or shorting out the EW modulator driver transistor Tr7533 proved that the EW modulator was working. The culprit turned out to be the 6.8 nF chip capacitor C2526 in the EW antibreathing bias circuit. It was leaky. We fitted a 100 V polyester replacement – the original capacitor was rated at 50 V.

Dead set: Two weeks after rebuilding the power supply in one of these sets we were called back because the set had again gone dead. When we switched it on there was silence for a few seconds then a purring noise came from the power supply. The cause of the trouble this time was the line output transformer, a replacement restoring normal operation.

Choke L5619 failing: Choke L5619 (3.9 μH) in this chassis is now tending to fail, with the result that the power supply self-destructs when put into standby. It's in the BUT18AF chopper transistor's base drive circuit. The parallel 10 Ω chip resistor R3619 is also worth checking, as it can be damaged. The coil is not included in the basic power supply repair kit, though it is in the extra kit ES7059. As L5619 can also become intermittent, it makes sense to replace it as part of a power supply rebuild.

Dead set, power supply OK: This set was dead although the power supply was OK when tested with a dummy load. So over to the line output stage, where we found that the BU508A output transistor was leaky. The root cause was C2546 (8.2 nF, 2 kV) in the EW diode modulator circuit – it had a large hole burnt in the side. After replacing these two items we confidently switched on. But there was nothing, still a dead set. As further checks failed to reveal anything amiss we disconnected the electronic trip (Tr7656/7655) by removing R3660. This time the set sprang to life when it was switched on, but there was no EW correction. The culprit turned out to be the 315 mA Wickman fuse T1534, which is in series with the collector of the EW modulator driver transistor Tr7533. The voltage build-up here is monitored by zener diode D6561, which provides one of the feeds to the electronic trip.

Dead apart from loud buzzing noise: After removing the line scan plug SG15 then connecting a meter and lamp across the 47 μF h.t. smoothing capacitor C2631 we switched on and were rewarded with the correct 148 V and a bright light. So the power supply was OK. We switched off and checked the resistance between the collector and emitter of the line output transistor. As there were no shorts here, a new transformer was fitted. We now had a working set, but there was a 1-inch band down the left-hand side of the screen as if the line centring was out. We went straight for the two surface mounted transistors Tr7594 (BC848) and Tr7593 (BC858). They were as short as short could be. No other component seemed to he faulty. So we fitted two new transistors, and also replaced the two 10 µF capacitors C2593/4 which appeared to be a little low in value. The set was now restored to full working order. On returning the set we were told that the 1-inch band down the left-hand side had been there for ages! It's odd that the two 1 Ω fusible resistors R3593/4 hadn't blown. Their values were spot on and there were no signs of distress.

Field collapse: The cause was simply a bad joint at the collector of the upper transistor (Tr7502) in the field output stage. Use of a multitude of miniature surface mounted components is a great space saver, but the majority of the PCBs used by manufacturers are very thin and flimsy. As a result, only slight flexing while manoeuvring a board into the service position can lead to cracked components. Surely with such a high density of delicate components the boards should be thicker?

Line output stage fault: If there is a line output stage short that's not caused by a faulty line output transistor or transformer, check the scan coupling/correction capacitor C2550 (390 nF) by replacement. We have known this capacitor to go both short- and open-circuit.

Sound but no picture: There was just a blank raster. The 200 V supply smoothing resistor R3375 (180 Ω) on the tube base panel was open-circuit because the TEA5101A/P RGB output chip IC7465 was faulty.

Philips G8 chassis

No colour, two faults: The first set was fitted with the combined signals panel. Overriding the colour-killer produced unlocked colour, but changing the TBA540 and TBA560C chips made no difference. We eventually found that there was no burst at pin 7 (burst output) of the TBA560C. The cause of the trouble was R3215 (1.8 MΩ) which was open-circuit, upsetting the bias at pin 10 (burst gate pulse input) of the TBA560C.

The second set had the BA00 combined signals panel. Adjusting the line discriminator balance control produced colour, but it was varying all the time. This variation could be seen on a meter as voltage variations at pin 9 (a.c.c. output) of the TBA540Q reference oscillator chip. The fact that the burst was distorted and varying led me to check the line pulse input. R212 (750 kΩ) in the pulse feed network was found to be almost open-circuit. Strange that the set seemed to be OK until a day or two previously, but as it has only just come into our area someone may have tweaked it up.

Hanover blinds: This set suffered from Hanover blinds that couldn't be cured by adjusting any of the controls in the decoder section. After a lot of soul searching and scoping we discovered that whoever had fitted a new line output transformer had made a wonderful dry-joint at pin three, as a result of which no line pulses were being fed back to the chroma circuitry. Resoldering put matters right.

H.t. was 250 V, couldn't be adjusted: The set employed the very early power supply panel with the zener diode in the BC147 control transistor's base circuit. R1365 had increased in value from 47 kΩ to 60 kΩ.

Philips G90 chassis

Tuning drift or no signals: If the green bar moves, check the ZTK33B 33 V stabilizer.

Field collapse: The Wickman fuse in the supply to the field output transistors had gone open-circuit. When a replacement was fitted the set worked but the 22 Ω resistors R3503 and R3507 in series with the two transistors were overheating. Resistance checks on the transistors showed that they were OK but R3501 (390 Ω) in the upper transistor's base drive network was open-circuit. Replacing this restored normal operation.

Action of contrast control limited: Check whether C2560 (33 nF), which is connected to pin 7 of the line output transformer, is leaky. This removes the standing bias for the beam limiter. Random failure of the BUT11AF chopper transistor is commonly due to the CNX83 optocoupler.

'Frilly' picture: When we tried the set we found that the picture was too large and had corrugated verticals. So we switched off quickly. On inspecting the main panel we saw that resistor R3668 and transistor Tr7652 in the chopper feedback circuit had burnt up. Replacing these items and fitting a new CNX83 optocoupler returned the 95 V h.t.

supply to normal, but the corrugated verticals were still present. Replacing the 47 μF h.t. reservoir capacitor C2630 cured that.

No teletext, 'F7' displayed: This set worked well apart from the fact that there was no teletext and F7 was displayed on-screen. Checks around IC7800 showed that its reset pulse was missing. This comes from a separate 5 V regulator/reset pulse generator circuit (transistor Tr7846 etc.) which checked out all right. Its input voltage was 1 V low at 7.3 V, however. C2843 (220 μF) had dried up. A replacement brought the voltage at the collector of Tr7846 back to 8.3 V and restored the teletext.

No line oscillator operation: Check the voltage at pin 16 of the TDA2579 timebase generator chip IC7470. With a normally operating set the reading should be 9 V. If the voltage is low, check the voltage across the 5.6 V zener diode D6455. If this is low, replace D6455 (BZX55FSV6). If it's correct, replace transistor Tr7455 (BC858).

Sound low or intermittent: Check the 20 V supply at pin 18 of the TDA8191 chip IC7220. If it's missing, check diodes D6272 and D6278 (both type BAS32) which can go open-circuit.

Dead set: Check for cracks in the print by the line output transistor's heatsink. The track most likely to break is the one that provides the line output stage's 95 V supply. To prevent the PCB being strained any more, make sure that the lower back retaining screws are fitted. Otherwise the cabinet can flex when the set is on its stand.

Power supply shut down after 10 minutes: The 95 V h.t. supply fell to 30 V and was pulsing to 40 V. A puff of freezer on the BC557C transistor in position Tr7652 restored normal operation. Everything was OK when this transistor had been replaced. If the power supply produces low outputs (less than 25 V on the 95 V line), check the BAS32 diode D6653 which is probably leaky or short-circuit.

Philips G90AE chassis

Set was tripping: A previous engineer had got in a right mess with this one. From an inspection of the PCB we could see from the soldering that someone had had a long and meaningful relationship with the power supply, the tuning and the teletext sections of the set. After a check for missing parts or chips fitted the wrong way round we switched on. The set was tripping. We disconnected the supply to the line output stage and connected a dummy load. The set still tripped, but this stopped when the set-h.t. control was turned down. In fact the supply could be set to 95 V as normal. Reconnecting the supply to the line output stage brought the set back to life but there was no sound and just a blank raster, though the on-screen channel display worked. When a signal was injected into the TDA5850 video switching chip there was activity on the screen, but there was no output from the TDA8341 i.f. chip. Resistance checks then showed that the crystal filter 1030 was short-circuit to chassis – there was a solder whisker across two of the pins! A touch with the soldering iron and all was well.

No colour: When the colour-killer was overridden by connecting a 470 Ω resistor between pins 1 and 6 of the decoder chip we were able to see that the reference oscillator was running through but couldn't be locked by adjusting C2352. The a.p.c. loop components are connected to pins 23 and 24 of the chip. Resistance checks here revealed that C2359 was leaky.

Set shuts down with a ticking noise from the power supply: We found that the PCB around R3668 (150 Ω) was scorched while Tr7652 (BC557C) was leaky. After fitting replacements we checked the set and found that although the h.t. was correct at 95 V R3668 was still under stress, while under certain conditions the verticals were ragged. A scope check on the h.t. line showed that there was a lot of noise. The h.t. reservoir capacitor C2630 (47 μF) was warm, a replacement finally putting everything right.

Power supply shut down: But the set would work when the mains input was reduced to about 90 V. This was not due to the protection circuit operating. We found that the supply to the optocoupler rose quite high as the mains input was increased. The only path is via D6653, which is normally reverse biased – it's forward biased in standby. A check showed that it was leaky, a replacement restoring normal operation.

Blank raster, but teletext OK: Checks around the TDA3561A colour decoder chip showed that the voltage at the contrast control pin (pin 7) was slightly negative instead of the 3.4 V specified in the manual. The negative voltage came from the beam limiter via D6326: C2560 (33 nF) was open-circuit.

Blank raster, no sound: There was also no on-screen display and the remote control received LED didn't flash in answer to a command. As all the supplies were present we carried out checks around the microcontroller chip and found that the reset line was stuck at 2 V.

Disconnecting the reset pin (IC7720, pin 33) proved that the chip had an internal leak. A new TMP47C434N3555 restored normal operation.

Intermittent tuning drift: We started by changing the tuner unit, then the microcomputer chip. As this made no difference it was time to think about the fault. A check on the varicap voltage showed that it varied when it shouldn't. Why don't we check the obvious things first? The ZTK33B regulator D6770 was the cause of the problem.

No red in display with RGB input: This set is used as a computer monitor in a local school. The complaint was of no red in the display with an RGB input. We found that the cause was the BAS32 surface mounted diode D6854 which is connected between the R input and chassis – it was short-circuit.

Reduced height with top fold-over: We eventually found that R3508 had risen in value from $24\,k\Omega$ to $70\,k\Omega$.

Parts would blow during standby: We've had two of these sets with the following serious problem. During production the junction leads of diodes D6624 and D6642 on the primary side of the chopper circuit weren't cut short enough. As a result, they tended to touch the solder-pad junction of resistors R3627 and R3625 intermittently. This doesn't affect the set in its working mode, but when the set is switched to standby the voltage across the 9V reservoir capacitor C2260 gradually rises from 9.2V to 25V, eventually shorting nearly all the diodes, transistors etc. in this part of the circuit with possible damage, if any attempt is made to switch out of standby, to most of the same items in the primary side of the chopper circuit. The first set drove me up the wall as it would behave perfectly for a few days before blowing all the above components. Now, before we do anything else when one of these sets comes in for repair, we check the connections to D6624 and D6622.

Intermittent fuse blowing: This set had been looked at by a dealer who, the fault being intermittent fuse blowing, had replaced many semiconductor devices – repeatedly by the look of the enclosed packet of defunct components. He hadn't fitted a Philips G90 SOPS repair kit, however. So the first job was to fit one. It consists of a chopper transistor and most of the other semiconductor devices on the primary side of the power supply, some of them surface mounted. This done we used a variac to wind up the input, with a dummy load connected across the 95V h.t. line. The set regulated beautifully. We then reconnected the line output stage, switched on and to my horror witnessed a flash from the mains fuse. A quick check across the BUT11AF chopper transistor showed that it had gone to a better place. After fitting another BUT11AF and checking other components on the primary side of the power supply we again disconnected the line output stage and wound the set up. Perfect! Full h.t. Rather than look for a fault in the line output stage we thought that it would be an idea to switch on the set with the full mains voltage applied, only to discover another BUT11AF was needed. So the set would operate when wound up, but certainly didn't like full mains voltage at switch on. A fault on the secondary side of the power supply was now suspected. D6649, D6615 and D6616 were all replaced as we've had trouble with them before. In fact we checked most if not all the semiconductor devices on both the primary and secondary sides of the power supply, a wise move with a fault condition like this. But nothing was amiss. Quite by accident I then discovered that D6645's anode didn't go anywhere. Sometimes empiricism succeeds where theory fails! It should be connected to pin 18 on the secondary side of the chopper transformer. But a check from this pin to the diode confirmed that there was an open-circuit. The print was checked and was fine. The only other item is a surface mounted jumper link in position 3645. Out it came and when checked was confirmed as being open-circuit. A replacement was fitted, also another SOPS kit in case the malfunctions had upset anything. The line output stage was reconnected and the set was then switched on. No bangs or flashes this time, just a full raster and sound. The set was finally soak tested, switched on and off repeatedly and pronounced fit.

Intermittent sound: Even the slightest attempt to move the chassis would cure the problem for months at a time. We eventually traced the cause to D6272 (BAS32), a surface mounted diode that's in series (along with another similar diode) with the 20 V supply to the TDA8191 sound chip IC7220.

'Drifts off tune': On test, no channels were tuned in. When retuning was tried a station was found but was lost again as soon as it was stored. A scope connected to the tuning PWM signal showed that it changed at the instant the store button was pressed, indicating that data was not being stored. A new X2402 EEPROM put matters right.

Line output transformer fizzing: This set led us a dance. The e.h.t. was low and the line output transformer was making a fizzing noise. Needless to say a new transformer made no difference. Scope checks on the supply lines showed that they all seemed to be noisy or oscillating. We then found that if pin 2 of the line output transformer was disconnected the fizzing stopped and the c.r.t. heaters lit up. Pin 2 is the source of the 14V supply. Various paths were disconnected. We eventually lifted pin 10 of the TDA2579A timebase generator chip IC7470. This brought the set on. A new TDA2579A chip put matters right.

No sound or raster: This fault had developed during a thunder storm. There was no sound or raster and the 95 V h.t. supply was low at 30 V. The h.t. was still at 30 V with the line output stage disconnected and a dummy load in its place, so there was a power supply fault. With the set fed via a variac, the h.t. was correct at 120 V a.c. input but fell back to 30 V as the mains input was increased. A check at the collector of Tr7656 showed that there was no protection active. The voltage at the cathode of D6600 (normally 8.2 V) was high, however. This should be the situation in standby, but checks in the standby circuit showed that it was not being triggered. Resistance checks in this area revealed that the BAS32 diode D6653 was short-circuit. Normal operation was restored when this diode had been replaced.

No sound or picture: The 95 V h.t. supply was missing because of a hairline crack in the print at the edge of the main PCB, where it slots into the cabinet. An insulated link put matters right.

Philips GR1-AX chassis

Tuning drift: When it was hot the tuning would be lost on channel change. As a bad batch of ZTK33 33 V stabilizers was used in some other sets the 33 V supply seemed a good place to start. It was stable, but we noticed that the feed resistors from the 95 V line were of lower value than given in the manual (it turned out that my set was a version 2, manual 4822 727 18084). The fault was brought on when the tuner was heated, a replacement U743 curing it.

Green screen: Sometimes a green raster with flyback lines was all that appeared when it was switched on. At other times there was no tuning and the channel wouldn't change. Yes, it was the tuning micro-controller chip, type TMP47C434–3559. Current replacements come with a metal shield to screen it and the RAM – this has to be connected to chassis.

Dead set, no 95 V from power supply: This set was dead with no 95 V output from the power supply. As no shorts could be found across the power supply outputs the set was fired up gradually via the variac.

Everything was fine until the 95V line reached 50V, then the overvoltage thyristor fired. One of the sensing zener diodes, D6638 (BZX79-B36), was short-circuit.

Dead set with 95 V supply low at 15 V: Checks in the chopper circuit showed that the BF487 driver transistor 7614 was conducting excessively – its collector voltage was much lower than the correct figure of 82 V. Resistance checks in its base circuit failed to reveal anything amiss, so we decided to try bridging the capacitors here. C2618 (27 nF) was open-circuit: it's of the blue plastic block type.

No picture or sound: However, the remote-command received LED flashed when the handset was tried. A quick check showed that the 95 V output from the chopper power supply was low at 10 V. We next made a few resistance checks on the power supply outputs and found that a reading of 60 Ω was obtained across the +9 (10 V) supply. The culprit was the sound chip, but although disconnecting the chip removed the short the symptoms remained the same. A quick read through the circuit description was called for. In order to start the power supply requires 15 V across D6613, which is a BYX79C15 zener diode. There was only 10 V because D6613 was leaky. A new zener diode and sound chip restored normal operation.

Sound but no picture: When the first anode control was turned up we found that the problem was in fact field collapse, brought about by failure of the TDA3653B chip IC7500. When this was replaced the set failed to come on at all. We discovered that there was a loose wire in the mains plug. Had this been the cause of the i.c. failure? Shades of the G11.

No colour and vertical black line: We've had this fault a couple of times now. The symptoms are no colour and a vertical black bar across the screen. The cause is a crack in the print by the line output transformer, where the PCB fits into the back cover. Check the track that runs from pin 10 of the transformer.

Set won't retune or reprogramme: If the set won't retune or reprogramme the volume, colour etc. and store the new values the store-lock facility has been accidentally set. To unlock it, select channel 38 then press the store and control buttons simultaneously.

Volume control would not turn down and stations could not be tuned in: This was an odd fault. As with most modern sets these functions are carried out by a microcontroller chip, but experience has taught us that failure of the associated RAM chip is much more common. So we replaced the ST24C02CP chip IC7785, which is also the cheaper of the two devices. This time we were wrong, however. The micro-controller chip IC7700 was the cause of the trouble. The original one was type TMP47C434N-3559 but the replacement supplied by Willow Vale was type TMP47C434N-3537, i.e. it had a different mask. In addition it came with a small tin shield. It's presumably a new, improved type.

Set was dead with just 5 V on 95 V line: The standby LED was alight but attempts to bring the set on using the front controls or the remote control unit failed. We disconnected the scan coil plug (you can't use a dummy load with this chassis) and tried again – with the same result. Next we tried powering the set via a variac. With a 100 V a.c. input the power supply produced an output of about 80 V, but when the a.c. input was increased to 110 V the power supply shut down and the standby line began to pulse. We left the a.c. input at this level and checked around the microcontroller chip where we found that the reset line was pulsing. Reset is generated by transistor Tr7673, whose base should be at 4.3 V. The voltage here was higher. Replacing the BZX79F4V3 zener diode D6671 solved the problem.

Dead set: Here's a good one! This 14 inch portable came in dead. Checks showed that the line output transistor was short-circuit, so a replacement was fitted. As there were no other obvious problems the set was then powered up. An odd squeak came from the line output transformer and the set promptly went dead. The line output transistor had failed again. We looked suspiciously at the line output transformer. We've had quite a few fail in this chassis. After fitting a new transistor and transformer the set worked. Feeling confident, we picked up the telephone to give the customer an estimate. But before finishing the dialling the set had once more gone dead. With a sinking feeling we found that yet another line output transistor had failed. Checks were made on various components in the line output stage, but the set could be kept on for only about 2 minutes otherwise the output transistor would fail. Scope checks showed that the line drive waveform at pin 27 of the TDA8305 sync/timebase generator chip IC7020 was very spiky and of incorrect shape. The chip itself was OK, as were C2519 and C2059 in the output coupling circuit (to the line driver stage). It turned out that the faulty component was the small 3.3 μH choke L5519, which is in series with the output from the chip. It presumably had a shorted turn. Anyway a choke salvaged from a scrap chassis restored the drive waveform to its normal shape and the set then ran with no further problems.

Dead set, line oscillator not working: Check whether the 12 V zener diode D6030 is open-circuit. For no sound or vision, just a blank screen, check whether C2044 (4 μF) is short-circuit.

H.t. supply not removed when the scan coil plug is disconnected: With most Philips chassis the h.t. supply to the line output stage is removed when the scan coils plug is disconnected. Beware! On some versions of the GR1-AX chassis this is not the case. We were caught out by a set that was dead with a whistling noise coming from the power supply. A resistance check across the 95 V supply reservoir capacitor C2660 produced a reading of 600 Ω. We figured that removing the scan coils plug should have disconnected the line output stage so, as the leak was still there, we chased around the power supply for a while. Finally we resorted to following the 600 Ω leak along the print and ended up at the line output transformer, not the scan coils plug as we would have expected. All that was wrong was a leaky BUT11AF line output transistor. Fitting a new one and attending to some dry-joints in the line driver stage got things working again.

No go, standby light flashes after a few seconds: If the cause is not the line output transistor it could be the 15 V zener diode D6613.

No sound or raster: The 95 V h.t. supply was present at the line output stage but there was no line drive. We found that the line drive stage was inactive as its supply was missing. Coil L5524 had gone open-circuit.

Low sound: We found that the 5.1 V zener diode D6715 was leaky. It's in the volume control line, being mounted adjacent to the TMP47C4343559 microcontroller chip IC7700.

Intermittent colour: There was also sometimes a dark, vertical bar about a third of the way across the screen. The fault was far more prevalent with VCR playback. This fact provided the clue. There was a break in the track that leads from pin 10 of the line output transformer around the corner of the PCB where it slides into the back cover (the fault had occurred following a move).

Sound slow to come back when channel changed: If search tuning was tried it continued when a channel was found. We discovered that the 'ident' signal to the microcontroller chip was permanently low because transistor Tr7046 (BC558) was short-circuit.

Dry-joints: We've had lots of these 14 inch sets in lately. Most of them have had dry-joint problems or a faulty 2SC3795 line output transistor.

In one of them, however, the 2 A mains fuse had blown because the posistor in the degaussing system had gone short-circuit. When this had been attended to there was no picture. The e.h.t. and heater supplies were present, also the RGB drives. When the first anode and focus controls were advanced there was dull, soft vision that could barely be discerned – and the adjustments interacted. A close look at the tube base showed that there had been spillage over the socket, which was very difficult to remove from the c.r.t. A new socket and a clean-up restored normal operation.

Weak sound: Check the disc ceramic capacitors around the sound i.f. chip. The main suspect is C2033 (22 nF).

Lines on picture: The complaint was of lines on the picture when the set had been on for about an hour. The fault was apparent on only some scenes. It consisted of a band of bright white lines, about 2 inches wide. The cause was C2523 (6.8 µF) which decouples the supply to the line driver transistor.

No picture: When the tube's first anode voltage was turned up a blank, unmodulated raster appeared. After much searching we discovered that a small, brown disc ceramic capacitor, C2045 (22 nF), was leaky. It's in the a.g.c. circuit, connected to pin 10 of the TDA8305 jungle chip IC7020. The readings we obtained were 3 kΩ both ways. A replacement restored the picture.

Excessive height, non-linearity: The symptoms were as follows: very excessive height, non-linearity and field blanking to the extent that only a quarter of the picture was displayed. They were all intermittent. Replacing the field output and timebase generator chips made no difference. Neither did replacing any relevant electrolytics. We removed and checked every resistor in the field output stage. Several hours later we traced the cause of the trouble to C2043 (82 pF), which was intermittently leaky. It feeds pulses from the line output trans-former to the field generator section of the TDA8305 timebase generator chip IC7020, at pin 2.

Would not start up: The h.t. was present and a slight grunt could be heard. The start-up pulse was present at the volume input of the TDA8305 chip, which contains the line oscillator, but the set wouldn't kick off. The chip wasn't to blame, however. Much investigation brought us to L5524 (1.5 mH) in the start-up feed to the line driver stage. It had gone high resistance, a replacement restoring the line drive. We've since had a second set with the identical fault.

No tuning voltage: Check whether R3703 is open-circuit.

Intermittently dead: The cause of this nasty fault turned out to be a small coil, L5524, which filters the supply to the line driver stage. It was going open-circuit intermittently.

Picture smeared: The picture was smeared and looked washed out, with bending verticals. The cause of the fault was traced to C2060 (680 μF) which is the smoothing capacitor in the 12 V supply to the TDA8305 timebase generator chip.

No sound or vision, raster OK: We see quite a few of these sets since entering into a deal with the operator of a local holiday site, where quite a number of them are in use. Suffice it to say that we buy line output transformers in bulk. But it was nice to get a different fault for a change. There was a raster, the graphics were OK, but there was no sound or vision. Checks showed that there were no outputs from the TDA8305A multipurpose chip IC7020, although its supplies were present and the tuning worked correctly. A new TDA8305A restored normal operation.

Intermittent tripping at switch-on: When we tested it we found that the overvoltage crowbar thyristor fired at switch-on, though the h.t. was correct at 100 V. So we removed and tested the three zener diodes D6638, D6639 and D6640 that are connected to the thyristor's gate, using a zener diode tester (see *Television*, July 1996, page 638). D6638 broke down at 30V, though the manual said that all three should be 36 V types, giving a trip threshold of 108 V. D6638 was marked 30 V, however. Maybe the wrong one had been fitted from new.

Won't stop searching: If the set won't stop searching even when a good strong signal is found, replace C2031 (22 nF) even if it reads OK when checked.

Dead, low power supply outputs: The set was stuck in standby, although the microcontroller chip sent the on/off signal to transistor T7631. Replacing R3610 and R3613 restored the set to life, but the fault recurred. The cause of the trouble was eventually traced to the 10 V zener diode D6610. It's in the chopper FET's gate circuit.

Would shut down every 2–3 hours: The set was put on test and sure enough shut down after the allotted time. Switching it off then on again restored normal operation. We removed the back and tried tapping around various parts, but there was no reaction to this. When a visual inspection was carried out, however, we found that the line scan coil connector was badly dry-jointed. The set worked perfectly once the connector had been cleaned and resoldered. Puzzles: why didn't it respond to tapping, and why didn't it fail more often?!

Dead, humming noise from speaker: The 95 V h.t. supply was present at the collector of the line output transistor, but it was without drive. The TDA8305 i.f./timebase generator chip IC7220 has a rather unusual start-up arrangement. A capacitor (C2058) that charges from the h.t. line supplies a pulse which is fed in at the volume control pin (11) to start the line oscillator! The pulse was present, and a line oscillator waveform was present for a second or so at the line drive output pin 26, but the drive output from this pin didn't get going and the 2.6 V normally present here was missing. The 12 V feed coil L5524, which is by the line output transformer, was open-circuit. Its part number is 4822 156 21293. P.B.

Philips GR2.1 chassis

Reluctance to start up from standby: Various faults had been found on this set and cured, except for this start-up problem. When the set was brought out of standby the red LED would go out then, a few seconds later, it would flicker at a very fast rate. If the set was left for a minute or so it would come on and work normally. Time was spent going through the power supply, but the cause of the fault was not here. In fact in the fault condition the h.t. was correct at 95 V. We found that the root of the trouble was that the line timebase didn't start up. Line drive is generated by a TDA2579B chip which had already been replaced. It seemed that this chip was faulty, however: the start-up supply, although about a volt low, was present at pin 16 but there was no line drive output at pin 11. We then noticed that a TDA2579 without the B suffix had been fitted. Could this be the cause of the problem? When a TDA2579B was fitted the set worked faultlessly. We proved the point by fitting a TDA2579, which brought the fault back. Moral: always fit exact replacement parts, and take note of suffixes.

Dead set: The set wouldn't turn on, although the power supply was running (phew!). Checks showed that the microcontroller chip wasn't working, but a replacement didn't help. We then did what we should have done initially, unplug the text panel. The faulty component was IC7880 (PCA84C81AP/098). If the chip from the earlier version is used, the Fastext keys won't work.

Thump from speaker then tripped at switch-on: This suggested failure of the audio output stage, and disconnecting the TDA2613 output chip brought back the picture. Unfortunately a replacement plus R3241 (8.2 Ω) failed to cure the fault. With the trip disconnected we found that the voltage at the chip's output pin started at 12 V then, after a few seconds, fell to 0 V. The set then worked normally. The soft-start circuit was clearly in trouble. Checks here soon showed that T7247 (BC848) was short-circuit.

Philips NC3 chassis

Intermittent picture failure: This was traced to a dry-joint on R608 (18 Ω) in this almost new set.

Shorted chopper transistor: After a fruitless check on the components that could possibly have led to the BUT11's demise we fitted a replacement, which immediately failed. The cause of the trouble was that R418 was open-circuit. This resistor is connected between pin 9 of the chopper transformer and the junction of L403/R409, not in parallel with R425 as in all the circuit diagrams we've seen.

Channel change switch problem: The channel change switch S300 has given trouble in a lot of these sets – symptoms are intermittently poor or even no picture and garbled sound. The only cure is to fit a new switch. Failure of the BUT11AF line output transistor Q501 is usually caused by bad joints around the line output transformer, especially at pin 10.

Dead set, C2483 exploded: We found that R3475 hadn't been fitted correctly at the factory, as a result of which C2483 which is rated at 25 V had some 50 V across it. Naturally C2483 had taken exception as had the 5 V regulator 7480 and diode 4480.

Sound and raster go after half an hour: The crowbar thyristor would then fire. We disabled the line output transistor and connected a dummy load across the output from the power supply. When the fault next occurred we found that the 100 V line had gone high. The cause of the problem was an intermittent CNX62 optocoupler.

Green raster: When this set was switched on the line timebase made a motorboating noise for the first few seconds then a green raster appeared. On removing the back we noticed that C2483 was dry-jointed, but resoldering it had no effect. With this capacitor out of circuit the voltage had risen, damaging the 78M05 5 V regulator which was giving out 6.8 V. A new regulator brought the voltage back to normal and cured the other problems as well.

Monochrome picture, false line lock: This set had a good fault to come back to after the Christmas holiday. There was a monochrome picture with false line lock and the brightness and contrast controls were inoperative. The cause of all these symptoms was a misshapen sandcastle pulse that confused the TDA4505 chip. The 680 kΩ line pulse feed resistor was OK, so a new chip was needed. As the chip had been damaged by an e.h.t. flashover from the line output transformer we also replaced the transformer.

Horizontal striations (NC3-CR): This set worked quite well but had horizontal striations over the screen. It turned out that C2413 (100 μF, 400 V) was completely open-circuit. We suspect that we shall see more of this due to the proximity of the chopper heatsink.

I.f. fault: If you have an i.f. fault with one of these sets – weak and buzzing sound or low gain and a ringing picture – it's a good idea to start by changing the i.f. filters. I've had examples of faulty SAWFs and ceramic filters – one set even had 5.5 MHz filters fitted!

No colour: Checks around the colour decoder chip revealed that the voltage from the colour control was slightly low at 0.6 V. We checked the control, the 10 kΩ series resistor and the 10 μF. decoupling capacitor. As we were getting nowhere we decided to consult the manual. To set up the reference oscillator frequency you link pins 5 and 1 of the chip. This supplies 12 V to the colour control pin to disable the colour-killer. When we did this the cause of the problem became obvious – the reference oscillator was slightly off frequency. A tweak to R635 was all that was required. The moral is that if you come across a set you've not met before it pays to read the book and think before you dive in!

Chopper transistor Q401 failure: Check for dry-joints at the connections to the chopper transformer T402 before fitting another BUT11AF. You often find that the rotary channel switch gets dirty. If you are careful it can be dismantled for cleaning.

Single green spot in screen centre: This set looked more like a defunct oscilloscope – it displayed a single green spot in the middle of the screen. The actual fault was no field scan: the picture was blanked out as usual, but not the vertical green tuning line which thus appeared as a spot! Resoldering dry-joints around the field output transistors Q551/2 restored a spotless picture.

Reduced height, horizontal ripples: The owner of this set had rescued it from being scrapped by a national retail/servicing company and brought it to us because he knew where to come . . . The problem was reduced height with vivid horizontal ripples over the picture. We noticed that a replacement Hitachi picture tube had been fitted, with scan coils that were different from the original ones. The ripples went when the missing 1.2 kΩ damping resistor was fitted across the Hitachi field scan coils. Full scan was restored when the parallel-connected scan coils were rewired in series.

Stuck in standby: This was because someone had fitted the wrong type of Preh on/off switch. These sets need low-voltage contacts that make when the set is on, not the momentary-make type. The part number is 4822 276 12503.

Stations drifted off tune: The customer complained that the stations drifted off tune every few seconds. We found that the cause was the 13 position rotary channel selector switch S300. A spray with switch cleaner immediately put matters right and improved the feel of the switch.

Set was dead but whistled: As there was no short-circuit across the line output transistor, pin 5 (h.t. input) of the transformer was isolated and the h.t. rail was loaded with a 60 W lamp. The power supply then ran happily, so where was the load? Pin 2 of the line output transformer supplies pulses to the green-line tuning circuit and a crowbar thyristor safety trip. Disconnecting this pin produced normal sound and picture. The crowbar thyristor's trip level is set by two series connected 8.2 V zener diodes (2581/2) in its gate circuit. One of these had a 200 Ω leak. We decided to replace them both.

Blank raster, but sound present: The on-screen display worked. A check on the waveforms around the TDA3565 colour decoder chip showed that the sandcastle pulses were present and a video signal went in, but nothing came out. Voltage checks then showed that the brightness control pin was high – 2.5 V instead of 0.6 V at the maximum brightness setting. A new TDA3565 was required.

Mains fuse open-circuit: If the mains fuse T400 (2 AT) is open-circuit and the BUT11AF chopper transistor Q401 is short-circuit, check the latter's 560 kΩ bias resistor R407. If this is high in value or open-circuit there will be no start-up.

No sound, picture lacks contrast: Check the 10 Ω safety resistor R416 in the 22 V feed to the audio output stage. This resistor also feeds a d.c.

offset to the beam-limiter circuit, via R637. Hence the lack of contrast.

Dead with blackened mains fuse: A quick check in the power supply revealed that the BUT11AF chopper transistor Q401 was short-circuit and the 4.7 Ω surge limiter resistor R401 open-circuit. There were also various dry-joints around the chopper transformer and the line output transformer. Resoldering these and replacing the failed parts brought the set back to life again.

Lack of height: It couldn't be cured by adjusting the height control – the raster wouldn't quite reach the top and bottom of the screen. We wasted a lot of time in the field driver and output stages before we moved back to the TDA4505 chip that contains the field oscillator. The timing components are connected to pin 2 of this chip. Checks here revealed that R581, which is connected to the 95 V rail, had risen in value from 3.9 MΩ to around 7 MΩ.

Rediffusion

REDIFFUSION MK 1 CHASSIS
REDIFFUSION MK 3 CHASSIS
REDIFFUSION MK 4 CHASSIS

Rediffusion MK 1 chassis

Thermal trip very 'touchy': Many of these excellent sets are still in use. Unfortunately the thermal trips fitted are becoming very 'touchy', tripping off at the slightest hint of trouble. Some have simply given up. Linking out the trip is all very well but can have unfortunate consequences – molten plastic from the tripler dropping on to the carpet. The correct course of action – the modification used by Rediffusion – is to link out the trip and insert a fusible 18 Ω resistor in the h.t. line. Details are shown in Figure 5 on the next page. Mount the fusible resistor on a tagstrip bolted to the screw that holds the strap for the l.t. smoothing capacitors. In this position it can be easily seen and resoldered and is away from combustible material. Note that with the rare 20 inch version of the chassis, which is physically quite different although electrically identical in most respects, block electrolytic C603/4/5 is mounted on the main frame. In these sets the fusible resistor should be mounted on choke L601, the red/grey lead being moved to the other end of the fusible resistor. Drill a hole in the plastic mounting for the choke's end tags and bolt on an extra tag for the new connections. The circuit is then as shown in Figure 5(d).

Here's a postscript to the above remarks on removing the thermal cutout in these sets. One of our regular customers moved to a large complex of apartments. After this her trusty set started to blow mains fuses violently. The resident handyman investigated, with the result that the plug fuse blew as well. What had happened was that while the set was in transit the thermal cutout, just about on its last legs, had received a blow. The contacts had disintegrated, shorting the live mains to chassis. Another good reason to modify any of these sets still in service where the cutout is an old one.

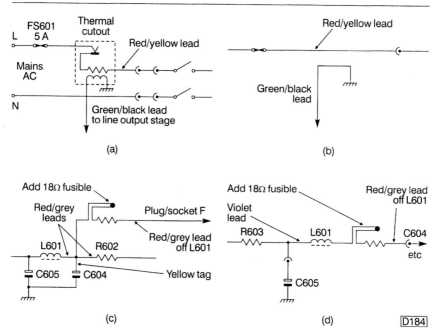

Figure 5 *The original thermal trip arrangement used in the Rediffusion Mk 1 chassis is shown at (a). Modifications to make it unnecessary are shown at (b) and (c). The 20 inch model is somewhat different – see (d) for modifications*

Line output stage fault?: The customer said that this set had a line output stage fault. When the brightness and contrast controls were turned up the sound and vision oscillated madly. At lower beam current levels all was stable. What had happened was that R101, the h.t. feed to the tuning voltage stabilizer, had risen in value almost threefold. When a new 33 kΩ resistor was fitted things quietened down. The giveaway was that there was no programme sound at switch-on, only noise, although the set was correctly tuned in and came on with sound, vision and locked colour as soon as the valves in the line output stage had warmed up. Programme sound normally comes on almost instantaneously with these sets. The owner had purchased new valves elsewhere but took it all in good heart – especially as the tube was still in excellent condition.

Rediffusion MK 3 chassis

Intermittent brightness variation: Under the fault condition the voltage at the emitters of the RGB output transistors went very high. The culprit was the 7.5 V zener diode 2D15 which returns the emitters to chassis. This diode is used to set the black level d.c. offset. It was going open-circuit intermittently. We've since had some cases where this diode goes short-circuit, blanking out the raster. This always seems to happen when there's trouble in one of the RGB output stages, the BF393/5 transistors go short-circuit. BF461/5 transistors are supplied as replacements. It would seem prudent to keep some 7.5 V zener diodes handy.

Blank display (white screen), intermittent sound: The 7.5 V zener diode 2D15 on the colour decoder panel turned out to be faulty.

Non-start and/or stopping: We've had an epidemic of intermittent non-start and/or stopping and staying dead with these sets. There were no signs of any distress and the sets were not tripping. We took one of the power boards for a two-day trip in the car before checking it over. The cause of the fault was then permanent and obvious – a dry-joint on one of the tags of the mains transformer 6T1's primary winding.

Dead set, fuse open-circuit: This was the first time we've had one of these sets in the workshop. The complaint was 'dead' and the 20 mm fuse on the power supply panel was open-circuit – but not blackened. On replacing this the set tripped and stayed off though the fuse remained intact. We disconnected the input to the tripler as this seemed as good a place as any to start, and when we switched on again the set burst into life minus a raster of course. A universal tripler was fitted, with the diode lead left disconnected to try it out. As the set then produced a picture the lead was cut off and insulated. But the picture was very weak: the raster was overbright and the contrast control had no noticeable effect. We checked around in the area of the tripler and decided that the trouble was probably in the beam-limiter circuit. It didn't take us long to discover that SR24 had risen substantially from its correct value of 220 Ω. When this had been replaced there was a normal picture apart from teletext lines over the top 4 inches of the picture – the field linearity was OK. We suspected that a capacitor in the field timebase circuit was leaky or open-circuit but eventually found that 8R26 had risen in value from 270 Ω to over 750 Ω. After replacing this the set produced a perfect picture. All this was done without a manual or the need for any special spares. Why isn't life always as rewarding?!

Tuning drift: For this problem don't strip the selector unit or suspect the tuner. Resolder pin 8 of socket 0PLA (at the top of the signals panel). This fault is becoming very common. It will save you time and trouble if you do this with every Mk 3 that comes into the workshop.

Tuning drift: This was particularly bad at the low end of the band, which displeased the customer as it affected the video channel. The set also produced an excruciating whistle. We cured the whistle by sealing the line linearity coil. Resoldering the chassis and input connections inside the U321 tuner and cleaning its case earthing contacts cured the drift.

Tuning problems: The very large number of these sets in our area are beginning to suffer from tuning problems. This particularly affects the six-button Rode–Stucky selector unit. It can be stripped and cleaned, clearing the problem, provided care is taken. To remove the unit from the set take out two self-tapping screws in the front panel. Next remove the buttons by inserting a small screwdriver in the holes on the underside of the surrounding moulding and prising them off. The cover over the switches and lamp is removed by prising off with a screwdriver in the four slots. You will find that the circuit board is attached to the plastic carriage by four screws, two of which are not immediately obvious as they are covered by plastic plugs. Remove these and the board is free. The selector unit can be unsoldered after which the tracks, on a subpanel riveted at one end of the main panel, are exposed. Cleaning is then a simple job, with no fear of damaging the plastic. Reassembly is also simple.

Would trip after a short while: We found that the set's h.t. rail was at 285 V instead of 270 V, posing a nasty threat to the thyristors in the line output stage and the tube's heaters. The 42 V feedback voltage from the line output stage to the power supply regulator, via 6R33 and 6D8, was low at 35 V as the 1 μF reservoir capacitor 6C25 had dried up. Replacing this increased the feedback voltage to 40 V and lowered the h.t. to 270 V.

No sound or blowing 6FS1 fuse: We get a lot of these sets ex-rental. No sound is usually caused by OR71 (820 Ω, 0.5 W) overheating and going open-circuit. This is now a very common fault. Fit a 1 W replacement. In the dual-standard versions (aerial/cable compatible i.f./s.f. boards) the component reference number is OR55. If the set tries to start up but the blowing of 6FS1 (1.6 AT) on the power panel puts a stop to the proceedings you need spare boards, a variac and patience. If a substitution check proves that the power supply panel is not responsible check the thyristors in the line output stage. On one occasion 5C4 (0.1 μF, 1 kV) was visibly bulged – a replacement restored normal operation. On two occasions recently the line output transformer has been the cause of the problem. If a variac is not available a neon tester held near the transformer will usually show 'pulsing' if the transformer is OK and the cause of the fault lies elsewhere. Even with a heavy leak in the line scan thyristor the neon lit.

Intermittent absence of signals: The fault was intermittent absence of the signals, curable by a good thump on the cabinet. Not an unusual fault, but we'd spent some time resoldering the power supply and line scan panels as the original report was that the set was dead. Plug and socket 0PLA on the signals panel was next accused. We cleaned and resoldered it but the owner still had to resort to a good thump occasionally. What finally seems to have cleared the fault is to have cleaned the pins and the holder of the TCA270S demodulator chip 0IC1. With another of these sets the cause of intermittent fuse blowing was traced to poor joints on the commutating coil in the thyristor line output stage.

Volume intermittently increases violently: We found that the control itself was as smooth as silk when rotated. The cause of the problem was traced to a high-impedance earth connection on the plug and socket that connects the volume control to the i.f. board.

Rediffusion MK 4 chassis

Line output transistor failure: These have been extremely reliable sets to date. We are now finding that the BU208A line output transistor is beginning to fail. What you find is that 4R2 (470 Ω, fusible) has sprung open due to the start-up circuitry being asked to run the set – unless the owner has switched off as soon as the fault occurred, which in our experience seldom seems to happen.

Failure of chopper transistor: These sets are now flooding the ex-rental market. A good few of them come off Granada maintenance contracts. The only troubles we've had in the past have been failure of the BU208A line output transistor and dry-joints on the chopper transformer – the latter often cause the former. You may also find that 4R2 (470 Ω) in the start-up circuit has sprung open. A new fault is beginning to appear, however. 4C16 (3300 pF) becomes dry-jointed at its bottom end with the result that the BU326A chopper transistor eventually fails and the mains fuse blows. When confronted with this situation check the BD433 chopper driver transistor 4TR2 which is sometimes leaky. If you miss this point and switch on after fitting a new chopper transistor the following will be destroyed: 4TR3 (BU326A), 4TR2 (BD433), 4TR1 (BC368), 4R22 (470 Ω), 4R23 (5.6 Ω), 4R26 (100 Ω), 4D15 (BA157) and 4IC1 (TDA1060). You will also be treated to a good bang and smoke. Guess how we found this out?! Going back to 4C16, this capacitor along with 4R32 and 4D18 form a snubber circuit across the chopper transistor, controlling the rate of voltage rise when it switches off.

Low width and burning smell: These sets have been very reliable over the years and still give good service. The symptoms with this one were low width and a burning smell. After resoldering the usual crop of dry-joints in the width and EW correction circuits the symptoms remained the same, with the parabola output transistor 3TR5 (a Darlington device, type BDW23C) getting extremely hot within seconds. Visual inspection showed that the culprit was 3C24 (3.3 nF, 1 kV) which had a pinprick hole through its casing. After replacing this capacitor the set performed normally.

Dead set: The start-up resistor 4R2 was getting hot. A check on the start supply showed that it was correct at 14.5 V, and no shorts could be found on the secondary side of the chopper transformer. A scope check at pin 15 of the TDA1060 chopper control chip showed that the drive output was at only about 0.1 V peak-to-peak – the manual gives the correct amplitude as being 0.4 V peak-to-peak. So we checked the supply voltage again: the reading was still 14.5 V. When the supply was scoped, however, we found that a 60 V peak-to-peak ripple was present on it. A check on 4C6 (1000 μF) showed that it had dried up, a replacement restoring the set to life.

Low h.t. (not caused by overload): Here's a fault that can catch you out. If there is little h.t. but the voltage comes up when the feed to the line output stage is disconnected don't go hunting around for an overload, because you won't find one. What you will find is that 4R5 (68 kΩ) on the power supply panel is open-circuit. You must test it when out of circuit.

Saisho

```
SAISHO CM16R
SAISHO CM250R
SAISHO CM260TT (FIDELITY ZX5000)
SAISHO CT141
SAISHO CT141X
```

Saisho CM16R

Dead set, standby indicator on: This was due to failure of the STK7308 chopper chip IC501. After checking all the components in the power supply – it takes only a few minutes – and also checking for shorts at the line output transistor a new chip was fitted and the set was switched on. A long soak test proved that all was OK.

Dead set: The 2 A fuse on the side unit was blown. A visual check showed that C510 (1 μF, 50 V) was weeping electrolyte. After replacing this and the fuse the set worked all right. The STK7305 regulator chip in these sets has a bad reputation for failure. Symptoms are standby indicator alight but set dead, set intermittently dead, bad power supply output fluctuations and D510 across the 103 V rail going short-circuit.

Saisho CM250R

No sound or raster: If you turned up the first anode preset you could see that there was field collapse. This was a bit of a red herring, however. The field drive comes from the TA7698 chip, which requires 12 V at pin 2. This supply was missing. We found that R408 was open-circuit because D408 read approximately 300 Ω both ways. Replacing these items cured the fault.

Sound buzz: Engineers at another repair shop had replaced every major item in the i.f. strip – the chip, the SAWF and the ceramic filters.

They had even tried to adjust the coils, all to no avail. We checked the work and found that all the replacement parts were correct. No matter how carefully i.f. adjustment was tried, it was impossible to delete the sound buzz. So we replaced the tuner. That cured the fault.

Low volume couldn't be adjusted: When we checked the set we discovered that none of the picture controls worked either, although channel changing was OK. I noticed that the set would switch on directly when the mains switch was used, although there's a standby button. The plastic standby button was found to be damaged, and was holding the standby touch switch in its closed position. When a new button was fitted the controls all worked correctly.

No sound, blank screen: It looked like a dead set, but when we checked the set we found it very much alive, as turning up the first anode control setting (it's on the line output transformer) proved by producing a field collapse display. Now we don't have the circuit diagram for this model, but confidently started to look for the field output chip. We were stopped in our tracks by the familiar smell of a cooking resistor. To our surprise the smouldering item (R428, 1.2 kΩ) seemed to have little to do with the field circuitry. As fellow colleagues out there know, apart from Toshiba, Aiwa and one or two others, you can forget about asking manufacturers for technical help – it just isn't there. Unsoldering the field output chip's supply pin made no difference: the resistor still cooked. We reached for paper and pen and started to draw the circuitry around R428. This approach can be tedious, with tracks that often divert all over the place, but with a little patience the drawing can start to make sense and a familiar circuit will begin to emerge. It seems that R428 is part of a start-up circuit based around Q403, which was also cooking. At switch-on Q403 supplies, via D401, about 12 V to the timebase generator circuitry. This continues until the line output stage comes into operation. Once this happens pin 2 of the line output transformer supplies D405 via R408 (1 Ω) and the running supply for the timebase generators is produced. D401 is then reverse biased. R408 and D405 had both failed. Hence the overloaded start-up circuit, which isn't designed to remain in operation. When R408 and D405 had been replaced all was well – proved by a long soak test.

Saisho CM260TT (fidelity ZX5000)

Chopper transistor short-circuit: This teletext set is fitted with the Fidelity ZX5000 digital chassis. The chopper transistor TR1 was found to be short-circuit between all its legs and the 1.6 AT mains fuse was open-circuit. Replacing the BU145A still left us with the dead set symptom, however – time to look a little deeper. The operation of the auxiliary 5 V supply was found to be all right, but the TIL111 optocoupler IC13 was providing a negative voltage at the input to the chopper control chip IC1 (type UC3844 fitted). Fitting a new optocoupler cured the dead set fault but bearing in mind the nature of the chopper transistor's failure we also replaced IC1.

Faulty line output transformer: This set had been got at before it arrived in our workshop. Supply resistors had been cut and lifted and a wire had been cut. When this damage had been repaired and the set was powered up it became obvious that the line output transformer was faulty. The set came to life when a new transformer had been fitted – but only for about a minute. If the set was switched off then on again it would do the same thing, run then stop. This final problem was caused by the 5 V regulator overheating as the clip which secures it to the heatsink was missing.

Chopper transistor short-circuit: If you find that the BUT11A chopper transistor is short-circuit, replace the following items as well: C5 (4.7 μF, 63 V); C7 (220 μF, 25 V); IC1 (UC3844N); and the CNY75B optocoupler IC13. The set will then switch on first time.

Saisho CT141

Power regulator fails to start: A stock fault that's beginning to appear on these very popular portables is that R503 (82 kΩ) goes high or open-circuit. As a result the power regulator (Q501, 2SC3158) fails to start up.

Sometimes fails to start: Once running the set was OK for the rest of the day. Replacing C508 (4.7 μF) provided a complete cure.

Intermittent, no results: When the set was checked on the bench, however, it was dead. After removing the chassis we found that there were various dry-joints around the chopper transformer T501. When these had been resoldered the set came to life, but when the back was refitted it went dead again. Further investigation showed that the h.t. supply rose to only 46 V. When the load was removed it reached 80 V slowly, 30 V less than the correct figure. We eventually found that C504 (0.047 μF) was virtually open-circuit.

Saisho CT141X

Low sound: The sound was so low that it was almost inaudible – just a very quiet hum. We'd no manual so we checked around in the sound i.f. section with a signal generator and found that C303 (100 pF) read in the region of 5 kΩ. It damped the i.f. coil so much that the sound was killed.

Dead set, usual faulty bits OK: This set was dead but the usual faulty bits were OK This time C508 (4.7 μF, 50 V) in the STR50103A's start circuit was found to be low in value. Replacing it restored normal operation. The fault can also be caused by R502 (100 kΩ) or R503 (82 kΩ) in the start supply.

No luminance: Check and replace as necessary D602, D603 and D613. Type 1N4148 diodes will do. Also check and/or replace the 2SA733 transistor Q601. If you have to replace the 5.6 Ω, 5 W surge limiter resistor R501, as you will have to do if the chopper chip IC501 and the overvoltage zener diode D508 have gone short-circuit, use the Amstrad type AM1422138 from CPC of Preston. It's the same but cheaper.

Low sound: Low sound was the complaint with this set – a simple fault for a change. Checks in the volume control circuit revealed that the hot end of the control was at 4 V instead of 5.9 V. The series resistor R145 (4.7 kΩ) had risen in value.

Salora

SALORA H CHASSIS (IPSALO 2)
SALORA J CHASSIS
SALORA J20 SERIES
SALORA K CHASSIS
SALORA L CHASSIS
SALORA M SERIES

Salora H chassis (IPSALO 2)

Tube looked soft, hum from loudspeaker and field fold-over when the aerial was disconnected: This set had a number of faults that at first sight didn't seem to be related. First the tube looked soft, secondly a low-frequency humming came from the loudspeaker, and finally and most curiously the field folded over when the aerial was disconnected. Where to start? Well, the tube is as good a place as any, so a check was made on its voltages. We found that the e.h.t. was low at 20 kV, so the tube was all right. We then checked the secondary supplies derived from the Ipsalo transformer and found that these were also low. The set-e.h.t. potentiometer RTB700 was adjusted to see whether we could get the correct voltages, but the slightest tweak was all that was required to send the e.h.t. sky high. A replacement preset didn't help matters at all. What did restore everything to normal operation was a new LF0034 Ipsalo control chip.

Both chopper transistors shorted: Also RB713 was open-circuit. Lots of bad joints were resoldered, including a very bad one at the line scan coupling capacitor CB532 (0.33 μF, 250 V). Further inspection revealed that this capacitor had failed – bulges were present. It's no longer available from NCS. Fortunately C831 (0.33 μF, 400 V) from the Ferguson TX10 chassis has the same pin spacing and will fit. It's available from Willow Vale.

Salora J chassis

Set would go to standby as soon as the picture appeared: A field engineer had tried disconnecting the tripler and the IR receiver module in case the latter had become noisy, sending random standby, channel change etc. signals to the remote control section. He'd also tried unplugging the teletext panel to eliminate it – it's quite a common cause of trouble. We found that by pushing the on/off switch right in (overriding the momentary contact) the set would run merrily, but you couldn't change channels either via the remote or the on-board controls. A handy feature of these sets is the switch at the back above the aerial socket. If you turn this the standby mode is overridden. By doing this we could check voltages and soon found that the SAA1251 chip (ICC9) was faulty. Any height and/or width twitching with these sets can usually be cured by replacing the LF0041 Ipsalo circuit control chip – it's a hybrid i.c., circuit reference HB1. It looked as if the static convergence was out on one of these sets – the reds were miles out, and of course there are no adjustments. So the only answer was a new c.r.t. plus yoke. Surprisingly the emission was good – these tubes (A51–590X) tend to go down prematurely.

Wouldn't start correctly from cold: If it was left to warm up it would start, but when running with the contrast turned up the width would fluctuate with picture content. The cause was CB712 (4.7 μF, 35 V electrolytic) in the base circuit of one of the power supply switching transistors (TB701) going open-circuit.

No colour: A check at pin 4 of the TDA3562A colour decoder chip revealed that the chroma signal was arriving from the filter circuit and was of the correct amplitude (500 mV p-p). After passing through the gated chroma amplifier in the chip, however, the burst only appeared at pin 28 (the output to the delay line). It seemed reasonable to replace the chip but the results were as before. We then checked the d.c. levels at the relevant pins and discovered that the voltage at the chroma input pin 4 was only 0.7 V instead of 3 V. The cause of the trouble was a leak in CB209 (10 nF) which when checked out of circuit measured around 250 Ω.

Dark patches on the screen (22J40): The picture was normal at switch-on, but within seconds dark patches smudged over the screen with irregularly large patches right across the screen, in varying bands; slowly at first but soon pulsating rapidly to give a juddering effect. This wasn't the same as another problem we've had, fast flutter on high contrast scenes and no contrast control action. That was due to D200

(1N4148) in the beam limiter circuit. This time the effect was more profound, and the controls worked normally. The chassis was lowered, with sinking heart, so that we could heat and freeze around the colour decoder chip etc. Before getting too involved we glanced here and there in the hope of seeing something burnt in a corner. There it was, R508, charred. We correctly deduced that the 12 V regulator IC500 was doomed, and after replacing these two items the display was once more correct.

Dead set: The h.t. filter resistor RB713 was open-circuit and the two Ipsalo transistors TB700/1 short-circuit. These items were replaced and after resoldering several suspect joints, including one that, at CB712, looked as if it could possibly have been responsible, we switched the set on. The on/off switch had to be pressed several times before the set would come on, then after about half an hour the two transistors failed again. We rather belatedly decided to consult Nick Beer's article on the J chassis (*Television*, December 1989, p.102). As a result CB712, CB726, RB703 and RB705 in the Ipsalo transistors' base drive circuits were replaced. RB705 was difficult to locate as it was designated CB705 on the PCB! Two new transistors and a filter resistor then had the set working normally.

Stuck on channel number 3: If you pressed the other buttons on the set the channels came up but didn't latch. Also some didn't do as they should – number 4 produced number 1, for example. There was a fairly obvious link in the lines between the SAA1251 control chip and the front keyboard. The switching transistors in these lines have been known to give trouble. The appropriate one, TC11, read OK, as did all the others. A replacement SAA1251 made no difference, and all the pull-up, -down and feed resistors had the correct values. Print continuity was OK, and the M293B1 programme chip was all right. For want of anything better to do we replaced TC11. This cured the problem, though the transistor still read right when tested out of circuit.

Sometimes didn't come on: Sure enough only the standby indicator stared at us. The problem was solved by replacing CB712 and CB726 (both 4.7 μF, 63 V).

Sound OK, no picture: The screen was completely dark. A check on the drive waveforms from the colour decoder chip, at pins 13, 15 and 17, showed that they were missing. In addition, the d.c. content was low. The brightness and contrast control voltages were present at the chip and altered appropriately depending on the function selected. The luminance input was also fine. Time to check the sandcastle pulses. These also looked fine. But wait a minute, something doesn't seem right. When the scope was set to the line frequency a perfect line blanking sandcastle waveform was seen, but at field frequency there was no blanking waveform present. We traced the pulses back to the point where the separate components are added together to give the distinctive waveform shape. The field component comes from the field output chip ICB400 via DB402 and TB400. ICB400 was delivering pulses but they were going AWOL at the base of TB400. The cause of this was that TB209 was short-circuit collector-to-emitter. When a new BC307 was fitted up came the picture.

Intermittent green faces: This chassis uses a TDA3562A colour decoder chip, with the PAL switching controlled by the sandcastle pulses. As the pulses didn't change when the fault was present we suspected the TDA3562A, but a replacement made no difference. Neither did replacement of the components connected to pin 2. While we were studying the fault we noticed that the field height increased when the ident dropped out. This was the clue that led us to the cause of the fault. The TDA2653A field timebase chip detects the frequency of the sync pulses, either 50 Hz or 60 Hz, putting out a voltage at pin 12 to adjust the height and switch the colour decoder chip to NTSC operation as appropriate. For PAL operation the voltage at this pin must be zero. When it rises above 12 V the set changes to NTSC operation. In this particular set the voltage at pin 12 varied intermittently. Fitting a new chip restored correct operation.

Excessive height, incorrect chroma: The fault report above highlights a familiar fault with this chassis – excessive height with incorrect chroma phasing (green faces). With this set, however, the field output chip was faulty. A far more common cause, however, is that the field hold control RT400 is noisy. Slight readjustment will usually provide a cure as long as the control is cleaned.

Teletext fault: The selected teletext lacked field sync, rolling through the screen. A check showed that the field sync pulses were missing at pin 13 of the SAA5030 VIP chip. Fortunately this chip plugs into a socket: a replacement cured the fault.

Peak brilliance at switch-on, then shut down: We've seen many of these sets over the years. They still seem to work well. This one had peak brilliance at switch-on, however, then shut down. We traced the cause to an open-circuit in the thin wiring loom, at connection 1 where it meets the main board.

Only top half of picture displayed: Only the top half of the picture was displayed, with the scan stretched at the top. Replacing CB400 (1000 μF, 40 V), the reservoir capacitor for the 25 V supply to the TDA2653A field timebase chip, cured this one.

Dead set, chopper short-circuited: The chopper transistors TB700 and TB701 were short-circuit while the mains fuse and RB713 were open-circuit. We replaced these items along with the 4.7 μF chopper drive coupling capacitors but the set was still dead. At least the mains fuse held, and there was 320 V at the collector of TB700. There were no start-up pulses, however, because the BR100 diac DB725 was short-circuit. A replacement got things going again.

No sound: So checks were carried out around the TDA2030 audio output chip ICB100. The 28 V supply was correct at pin 5, the speaker was linked to pin 4, and audio could be seen when a scope check was made at the input coupling capacitor CB108. But the feedback/input pins 1 and 2 were both at 26 V. Potential divider RB124/5 provides a bias voltage for these pins. RB124 (1.8 kΩ) had risen in value to around 6 kΩ, upsetting the d.c. conditions with the result that the chip was cut off.

Dead set: We seem to have had more than our fair share of these sets in lately. In the past the cause of a dead set has generally been failure of the 4.7 μF base drive coupling capacitors in the Ipsalo circuit. Nick Beer told us about this in the pages of *Television* (February 1990, page 274). More recently the cause of this fault has in a number of cases been the Ipsalo circuit control chip HB1, where pin 2 has been at a much higher voltage than the stipulated 7 V. Replacement is the only remedy.

Randomly produces dead set symptom: In fact this set produced the sort of half working/low output state that's characteristic of one of these sets in trouble. DB720, a 15 V zener diode connected to pin 5 of the LF0041 chopper control chip HB1, was found to be intermittently leaky. The cause of a recent case of a dead set with no 15 V input to the 12 V regulator chip ICB500 was traced to rectifier diode DB507 (EGP20D) being open-circuit.

Dead, power supply trying to start: We found that there was about 50 V at the collector of the line output transistor. A few resistance checks soon revealed that the BY133 diode DB510 in the EW modulator circuit was leaky.

Salora J20 series

No line lock: A scope check at pin 11 of the TDA2594 sync/line oscillator chip ICB501 showed that a suitable video waveform was present – and also cleared the fault! The scope had been set to d.c. input: when switched to a.c. the fault was still present. A look at the circuit suggested that RB514 (2.7 MΩ) could be responsible for the trouble and when checked it turned out to be open-circuit.

Degaussing problem (16J20): This set had an odd cyclic fault. Every 10–15 seconds the picture would gradually darken then, with a ping from the degaussing circuit, it would jump back to the correct contrast level, usually with a bad purity error. The noise and purity problems would go if the degaussing coils were disconnected, but the picture contrast would continue to vary as before. A check on the beam current limiter voltage (across CB503) showed that this dropped as the picture darkened. This is the opposite of how it should work, i.e. when the beam current rises (high contrast/brightness) the voltage should fall in order to limit the drive from the TDA3562A colour decoder chip, using transistor TB200 to hold down the d.c. brightness and contrast control inputs. The cause of the fault turned out to be a dry-joint on the upper connection of the screening 'wall' between the colour decoder and the RGB output stages. This wall is used to link a chassis return from the c.r.t. base to the main panel. Loss of this connection meant that the grid supply potential divider had no link to chassis.

No picture (16J20): The problem with this set was no picture, but the sound was OK. Knowledge of the TDA3562 colour decoder chip was the key to discovering the cause of the fault. A scope check on the sandcastle pulse input at pin 7 of this chip showed that the field component of the pulse was incorrect. Moving back through the circuit brought us to transistor TB400 (BC237B) which proved to be defective. Replacing this brought everything back perfectly. When scoping the field component of the sandcastle pulse it helps to remove the other components by isolating pin 7 of the TDA2594 chip – clearing the decks so to speak.

Wouldn't change channel: At switch-on this set would start up on channel 1 as normal. It couldn't be shifted by using either the remote control unit or the on-board buttons, however. We found that if the standby button on the remote control unit was held down the channels could be changed normally. When this button was released the set stuck on the selected channel. The remote control decoder/ set keypad encoder chip ICC9 (SAA1251) was suspected and when it was replaced the fault had cleared.

Salora K chassis

Wouldn't switch on from cold: As h.t. was present at the collector of the chopper transistor TB701 in the fault condition we checked for shorted diodes around the power supply and that the 8.5 V rectifier connected to pin 18 of the Ipsalo transformer wasn't open-circuit. A replacement hybrid chopper control chip was then fitted, but still no go. As the 8.5 V supply is very critical we checked the 1000 μF reservoir capacitor CB604. It had gone low in value, the result being ripple on the 8.5 V line. This was upsetting the hybrid chip.

Tripping out when changing channel: Also for the field output chip IC501 having a very short life, check choke L601 in the line output stage. In the set we had in L601 looked as if it had got very hot and clearly bad shorted turns.

Thick black horizontal lines: This chassis seems to crop up under many guises – in this case the set was a Finlandia C59C27. The intriguing symptom was of thick, black horizontal bars that almost obliterated the picture when the set was switched on from cold. After several minutes the bars would gradually disappear, revealing a perfect picture. We didn't have a manual but did have knowledge of a similar fault some years previously. So we proceeded to the field timebase, armed with a can of freezer. Our efforts here revealed that the culprit was C574 (100 μF, 40 V) which is next to the field output chip.

Severe blanking fault for first 10 seconds: A large horizontal band was blanked from the picture. The band gradually shrank, leaving a normal picture. We had the feeling that this was a capacitor fault; we were right. When the super-sandcastle pulse was monitored we could see the blanked portion varying in size. Use of freezer soon narrowed the cause to C574, which is beside the field output chip. A new 100 μF capacitor put matters right.

Intermittently goes dead: If one of these sets comes on then intermittently goes dead, working all right when switched off and on again, check the mains on/off switch. We've found a number of these to be intermittent.

Salora L chassis

White across picture 2 inches from the top: We've found Salora/Luxor sets to be very reliable, so when this one (a 20L30) arrived in the workshop we hardly knew where to begin. The symptom was a white line across the picture, 2 inches from the top. It looked just like a linearity fault, but we couldn't find anything wrong with the field timebase circuit. So we checked the power supply voltages, which we should have done first, and found that they were on the low side. The chopper circuit is controlled by the LF0059 thick-film chip HB600 and when this was replaced our field problem had been cured.

Intermittent going off or standby: The Salora L chassis is also used in Hitachi sets such as the C17-P118. It's prone to going off or to standby intermittently. The problem area is around the diode split Ipsalo transformer M600. It's a large, heavy component and is mounted solely by means of its leadout wires, with no other support for the core or windings. This seems to be the root of the trouble. You get an assortment of dry-joints at the leadout points to the print lands, also cracked print. It would seem that these are caused by movement of the transformer with respect to the PCB during transit. Any of the solder points can be dry or cracked, and the print track from pin 1 to D709 is frequently the site of a hairline crack. The usual precautions apply. Remove the transformer completely, check all the leadout wires carefully, scrape and re-tin if necessary, and check the soldering of the winding ends to the leadouts. Use a magnifying glass to examine the print in the area of the transformer mountings and remedy any suspect cracks etc. Clean up and re-tin the solder pads and prepare to remount the transformer. We prefer to provide some extra support for the transformer either by fitting double-thickness 'Pritt' self-adhesive pads between the core and the PCB or by slipping pairs of large ceramic beads (as used for insulation in electric fires etc.) over the leadout wires at the corners of the transformer connection panel, again to support the weight of the transformer. Then remount the transformer, paying particular attention to the soldering. When testing afterwards, make a point of applying pressure to the transformer while the set is on, to confirm that the repair is secure. Treat other large components in the area in the same way. It also pays to remove the driver transformer MB601 and wirewound resistors RB702 and RB703, then clean, re-tin and refit as above.

Loud crack from speaker, collapsed picture: The cause was dry-joints on C511, C512 and C507 in the line output stage.

No sound or a blank raster: This chassis is used in the Hitachi Model CPT2658. We've had two of them in recently with the TDA4505 chip faulty, the symptom being no sound or a blank raster. Depending on the make of the chip, adjustments to the line and field hold presets may be necessary after fitting a replacement. Before condemning this chip, try disconnecting the text panel (if fitted). If the picture then returns, replace the wire links (about eight in all) that connect the upper and lower tracks on the panel and resolder the other components that link the two print patterns. If the text panel is still faulty, the DPU2540 chip is usually the culprit.

No sound: This set was actually a Tashiko Model 51E941. The fault was no sound. We soon discovered that there was no audio output from the TDA4505 chip, where voltage checks showed that pin 14 was at 0.2 V instead of 1.5 V. The 22 nF decoupling capacitor CB119 was faulty, a replacement restoring perfect sound.

Salora M series

Dead set, fuse and RB701 blown: This set was completely dead, with the 2 AT mains fuse and the 2.2 Ω, 5 W surge limiter resistor RB701 both open-circuit. We spotted the cause of the breakdown as soon as the back cover was removed – CB705 (680 pF) hadn't been fitted correctly during manufacture. Only one end was soldered through the board, the other end resting on the surface of the PCB. As this capacitor is part of the snubber network across the primary winding of the combi-transformer it wasn't surprising to find that the chopper transistor TB701 (BU603) was short-circuit, hence the blown fuse and surge limiter. Zener diode DB709 (12 V) in TB701's base circuit had also failed. TB701 was in fact shorted between all its connections. When these various items had been replaced, with the capacitor fitted correctly, the set was restored to correct working order.

No picture: Occasionally a 2-inch wide band of illumination would appear on the screen. The cause was traced to the BS208 transistor TB526 in the EW modulator drive circuit. It's a field-effect transistor.

Came on with high volume: When the sound was turned down it went higher then muted, then we had only one channel and after that the other. When the cursor reached the minimum setting the sound muted and could then be turned up in the normal manner. Turning it down produced the odd effect. All this was caused by the X2404P memory chip playing tricks.

Blown line output transistor: The transistor is TB525 (2SD1577). There are several causes of this, commonly dry-joints around the LF0070 hybrid chip HB701 in the power supply, often C523, C624 (both 220 μF, 16 V) and C622 (470 μF, 16 V). It's advisable to uprate these electrolytics to 25 V types.

No picture: Where to start? All those huge chips! As the sound was present and the tube's heaters were alight it seemed that the combined power supply/line output stage was working normally. Then, for the briefest moment, we noticed a narrow band up the screen. This led me to the BS208 EW modulator driver transistor TB526, which is a FET type. Replacing it restored normal operation. All we then needed was a few yards of solder to stick all the dry-joints together. . .

Line output transformer short-circuit: This was actually an Hitachi set, Model CPT6608. We found that the line output transistor was short-circuit but, after replacing it and checking for the usual dry-joints on the large capacitors, the line output stage was still clearly in distress. The new transistor was overheating, there was a defocused picture and hum on the audio. Time to ask Mr Hitachi. He suggested replacing RB532, RB533, RB535 and RB536 (all 1 Ω) in the line driver stage. This improved matters, but in the end TB522 (BC307), TB523 (BC368) and TB524 (BC369) all had to be replaced, as well as the usual BS208 FET in the EW circuit. The line driver stage in this chassis doesn't use a transformer. If all is well in this stage, the line output transistor will have between −1.4 V and −1.7 V at its base. When resoldering dry-joints on the polystyrene capacitors in this part of the set, the print must be scraped shiny and the leads bent over before soldering. This will prevent recurrence and possibly severe PCB damage.

Dead set: In this event the thing to do is to switch on, remove the mains power, then plug back in to see if the set is in standby. If it is, you can be fairly sure that the power supply is OK. A short-circuit line output transistor was the cause of the trouble with this set. No dry-joints, which are often the cause of this failure, could be found. But experience has taught us that the line output transistor seldom fails without cause. We found that C527 (8.2 nF) was low in value, reading about 1 nF when checked with our tester. Replacing the transistor and capacitor restored reliable working.

Samsung

SAMSUNG CI3312Z (P58SC CHASSIS)
SAMSUNG CI3351A (P68SC CHASSIS)
SAMSUNG CI5012Z (P58 CHASSIS)
SAMSUNG CI5013T (P58SC CHASSIS)
SAMSUNG CI537V

Samsung CI3312Z (P58SC chassis)

Intermittent loss of colour on changing channel: This was cured by replacing the 8.8 MHz reference oscillator crystal X501. Unfortunately the manual gives no information on how to set up this circuit. The correct procedure is as follows. Short together pins 1 and 6 and pins 21 and 22 of IC501, then set CV01 for floating, locked colour. Finally remove the shorts.

Lack of height: The obvious thing to do was to check the d.c. voltages around the field output chip, IC901. These were all OK, but a new chip was tried in case. This made no difference so we moved back to the TDA8305 chip IC101. The d.c. voltage at the ramp generator pin 2 was found to be low. Replacing R302 cured the problem. It had risen in value from 470 kΩ to about 1.5 MΩ.

Requirement for repeated degaussing: We've sold large quantities of these excellent sets and have had very few faults. One problem we've had, however, has been a need for repeated degaussing. The cause of this is a poor crimp connection in the degaussing coil plug, as a result of which you get slight burning and carbonizing inside the plug. We think that this produces a diode action, passing d.c. into the degaussing coils. Hence the green faces. In the past we've cured the fault by fitting a new plug – old Sony sets are a good source of these two-pin plugs, though releasing the pins is sometimes difficult. As the sets are still under guarantee, however, we now have a stock of replacement degaussing coil assemblies – Samsung part number. 32479–029–100.

Set wouldn't come off standby: The complaint with this set, which we'd sold four months previously, was that it wouldn't come off standby. Checks showed that the power supply was running but the 2SD288 12 V supply switch transistor Q802 wasn't being turned on. We traced the circuit back to RQ11 which was OK, then pin 41 of the PCA84C640 microcontroller chip RIC01. The voltage here wouldn't go low to switch the set on. By shorting this pin to chassis we were able to switch the set on but there was only a snowy raster, no on-screen display and no tuning. Clearly the microcontroller was in trouble. Pin 39 (serial clock) and pin 40 (serial data) were both dead. Disconnecting them from the EPROM RIC02 made no difference, neither did replacing RIC01. When the 10 MHz clock crystal RX01 was replaced, however, the set sprang to life.

Signs of burn-up in line o/p stage: After replacing various components here we switched on. A loud bang and a bright flash from two of the resistors we'd replaced woke us up. The set continued to work, but the picture was oversized and there would occasionally be arcs from the two resistors previously mentioned. A check on the h.t. produced a reading of 157 V at the minimum setting. The manual specifies 125 V at D821. The cause of this high voltage was that R807 had increased in value from 11 kΩ to 19 kΩ. A replacement restored normal operation.

Common dry-joints: A couple of dry-joints are becoming common in these sets. Check R823 in the feed to the 12 V regulator and R411 which is connected across the primary winding of the line driver transformer.

Wouldn't tune: The on-screen display indicated, however, that the tuning was OK. A check showed that the tuning voltage at pin VT of the tuner was zero. Tracing this back brought me to the integrating transistor RQ01 which was without a supply at its collector because its 10 kΩ load resistor RR05 was open-circuit.

Refusal to search-tune: A check showed that the tuning line voltage BT was stuck at 29.75 V. The cause was resistor RQ01 which was dry-jointed.

'Arcing then dead': When we switched on we found that the line output stage was in distress, with the h.t. low at 70 V. We then saw that R412 was overheating. Disconnecting it brought the line output stage to life. The KA2131 field output chip was found to be leaky, probably as a result of the e.h.t. arcing. Replacing it brought the set to life, but with the h.t. at 180 V! R807 (11 kΩ) which is in series with the h.t. preset control had risen in value.

Mains fuse or R601 open: If you get one of these sets with either the 3.15 A mains fuse or the 5.6 Ω, 7.5 W surge limiter resistor R601 open-circuit, but with no obvious short-circuits, check C816 (2.2 nF, 2 kV) which may be dead short or leaky. This capacitor is connected in parallel with the chopper transistor: you will often see a line around its case. It's so annoying to blow another surge limiter resistor, especially when it's your last one!

No sound: The speaker and the audio output chip appeared to be in order, as were the d.c. supplies to the audio section. But the ident level (approximately 0.5 V) at pin 29 of RIC01 was missing because R709 (27 kΩ) was open-circuit. A replacement resistor restored the sound.

Samsung CI3351A (P68SC chassis)

Dead set, start-up voltage low: A check on the start-up voltage at pin 4 of the power supply chip IC801 produced a very low reading (0.6 V). The feed resistors were OK and there were no shorts, the fault being within the chip itself. It's a special hybrid device, type SDH209B.

Intermittently dead: This would usually happen when the set was hot. There would be no standby LED illumination, although it would sometimes come on later. The cause of the problem was obviously on the primary side of the power supply, but the symptoms were rather illogical. We traced the cause to C808, a 330 pF 2 kV disc capacitor, in the snubber circuit. When it was removed and checked it read perfectly!

Low brightness picture: Even turning up the first anode control on the line output transformer failed to produce adequate brightness for normal viewing. A check showed that the first anode voltage at the tube base was lower than expected, and varied over a period of time – as did the brightness. The cause of the trouble was a varying leak within the 1 nF, 2 kV decoupling capacitor C907 on the tube's base panel. We were able to lift it to prove the point.

Samsung CI5012Z (P58 chassis)

No signals: The tuning voltage showed that it swept, but no station was ever found. A check at the tuner showed that the tuning voltage was missing. There was 33 V at the cathode of DZ824 but nothing at the collector of the integrating transistor RQ04. Its tiny 0.25 W feed resistor RR05 (10 kΩ) was open-circuit.

Field linearity problem: The field was up at the bottom and stretched at the top. When it was adjusted the field linearity control VR303 had little or no effect. The culprit turned out to be C306 (2.2 μF, 50 V), which is between the linearity and height potentiometers.

Dead set: In this case it's worth checking C816, which is connected in parallel with the chopper transistor. We quite often find that it's leaky. You will also find that the 5.6 Ω, 7 W surge limiter resistor R801 is open-circuit. Another common fault with this chassis is failure of the mains rectifier diodes, again blowing R801. Why is a fuse included? It never seems to fail!

Dead set: After a good clean out with the vacuum we could see the PCB and start work! We found that the 5.6 Ω, 7 W surge limiter resistor R801 was open-circuit, the cause being failure of C816 which was short-circuit. This 222 pF, 1 kV capacitor is connected in parallel with the chopper transistor Q801.

Samsung CI5013T (P58SC chassis)

Drifts off tune: This set would drift off tune about 5 seconds after selecting a channel stored in the memory. If sweep tuning was tried the set wouldn't stop when a channel was found. There's precious little by way of voltages or waveforms in the manual, but we decided to check the voltage at pin 9, the a.f.c. input, of the microcontroller chip RIC01. There was nothing much by way of d.c. here. The input comes from pin 18 of IC101 via a potential divider that consists of two 120 kΩ resistors, RR26 and R116. The latter was open-circuit.

Won't start, no standby light or possibly trips: Check R811 (100 kΩ) which tends to go high in value. It's connected to pin 8 of the TDA4601 chopper control chip.

At random intervals picture goes blank: It would leave only sound and a dark raster. Pressing the text button would bring up a text display, then pressing the return to TV button would restore the picture. The cause of the trouble was traced to the 27 MHz crystal on the text panel.

Stuck in standby: Checks in the power supply revealed that the 16.5 V and 12 V outputs were missing. As these are switched on by the system microcontroller chip RIC01 we checked here and found that the required switch-on voltage was missing. Unfortunately the service manual is of little help in this connection as there are no voltages or waveforms. Since microcontroller chips are expensive and seldom fail, we decided to persevere with our blind investigation in this area. The two most important things for the operation of such a chip are its 5 V supply and the clock signal derived from an associated crystal. The 5 V supply was OK. Replacing the 10 MHz crystal put matters right.

Samsung CI537V

Failure of field output chip: The only stock fault we've had with these sets, which have been popular for rentals, is failure of the TDA2653A field output chip. As you'd expect, the result is a blank screen. In each case we've found that almost no heatsink compound was used. After fitting a new chip with adequate compound we've had no further trouble.

No vision, blank raster: The set was permanently taking its vision input from the AV phono input sockets. Inject a video signal here and up comes the picture. The cause of the fault is in the tuner, where the video switching is carried out. A different tuner is supplied as a replacement. Two of its pins have to be cut off before insertion.

Intermittently dead, chattering relay: REL901 frequently suffers from burnt contacts, but the field engineer had cleared this possibility. When the fault occurred on the bench we found that the relay coil wasn't being earthed. Tracing back we discovered that there was a broken pin in connector PHO1 (pin 3, the pink lead).

Sanyo

Sanyo CBP2145 (E2 chassis)

Loud arcing noise, flashing lines across picture: When this set was switched on from cold it would emit a loud arcing noise with flashing lines across the picture, almost as though the focus spark gap was arcing. The fault would clear after 10 minutes or so, making fault-finding difficult. To cut a long story short, after using three-quarters of a can of freezer we found that the culprit was C364 (100 μF, 16 V) which smooths the 12 V supply. When we had removed this capacitor we plugged it into the bench digital capacitance meter and then heated and cooled it to see what was happening. The more it was cooled the further its value dropped towards zero.

For first 10 minutes no signals: Only channels 3 and 4 could be selected. During the period of the fault there was ripple on the 5 V line. Replacing C395 (100 μF, 16 V) restored the signals but they were noisy with slight lack of height. The fault again cleared after 10 minutes. This time replacement of C397 (100 μF, 16 V) provided a cure. These sets seem to be becoming noted for the failure of 100 μF, 16 V capacitors to work from cold.

One colour drops out intermittently: Its cause was on the c.r.t. base panel, where the RGB output transistors are mounted in the same well-known way as the field output transistors in older Hitachi sets. The cure is to resolder the transistors and edge connectors on the board.

No picture, just snow at switch-on: This fault is becoming quite common with this model. A normal-quality picture may appear after a while, but when the set is switched off and on again the fault symptoms are repeated. The cause of the problem is the two 10 μF, 16 V electrolytics C397 and C364. They are near a stand-off resistor that gets quite warm and dries them out.

Picture break-up and bad 'twitching': When this set was switched on from cold the picture would break up into horizontal lines, with bad 'twitching'. The sound would buzz and pop. If you've not come across these symptoms before, you would be forgiven for switching the set off and reaching for another set to repair. But don't panic! Just replace C364 (10 μF, 16 V) which decouples the 12 V supply. It's mounted close to a high-wattage resistor that runs warm, thus drying out the capacitor.

Complete loss of signals: This is often caused by failure of choke L391, which supplies 5 V to the u.h.f. tuner. It goes open-circuit, but we've known the break to heal itself as the set warms up, adding a bit of interest to the diagnosis.

Wouldn't come on for 15 minutes: The set would then display a snowy picture, and the channel change was sluggish. This kind of fault always leads me to suspect low supply voltages and poor smoothing. Sure enough the culprits were C397 and C364 (both 100 μF, 16 V). Replacing them restored correct operation. Apparently this is a common problem with these sets, but it's not when you don't see many of them!

Line tearing at high brightness: There was a good picture when this set was switched on, but when the brightness was increased there was line tearing across the screen. I went for the electrolytics, which is always a good bet with these sets. C364 (100 μF, 16 V) was open-circuit, a replacement restoring normal operation.

Partial field collapse, no sound and a shaky picture: Replace C364 and C398 in the power supply. They are both 100 μF capacitors rated at 25 V.

Patterning and lack of height: These two faults showed up only after about an hour. When freezer was sprayed near the 12 V and 5 V regulators IC361 and IC365 the faults cleared. Both regulators have 100 μF, 25 V decoupling electrolytics – C364 and C398. Replacing them cleared the two faults.

Sanyo CBP2152 (E4-A21 chassis)

The text TV handset didn't function: The cause of the fault was a fracture in one of the legs of the ceramic resonator.

Occasional failure to come out of standby: The set also suffered from disturbance on the picture and random channel changing. After fruitless checks in the power supply we decided to take a look at the army of zener diodes between the front panel 'tac' switches and the system microcontroller chip. By disconnecting each one it turn and watching the results we found that D746 was the culprit. A new 6.2 V zener diode restored correct operation.

Dead set: The set would come out of standby for a few seconds, then go back to standby again. The h.t. supply to the line output stage was present, but the drive was missing at the line output and driver transistors. Checks around the TDA4505 chip showed that there was no line drive output at pin 26, though there was oscillation at pin 23. In addition the 12 V supply at pin 7 was low. Tracing the source of this back to the power supply, we found that the output from the L78M12 regulator chip IC362 was also low. A new L78M12 restored the set to life.

Sanyo CBP2180 (A5 chassis)

'Went bang with a puff of smoke': So said the customer. We found that the degaussing posistor had burnt out and melted, and that the main fuse had blown. Replacing these items put matters right.

Set stuck in standby: If the remote control unit's standby button was pressed the standby light would go off then come back on again a few seconds later. During this time all the supplies except the 12 V line came up. The reason for the failure of the 12 V supply to appear was that the 78M12 regulator IC552 had gone open-circuit.

Couldn't be tuned: This was because the set's tuning voltage was missing. R714 (10 kΩ) on the front control PCB had gone open-circuit.

Field fold-over: This set suffered from field fold-over at the top of the screen. The cause was a dry-jointed chassis connection at the 12 V regulator IC552.

Sanyo CBP2572 (ED1 chassis)

Surge limiter open-circuit: The 3.9 Ω, 15 W surge limiter resistor R301 was open-circuit and the 2SC4429 chopper transistor Q303 short-circuit. The cause of the chopper transistor's failure was the fact that the resistor in the snubber circuit (R315) was dry-jointed. After attending to these items the set still failed to come on. The TEA2260 chopper control chip IC301 had also failed, a replacement finally restoring normal operation.

Strange EW fault: Scope checks showed that only some information was coming out of the digital signal processor unit. The DPU2553/75 chip was responsible.

Blank raster, no sound: If the set comes on with a blank raster, no sound and no on-screen displays it's likely that the two non-volatile memory chips on the signal's PCB have become corrupted. They can be replaced or reprogrammed, as the set has a built-in facility for reprogramming corrupt memories. To gain access to this, press the service button and the volume plus button simultaneously – the service button is accessible through a small hole in the front panel. Care should be exercised when using this feature, as all the previously stored data will be lost. It's quite time-consuming to re-enter this.

Intermittent loss of off-air signal: For once the fault wasn't on the digital board! There was simply a dry-joint at the emitter of Q173 (2SA608).

Sanyo CTP6144

Power supply pulsing: Check whether the l.t. transformer T391 (part number PT0144) has a high-resistance or open-circuit winding.

Set went off, dry-joints around the chopper transformer: The report on the card said something about the channel indicator flashing 88 or 99, and the engineer had added that the raster became pear shaped. The vital piece of missing information was that the set went off, the reported symptoms being ancillary ones. The cause of the trouble was evident as soon as the back had been removed and the set had been upended – there were dry-joints on all pins of the chopper transformer T301.

Excessive brightness: We found that R478 was open-circuit.

No colour: Replacing the TDA3565 colour decoder chip restored normal operation.

Sanyo CTP7132 (80P chassis)

Dead set: If 320 V is present at the collector of the chopper transistor Q304 first check its 470 kΩ base bias resistor R302. If this is OK, check or better replace the 10 μF drive coupling capacitor C312. It's just below Q304's heatsink.

EW fault: R4012 (2.2 Ω) was burnt, transistor Q4005 was short-circuit and R4011 was also cooked. When these three items had been replaced the set was switched on and R4011 immediately began to smoke. All three transistors in the circuit were checked and found to be OK. It turned out that the EW coil L4001 (green) was short-circuit. The type used in the Philips G11 chassis was tried as a replacement and worked well.

Switch-mode PSU wouldn't start: We disconnected the feed to the line timebase and checked the power supply with a bulb across its output. The bulb lit and the output voltage was correct. There was no apparent fault in the line output stage in terms of a short-circuit, however – the transistor, transformer etc. were all checked. So we tried reconnection and testing with the bulb still attached. Lo and behold, a fully operational TV set! Well, apart from the idea of fitting a table lamp on top of the set and wiring it to the power supply, what were we to do? We checked around the switch-mode control circuit and found that the waveforms didn't match those shown in the manual, despite full regulation. We decided to change the two electrolytics C312 (10 μF) and C314 (47 μF) and after that there was no further trouble. A similar fault with an Hitachi CBP260 (NP9A chassis) produced tripping at switch on, though the set would work normally after a few attempts. We decided to change C918 (220 μF), C919 (22 μF) and C909 (10 μF), which put matters right.

Dead, short-circuit line output transistor: While fitting the replacement we resoldered the heatsink plate connection to the main PCB as this is often the cause of line output transistor failure in this chassis. Two weeks later the same fault occurred. The set worked normally when another BU208D transistor had been fitted, but after 10 minutes the line output transformer started flashing. Assuming that the trouble was insulation breakdown, we replaced the transformer. Yes, you guessed right, the set came back a week later. This time we were in luck. While checking the power supply, with its 110 V output disconnected and a 100 W bulb as a dummy load across C321, the phone rang. On returning 20 minutes later the bulb was noticeably brighter than before. The h.t. had risen to 160 V. The cause of the power supply instability was traced to zener diode D305 in the error detector circuit. From time to time the voltage across it was as low as 5 V instead of 7.6 V. Its replacement restored normality to both the set and us!

Short-circuit LOP transistor: If the 2SD871 line output transistor has gone short-circuit, replace C312 (10 µF) in the power supply and check the other two small electrolytics before switching on again. Also clean the line hold control (under the control flap), as little fingers twiddling this often lead to failure of the line output transistor. If the latter fails again at switch-on, the line output transformer is suspect. If a picture does appear it will probably be slightly out of focus. The transformer is again to blame – the tubes have a long life in these sets. The cause of field collapse is usually the 2200 µF field scan coupling capacitor C445 going open-circuit.

Power supply tripped slowly at switch-on: If the set was left on the fault would correct itself. Replacing C312 (10 µF) restored normal operation.

Sharp

Sharp C1421

Dead set, standby relay not energized: Linking across the relay produced a snowy raster with no tuning control. Checks around the microcomputer chip IC1002 showed that there was no voltage at the reset pin 11. This was due to operation of the overload protection transistor Q604. By isolating the various protected circuits we found that the basic cause of the trouble was zener diode D607, which was short-circuit.

Stuck in standby: When we switched the set on the e.h.t. rustled up for a second then the set went into the standby mode. As the e.h.t. rustle seemed to be rather violent we tried the set with a 110 V mains supply. It then came on, but if the mains supply was raised the 115 V supply to the line output stage rose as well. The STR40090 chip was faulty.

Sound is low: L301 in the i.f. unit could be faulty. When this coil is defective the action of the volume control changes – as you turn the control up above the half-way point the sound actually becomes quieter.

Sharp C3720H

Weak blue drive: The symptom with this set was weak blue drive. We found that D807 on the small subpanel connected to the tube's base panel was leaky. A 1N4148 diode fitted the bill, but the fault remained as before. The TDA3566 decoder chip had to be replaced as well. Perhaps a c.r.t. flashover had been responsible for the failures.

Would trip at switch-on: Because of the comprehensive safety/trip circuitry incorporated in many sets, such as this one, fault-finding is impossible without a manual. At switch-on the e.h.t. came up then the set tripped back into standby. After checking on many possibilities we found that there was no supply at the field output chip. Both D502 and R521 in the 24V supply were faulty. Thus in the event of field collapse the set is switched back to standby in approximately 2 seconds.

Would trip after a few seconds: We traced the cause to R521 ($3.9\,\Omega$ fusible) in the feed to the field output stage. To prevent tube damage the protection circuit puts the set into the standby mode in the event of field collapse.

Sharp CV2123H (7P-SR1 chassis)

Wouldn't come out of standby: Although the power relay clicked on and off in sympathy with the standby command this set wouldn't come out of standby. A quick check showed that there was h.t. across the mains rectifier's reservoir capacitor, but the STR41090 switch-mode power chip wasn't performing. After eliminating the possibility of shorts in the line output stage and across the secondary supplies a new STR41090 was fitted. This restored normal operation.

Appeared as if tube was defective: On further investigation, however, we found that the c.r.t.'s heaters were not being fully supplied because R621 ($1.2\,\Omega$) had increased in value to about $5.6\,\Omega$.

Trips at switch-on: First check for dry-joints in the chopper power supply. Then bring the mains input up to 40 per cent via your variac. The set should switch on. If not, the STR41090 chopper chip is probably faulty (in the overvoltage mode). Replacing it should enable the set to work happily with the full mains supply restored.

Would not come out of standby: When power was applied to this set the red LED came on and the standby relay operated when asked, but the set remained dead. The chopper power supply was inoperative because the STR41090 chip had failed.

Sharp DV5103H

Blank raster and no on-screen display: After replacing the line output transistor and transformer in this set (by arrangement with the customer's bank manager!) we were left with a blank raster and no on-screen display. There was good sound, however. Panic set in, as all the processing is carried out on the digital board. We've had the non-volatile RAM chip cause this fault, but replacing the PCB with a test one made no difference. As the line output transformer had failed, we decided to investigate the beam limiting circuit where we found that C612 had burnt out. Phew!

Luminance and chrominance out of step: The chrominance information was displayed about one and a half inches to the right of the luminance information – just as if there was a luminance delay-line fault. The cause of the problem was the small plug-in non-volatile memory chip on the digital PCB.

Dead, fuse and surge limiter open: These sets sometimes come in dead with the 2 A fuse and the 8.2 Ω, 7 W surge limiter resistor open-circuit. The cause can be a short-circuit in the chopper transformer, between the primary winding and earth.

Appeared to be dead, h.t. was low: A check showed that the h.t. voltage was low at about 45 V instead of 113 V. As there appeared to be no short-circuits we checked around the feedback loop in the power supply and found that the 6.2 V zener diode D754 had a 62 Ω leak both ways. Replacing D754 and the CNX82A optocoupler cured the fault.

Stuck in standby: The set's bicolour LED stubbornly remained red. A check showed that the 113 V h.t. supply was at 185 V! The cause was traced to R753 and R781, which are both 39 kΩ and were both open-circuit. Replacement resistors cured the fault, but we then found that there was a pinhole failure in the line output transformer's insulation.

Sharp DV5132H (DECO 5 chassis)

Goes into protection mode: When one of these sets goes into the protection mode you have approximately 5 seconds before it reverts to standby. Fortunately all that was wrong with this particular set was that R612, the fusible resistor that feeds the supply to the field output chip, was open-circuit. A replacement restored normal operation.

Lack of height: The problem became progressively worse as the set warmed up. Replacing the RH-IX1426BMNO chip IC1400 on the digital PCB cleared the fault.

R705 slow to go open-circuit: If the $3.3\,k\Omega$ fusible resistor R705 in the start-up circuit is slow to go open-circuit, check R725 $(39\,\Omega)$ which tends to go high in value.

Solavox

Solavox 140

Field collapse: This was simply a case of the LA7830 output chip going short-circuit. Its 3.3 Ω feed resistor R122 had also failed. Replacements produced some scan, but with poor top linearity and some top compression. C108 (1000 μF) was responsible for this. We've seen these sets in Nikkai cabinets.

Blank raster: There was no snow, and it looked as if the set was stuck in the AV mode. Checks around the TDA4505 chip IC101 suggested that it might be faulty, so a replacement was fitted. As this seemed to cure the problem the set was returned to the customer. Within a month it was back with the same fault. This time we found that the area around IC101 was very sensitive to heat, producing all kinds of symptoms when either a hair dryer or freezer was pointed in its direction. We felt it unlikely that the new chip had failed, but a replacement had to be tried. No luck. Plying the panel eventually led us to the cause of the problem: after much soul searching and prodding around the chip we found that L105 in the video output feed was intermittent. A replacement salvaged from a scrap panel cured the fault.

Severely crinkled verticals: The picture produced by this 14 inch. colour portable was marred by severely crinkled verticals. On investigation we found that C116 (1 μF, 63 V) and C117 (4.7 μF, 63 V) had both fallen in value. Replacements straightened the picture.

Solavox 141

Dead set or persistent field collapse: These sets always seem to come in dead. Here are various faults we've had. R109 (180 Ω, 0.5 W) goes open-circuit. This resistor's body colouring makes it look as though the value is 1.8 kΩ – we've even had a faulty one measure 1.8 kΩ! Q117 (2SC1573A) often goes open-circuit. It's an npn transistor rated like a video output device. Another regular failure is the remote standby transformer whose primary winding goes open-circuit. The 12 V supply filter resistor R104 (5 Ω, 1 W) can and does go high in value. This usually results in a dead set although in one case the symptom was persistent field collapse because the low 12 V supply upset the TDA4503 chip, removing the field drive.

No signals, just snow: A check showed that the tuning voltage was missing at pin VT of the tuner. We didn't have a circuit diagram, but as the tuning and channel preset circuitry is on the front vertical panel this seemed to be a good place to start. As luck would have it there was a very sad looking 470 kΩ resistor (R010) – the colour bands could just be recognized. It read open-circuit and on fitting a replacement the set's tuning was restored.

No tuning: There was just a snowy screen. A check showed that the tuner's VT pin was deprived of its tuning voltage. We noticed that R1010 (470 kΩ) on the front tuning panel was very discoloured. When we tested it the reading was well above 5 MΩ. Normal tuning was restored after replacing it.

Snowy black and white picture, lack of height: It seemed reasonable to suspect problems with the multifunction TDA4505 chip IC101, so we started off by checking its supply which was low at 9 V. The 12 V supply was OK, however. The cause of the problem was R104 in the feed to IC101. This 3.3 Ω, fusible resistor had gone high in value.

Lack of height: Check whether the 12 V supply is low. In a recent case I found that R104 (3.3 Ω, 1 W) had gone high in value.

Sony

SONY AE1 CHASSIS
SONY KV1412
SONY KV2000 MK 2
SONY KV2052
SONY KV2062
SONY KV2090 (XE4 CHASSIS)
SONY KV2090
SONY KV2092 (XE4 CHASSIS)
SONY KV2096 (XE4 CHASSIS)
SONY KV211XMTU (AE1 CHASSIS)
SONY KV21XRTU
SONY KV2212 (YE2 CHASSIS)
SONY KV2217
SONY KV2704
SONY KVM14TU
SONY KVM14U
SONY KVM2131U
SONY KVX2121U (AE1 CHASSIS)
SONY KVX2521U (AE1 CHASSIS)
SONY KVX25TU (AE1 CHASSIS)

Sony AE1 chassis

No channel storage: This set could be tuned in perfectly but you couldn't store the channel. As soon as the preset button was pressed the picture would be lost. Replacing the M58655P chip IC003 made no difference. We then discovered that there was only $-14\,V$ instead of $-33\,V$ at pin 2. Coil L807 was open-circuit (the $-14\,V$ was coming via a $10\,k\Omega$ resistor in parallel with the coil). Incidentally the M58655P is an expensive device.

Intermittent picture and sound: We found that all the connections to the 2SD1548LB chopper transistor Q608 were dry-jointed. When the surge limiting resistor R805 $(1.2\,\Omega)$ in the $200\,V$ supply goes open-circuit you get the blank raster with flyback lines symptom.

Lack of height: For lack of height, top fold-over and severe cramping at the centre of the picture replace R802 (0.47 Ω).

No sound or vision: Check for 12 V at L606 in the power supply. If the reading is low, disconnect L606. If the 12 V output from the regulator Q608 then appears check for h.t. at the collector of the line driver transistor Q805. No voltage here means that R822 (1 kΩ, 1 W, 5%) is open-circuit.

Various symptoms: For drifting off tune, white streaks on the picture, will search-tune up but not down, will not visually lock on a channel – find board A, remove the screening can and inspect the soldered connections on the can mounting on the PCB. These connections form an earthing band and are usually cracked. Also check and resolder as necessary the connections to T101 and T102.

Intermittent loss of luminance: The cause was traced to the delay line (DL332) on PCB B. Its part number is 1–236–062–11.

Teletext contrast variations: The cause of intermittent teletext contrast variations was traced to a dry-joint at the emitter of Q02 on board V.

Dead after water spillage: Fortunately the damage was restricted to the 2.7 Ω, 7 W surge limiting resistor, the TEA2164 power supply control chip and the 2SD1548 chopper transistor. We replaced these items then wound the set up using a variac. All was well.

Motorboating in standby: When this set was switched to standby it produced a motorboating groan. This was caused by the main metal plate across the PCB. It was dry-jointed. Resoldering the lugs put matters right.

Bad joints in the i.f. section: Bad joints in the i.f. section on board A give a lot of problems. We've had intermittent loss of colour, poor picture or sound, picture rolling and also intermittent failure to stop in the search-tuning mode. In all cases the bad joints have been either on the earthing lands that anchor the screening plate mounting pins or around T101 and T102.

Sony KV1412

Wouldn't start: Having had a similar occurrence previously we condemned IC601 (μPC1394C), but a replacement made no difference. But we hadn't checked the start-up supply resistor R602 (2.2M Ω) which was open-circuit.

Wouldn't start: The fuse was intact and there was 115 V at IC601 (μPC1394C). While checking voltages the set started, then went off. After resoldering a couple of suspect joints on IC601 the set worked for a couple of days, then on a particularly cold morning it refused to start. Heating IC601 got it going once more and after fitting a replacement we had a lasting cure.

Took 20 to 30 minutes to start up from cold: Once it was on the picture was perfect. With the help of a hairdryer and freezer we established that the fault was in the power supply, around the μPC1394C chip. Replacing the chip made no difference and we then found that R602 had increased in value to 5 MΩ. Fitting a replacement cured the trouble.

Dead set: A scope check showed that there were no output pulses from the chopper control chip. Pin 4 read 2.5 V but the circuit diagram said 2.9 V. Was this 0.4 V difference enough to stop the power supply working? The answer was yes. R602 (2.2 MΩ) was open-circuit.

Tuner out of action: The tuner in this set was out of action, but why? The 33 V regulator was OK, and Q10 was supplying a control voltage. But by the time this reached the tuner's VC pin it was only a fraction of a volt. The almost complete short-circuit was traced to C134 within the tuner. It had an 80 Ω leak.

Sony KV2000 MK 2

Whistle from back and corrugated verticals: These symptoms indicate a problem in the power supply. It didn't take us long to find that C612 (3.3 μF, 25 V) was the cause of the problem. C609 (0.47 μF) and C610 (220 μF) are worth checking as well as these also tend to dry up.

Eratic width and whining from power supply: The fault clears after about 10 minutes. The cause is C612 (3.3 μF, 15 V) on the power supply PCB.

Dead set: The mains fuse had blown and the chopper transistor Q607 was short-circuit. When these parts had been replaced the power supply worked with a dummy load, but there was a 50 Ω resistance to chassis at pin 2 of plug F3 – the 135 V feed to the line output stage. C901 (330 pF, 2 kV) was leaky.

Appeared to be a blanking fault: The top of the picture, to about 4 inches down, seemed to be dark: then it began to brighten until it became normal. Actually the problem was that the lower part of the picture was too bright, the top part being correct. The cause of the fault was the reservoir capacitor for the supply to the RGB output stages, C827 (4.7 µF, 250 V). Note that there were different versions of this set with variations in this area.

Sony KV2052

Picture fold-over: In the event of picture fold-over down the centre of the screen, check for a high-resistance connection on the earth rail around the line output transistor. This fault can be intermittent and can destroy the line output transistor.

Blown up power supply: A major rebuild cured the problem for two days after which it blew up again. The cause of the trouble turned out to be a pinhole in the chopper transistor's mica insulating washer.

Intermittently dead: When the fault was present the e.h.t. would come up then die away. When the set did come on there was picture tearing. In the fault condition the set was in a semi-standby state. The cause of this trouble was Q001 in the standby circuit – it was leaky collector-to-emitter. A BC237 made a suitable replacement. The set would work if the standby switching transistor Q501 was removed.

Sony KV2062

H.t. appeared then set shut down: The set then went completely dead. With a dummy load connected the power supply ran perfectly, so it seemed likely that there was a line output stage fault. We wound the set up via a variac, and at about 80 V the set tried to start. It was squealing, however, and R855 began to burn up. We concluded that one of the e.h.t. rectifiers in the line output transformer was faulty and fitted a replacement. To our dismay this didn't cure the fault. Now one of the differences with the Trinitron system is that the e.h.t. lead from the line output transformer doesn't go to the tube directly: it goes to the H stat unit. When this item was unplugged from the transformer the set started up. A replacement put matters right. We've had this fault on a couple of occasions now.

Dead set: The 3.15 A mains fuse was blown. A check on the bridge rectifier showed that D609 was short-circuit. Fitting a replacement and giving the set a long soak test proved that all was now well.

Screen blank: There was sound but the screen remained blank although e.h.t. was present. Checks at the cathodes of the c.r.t. showed that it was cut off. We traced back to the CX109 colour decoder chip where the voltages at the colour-difference signal output pins were all low. We next found that the pulse input at pin 22 was missing. It comes from pin 20 of the μPC1377 line oscillator chip, where the voltage was at almost 0 V instead of 2 V. A new chip brought the picture back.

Sony KV2090 (XE4 chassis)

Chopper/line output transistor blowing problems: Replace D605 and D614 (both type RGP15J) which are connected in series with the chopper transistor and C513 (2700 pF) in the line oscillator circuit. Then resolder the stand-off resistors, using high melting point solder.

Dead set: When you find one of these sets dead with the usual dry-joints at R621/622 (which quite often have to be replaced because arcing has eaten the centre leg away) and a failed 2SD1497 chopper transistor (expensive! – use a BU508), don't switch on until you've checked D605 and D614. They can be short-circuit or leaky – take them out to check.

Intermittently dead: The power supply continued to work with the h.t., 7 V and 4 V rails OK. The line drive was disappearing, and once this had happened the set wouldn't come on again until it was switched off then on. Tapping the PCB gave the impression that dry-joints were present, but a blanket resolder job didn't improve the situation. Eventually gentle probing revealed that C513 (0.0027 μF) had an internal connection problem. It's connected to pin 15 (line oscillator) of the TDA2579 timebase generator chip IC501.

Sony KV2090/2092/2096 (XE4 chassis)

Dry-joint problems: A weakness with these sets seems to be dry-joints on the high-wattage resistors on stand-off pillars. Sony recommends resoldering with RS high-melt solder and checking for this whenever you see one of these sets. Models that have this trouble are also prone to mains switch failure. Sony have issued an official modification kit covering both problems.

No results: The owner reported that prior to this the set had for several days been reluctant to start when first switched on and would suddenly go off apart from a fast, loud ticking noise from the back. We found that the 1.25 A h.t. fuse F602 had failed because the 2SD1398 line output transistor was short-circuit. There was a good picture when these two items had been replaced. But after a short time the set would, just as the owner had described, shut down intermittently. We checked the usual dry-joints for which these sets are noted but there were no problems here. The cause of the fault was eventually traced to intermittent breakdown of C513 (2700 pF) in the line oscillator circuit.

Sony KV2092 (XE4 chassis)

Volume would alter or not adjust: Unfortunately this was an intermittent fault. While the set was playing up we carried out a check at the base of Q015, the volume buffer transistor in the line from the microcomputer and DAC chips to see whether the d.c. level changed when the volume was adjusted. It did, so this eliminated the circuitry prior to Q015, putting us in a much better position when the fault next appeared for a reasonable time. When the set faulted again we carried out checks around Q019, Q021 and Q017, which process the volume on-screen display information. Q017 turned out to be faulty. As we didn't have a Sony replacement we fitted a BC546 which did the trick.

Dead set, mains fuse blown: The chopper transistor Q602 was short-circuit. As no dry-joints or obvious causes of the failure could be seen a BU508 was fitted temporarily and the set was tried again. The BU508 failed immediately. Further checks showed that both D605 and D614 were short-circuit. Fitting new diodes and the correct 2SD1497 in position Q602 restored normal operation.

Sound intermittent: Also the volume control wouldn't operate correctly, sometimes with the volume bar, the sound jumping in fits and starts. It took us a time to pin the cause of this fault down: D016 was eventually found to be leaky.

Line output transistor failure: If the problem with one of these sets is intermittent failure of the 2SD1398 line output transistor, desolder the base drive coupling coil L801 then re-tin and refit it. It's also a good idea to go over the connections to the line driver transformer.

Sony KV2096 (XE4 chassis)

Intermittent loss of sound and vision: This fault seemed due to a mechanical problem – if the main panel was touched almost anywhere the fault would occur. After some careful flexing and tapping around the cause was found. There was a dry-joint at one end of coil L852 which couples the 12 V supply to much of the set.

Classic dry-joint: This was a panic job – we were called to the house on the eve of a holiday and had no service information and didn't know the set. The firm that supplied it was shut for the duration. When we found out how to slide out the chassis we discovered a classic dry-joint. There are three wirewound resistors mounted on end next to each other near the fuses, just asking for overheated joints. We resoldered them all and the set worked. Phew! The cabinet is frighteningly unstable when the chassis is slid out but the main thing to watch if you meet one of these beasts is those wirewounds.

Field fold-over at bottom of screen: There was excessive height at the top of the screen and the bright-up at the bottom was about 2 inches. high. The voltages around the field output chip were more or less correct except for that at pin 4 which was low. This is associated with the linearity feedback between the output and generator chips. An investigation of this circuit revealed a heavy leak in C527 (470 μF). Replacement provided a complete cure.

Field fold-over and red shading: There were two faults with this set. First, field fold-over at the bottom of the picture with bent verticals at the edge. Secondly, red shading around white verticals. The cause of the first problem was C527 (470 μF) which had become leaky. The second problem was due to loss of the focus voltage as R853 (1 kΩ) had gone open-circuit. Though diode D852 measured OK, when we removed it from the panel we saw that there was a split down its side where it was going short with the high voltage across it.

Dead set, EW correction poor: This set came in with a report saying that it was dead, also that the EW correction was poor. We soon found that the chopper transistor was short-circuit due to dry-joints on the 680 Ω snubber network resistors. These should have been resoldered using the special solder that Sony supply with the mains switch modification kit. We were told that the switch modification had been carried out, but whoever had done it had fitted the wrong switch. We obtained the correct switch kit from Sony, fitted it and repaired the power supply. The result was a good picture, but we couldn't obtain straight verticals even with the pincushion amplitude control at one end of its travel. This was eventually traced to C527 having gone low in value. It should be 470 μF but read 350 μF when checked with our capacitance meter.

Intermittent picture failure: Several calls had been made to this set because the picture went off, but the fault never occurred when an engineer was present. We took it back to the workshop for a soak test and after a couple of days the fault developed: instead of a picture there was a dark, blank raster. Unfortunately as soon as the back was removed the picture returned. No amount of tapping around or heating/ freezing would bring on the fault. So we waited. After a couple of brief appearances we were still clueless as to the cause of the fault. Then on its third return we were provided with a major clue as to where the cause of the fault might lie. The screen went black as before, but a slight coloured shimmering and, occasionally, a few teletext characters were seen. This time the fault remained when the back of the set was removed. By this time everyone was walking on tiptoe. Several components were lightly pressed or tapped. Only when the SAA5050 text chip was lightly touched did the fault clear. As this chip is in a holder we unplugged it and soldered it directly to the PCB. We haven't seen the set since.

LOPT short-circuit and 1.25 A fuse blown: The line output transistor was short-circuit and the 1.25A fuse had blown. As no obvious faults were found these items were replaced. At switch-on the transistor instantaneously went short-circuit, the power supply making no attempt to trip. I removed the transistor and checked the h.t. voltage, using a 60 W bulb as a load. It was correct at 115 V. The line drive was then checked. There seemed to be plenty of power, although the waveform was misshapen due to the absence of the line output transistor. The flyback tuning capacitor and the various protection capacitors around the line output stage were next checked but no faults could be found. Changing the line output transformer finally provided a cure.

Sometimes stuck on AV at switch-on: This set had an odd intermittent fault. Sometimes it would be stuck on AV when switched on, with AV displayed on the screen. At other times the IR receiving diode would flash as if a remote control command had been received and the channel number would appear on the screen. The cause of the fault turned out to be dry-joints on the screening can around the main microcontroller chip. This can is also used to earth print lands around the chip. Also check the main PCB where the subpanel is plugged in.

Picture shimmer on verticals: If this gets worse as the set warms up, it can be cured by resoldering the connections to the field output heatsink. They are used to make an earth path connection.

Field fold-over at the bottom: Go straight to the 470 μF, 25 V scan coupling capacitor C527. It dries up with age.

Print damaged by dry-joints around the snubber resistors in the chopper circuit: A first-class repair can be achieved by fitting the Sony power supply modification kit part number X-4377–097–2. It contains replacements for all the parts that have been damaged and a new piece of circuit board which you bolt on to the top of the main PCB.

Ragged verticals in screen middle: The symptom was present when the set was cold, disappearing after 10 minutes. The cure was to resolder the heatsink on the field output chip – it acts as a link between earth lands.

Sony KV211XMTU (AE1 chassis)

Failure of rectifier and feed resistor in tube's first anode supply: An extremely common problem with these sets is failure of D803 and R807 respectively. The result of course is no picture. Evidence that the resistor has failed is clear to see, but the diode can check OK. If it's taken out, however, you can see signs of arcing on the under side. It seems that the problem has now been recognized. Sony supply an uprated diode, type RGP02–17, instead of an ES1F. Part numbers for the diode and resistor are 8–919–300–65 and 1–218–025–51.

Bowed raster: This was a weird one: the raster had bowed sides, with only the left-hand quarter of the picture showing. It was a strange sight indeed, with three-quarters of the screen completely black. Curiously the on-screen display was perfect. The EW fault was the give-away and, believe it or not, the pincushion correction processing is carried out on the AV interface board! All became clear when the board was hinged down: two large dry-joints were obvious at the earthing pins 17 and 18 of the PCB plug and socket CNJ51. Resoldering these restored full scan.

Intermittent bright green picture: The fault was heat related, so it was time for the hair dryer. We found that the cause of the fault was definitely on the video panel, which was a bit of a surprise as we've had a few faulty transistors on the tube base PCB in these sets. The fault could be made to appear by heating around the colour decoder chip. C302, the green sample-and-hold capacitor, was suspect but proved its innocence on being replaced. It didn't take long though to trace the source of the fault to the green on-screen display buffer transistor

Q311. The original type is JC501, but the 2SC2785 is a direct, Sony official replacement. A new one cleared the trouble.

Fine black lines on screen: The whole picture was covered by fine, shimmering horizontal black lines. A scope check on the field output waveform showed that there was a lot of h.f. oscillation present. The 330 Ω scan coil damping resistor R544 was found to be open-circuit.

Red heater-cathode fault: We'd fitted a new tube in this set a couple of months previously because the original one had a red heater-cathode short. Here it was back with a ticket that read 'same fault as before'. On test we found that the picture was covered with flyback lines and was too bright. Excessive first anode voltage was the cause of the trouble. R722 (680 kΩ) at the chassis side of the first anode potentiometer was open-circuit.

Sony KV21XRTU

Rope pattern on left side of screen: A number of these sets seem to have the same fault when unboxed – a rope pattern about 1/8 inch wide one-third of the way across the screen from the left-hand side. In each case the pattern has been more noticeable on BBC-1 (channel. 55). Having had similar troubles with earlier Sony sets we checked that all leads are dressed correctly and for dry-joints on heavy legged components in the line timebase and power supply areas. In most cases the suspect joints have been around C715 and the scan coil connection plug.

Poor picture, reverting to monochrome: We ran the set for several days in the workshop before the fault appeared. As the tuning had shifted slightly there was a ringing picture. We discovered that switching the a.f.c. on made no difference and that when the sweep tuning was started it didn't stop when a channel was found. So checks were made around the a.f.c. detector chip IC102 on the i.f. panel. A dry-joint was spotted at one end of the detector coil L105: when this had been resoldered the set worked perfectly.

Dead set: This was was dead and a quick check showed that the 135 V h.t. supply was missing. When pin 5 of the line output transformer was disconnected the power supply returned to life. A new transformer cured the fault.

Teletext selection problem: When teletext was selected we obtained only TV chroma. When mix was selected there was TV vision but no text. When TV was selected all was well. The cause of the problem was a high-impedance connection at pin 5 of connector D3, the blanking output from the text decoder. Cleaning cured the fault.

Sony KV2212 (YE2 chassis)

No tuner memory: It would scan through the channels but wouldn't stop. To wade in feet first seemed to be rather a daunting proposition, so we decided to check the miniature pushbutton switches under the tuning panel escutcheon. Out of a total of ten switches four had broken springs internally. A complete set was ordered from Sony and when these were fitted the fault was completely cured.

Sides of the picture 'went funny' after about half an hour: Since we expected to see an east/west fault we were quite surprised to find that the red/green convergence at the sides of the screen jumped out then flicked back to normal. It happened again 5 minutes later and soon became permanent until the set was switched off and allowed to cool. The back was removed and, armed with a can of freezer and the hot-air gun, in we delved. After much freezing and frying it seemed that the fault was around Q551/552. Using a piece of paper as a shield we then froze a few individual components in this area. After some effort the culprit turned out to be coil L551. It should be 27 mH but the nearest we could find, from a scrap chassis, was 10 mH. This restored normal operation.

No teletext: When text was selected there was just a blank screen. If mix was selected the text came up in monochrome. An empty box was displayed when time was selected. Since monochrome text was present in the mix mode the decoder was obviously working. It seemed likely that the RGB drives were incorrect and a scope check showed that there was no activity at any of the outputs. The manual shows the collectors of the three buffer transistors Q5/6/7 connected to chassis, which caused us some confusion until we followed the print paths and found that they are supplied from a 5 V line via the level set potentiometer RV3. This was open-circuit.

Set permanently in search tune mode: When we removed the back we saw that the channel search button was stuck down. A call to our Sony dealer friend proved to be fruitful. He didn't have the switch but provided us with the part number and recommended replacement of all ten pushbuttons on this panel to prevent further trouble. When we removed the panel we found that six of the buttons were stuck down. The Sony part number is 1–552–774–00.

Narrow picture with crinkle-cut edges: When the set had warmed up a bit the edges of the picture straightened and then the width popped out. Application of freezer to the SG246A SCS on the scan board proved its guilt.

Field collapse: The cause was traced to R851 (1.2 Ω) which is connected to pin 9 of the line output transformer. We could find no reason for its failure.

No picture: We found that the h.t. voltage was low at only 83 V. C652 and C653 (both 33 uF, 250 V) in the power supply were both faulty. Replacements restored the picture.

Narrow band of picture on screen: This set had a narrow band of picture across the screen, similar to field collapse. As the field output stage's 40 V supply was present and correct, we started to check the electrolytic capacitors in the area. When we came to the field scan coupling capacitor C522 (470 μF, 63 V) one leg came out as we lifted it off the board. It was in a very sorry state. A replacement cured the fault.

Sony KV2217

When text called up screen went blank: The sound and picture were normal but when text was called up the screen went blank. There was a normal picture and black text in the mix mode. The cause of the problem was the preset RV3 (100 Ω) on the text panel: it read over 10 kΩ.

Set stuck on channel 6: There was no sound or picture. All the supplies were present and correct, the line output stage was working and the tube's heaters were alight. Checks in the control section at the top of the cabinet revealed that the volume minus button was short-circuit. Replacing this restored all functions.

EW fault: Our field service engineer had replaced a number of capacitors before giving up. The first thing we noticed was that the tube was as flat as a pancake. It was an insurance job, however, so we delved into the EW correction circuitry and found that coils L552 and L554 were open-circuit. After replacing these and the tube, then setting up, we had an excellent picture.

Sony KV2704

Narrow picture, bowed at the sides: Adjusting the width and pincushion amplitude controls had no effect. As the SG264A pincushion output thyristor can cause this trouble a replacement was tried but made no difference. When voltage checks were carried out in the pincushion correction circuitry we found that Q510 was biased off. A look around the timebase panel revealed that the soldering to the flyback transformer was extremely messy. A blob of solder hung from one of the pins and rested on an adjacent track, the green varnish providing insulation between the two. It would appear that flyback pulses had eventually punctured this barrier taking about seven years to do so – thus shorting together the transformer's ABL and 14 V pins. Removing the solder bridge and tidying up the soldering on the other pins cured the fault – once the width and pincushion amplitude controls had been returned to their original settings. Incidentally the e.h.t. lead was nicked where it rests against the degaussing shield. We added sleeving to prevent possible flashover problems. This is something that's worth checking whenever one of these sets comes in.

Works for 5 minutes and goes to standby: We noticed that the width was excessive at switch-on – in fact there was an EW fault. As soon as we tapped the line panel the set went into standby. We suspected a dry joint but decided to deal with the EW fault first. A check on the SG264A gate-controlled switch which drives the EW modulator diodes revealed that it was leaky. When it was replaced the set worked perfectly and no longer went into the standby mode.

During the same week we had a call from another customer who reported a similar fault on one of these sets. This time the SG264A GCS was OK, the problem being due to one of the EW modulator diodes. We fitted a BYW96E and it worked very well.

Set didn't always start at switch-on: The switch would have to be operated several times before the set would come on. Resistors R605 and R606 were both OK, but the h.t. was low at only 80 V. The cause of the fault was traced to C623 (33 μF, 250 V).

Intermittent operation: The basic complaint with this set was of intermittent operation – intermittent starting, going off and coming back on at random, and brightness and volume variations. Apparently it would sometimes work for weeks with no problems. Luck was on our side (for a change) when we soon found that pins 1 and 2 of the chopper transformer T602 on panel F2 were badly dry-jointed. We then found that three of the eight front panel function switches (board H1) had seized. This is quite a common fault now, accounting for various function fault symptoms. For reliability all eight switches were renewed – part number. 1–553–363–11. To complete the job we gave the power supply panel a good going over with the soldering iron.

Needs several attempts to switch on: Originally it would require several attempts to switch this set on. It now remained dead, although a noticeable fast ticking noise came from the chopper transformer T5. This was struggling to provide a 135 V h.t. supply but managed only 39 V. The h.t. reservoir capacitor C623 (33 μF, 250 V) was open-circuit, a replacement restoring normal operation.

Wouldn't start up: Originally several attempts were required to switch this set on. Subsequently it became lifeless. Though the power supply was trying to deliver the h.t. (135 V) it managed only 39 V. The cause of the trouble was the h.t. reservoir capacitor C623 (33 μF, 250 V), which was open-circuit.

Sony KVM14TU

Various symptoms: The symptoms with this set were as follows: the search tuning didn't stop at stations, there was no sound and the picture came only when the 'preset' button was pressed. The microcontroller chip IC001 needs a 7 V line-frequency pulse at pin 51, a 6 V sync pulse at pin 36 and an a.f.t. 'dip' at pin 35. We found that the sync pulses were missing because the sync generator transistor Q071 was short-circuit base-to-collector. When the set is working correctly the voltage at the a.f.t. pin 35 is about 2.3 V. If you find that it's 0 V then C012 (0.01 μF) is probably short-circuit.

Reds flared: This gave the impression that the tube was soft. Checks at the tube base showed that the first anode (G2) voltage was low at only 190 V. D852 was short-circuit and R852 (680 Ω) burnt out. Replacing these items restored the first anode voltage to 880 V, producing a normal picture.

Intermittent a.g.c. overload: Check for dry-joints inside the i.f. module.

Sony KVM14U

Grey-scale drift: There were varying symptoms due to grey-scale drift. The auto-clamping circuit didn't work well although it was all right itself. The main symptom was that the green level varied from fairly acceptable to very high. When adjusted the first anode voltage control had a very odd effect on the picture. The cause of the trouble was a faulty tube.

Intermittently going off tune: Here's a fault that is becoming as common as the defective first anode rectifier in these portable sets. The symptom is intermittently going off tune. A slight tap on the side of the i.f. can will produce the fault, which is caused by dry-joints on either T01 or T02 – the vision detector and a.f.c. coils.

Poor picture, then set smoked: The customer who brought this set in said that the picture had been very poor, then the set had started to smoke. When the back was removed R852 and D852 in the tube's first anode supply were seen to be burnt out. Replacements brought back a perfect picture.

Burning smell, dim blurred picture: Once the back had been removed we saw that R852 ($1\,k\Omega$) in the first anode (G2) supply had fried to a crisp. The cause was D852, which was leaky. As diode types seem to be very critical with Sony sets, we obtained the correct type from CPC. This and a new $1\,k\Omega$ resistor restored a really excellent picture.

Sony KVM2131U

Dead set: There was no h.t. from the power supply. A resistance check across the h.t. line confirmed our suspicions that a short-circuit was shutting the power supply down, and in fact the BU506 line output transistor was very much short-circuit. When a new one was fitted and the set was switched on the e.h.t. crackled up for an instant then everything was as dead as before. This time the protection diode across the h.t. line, D611, was dead in addition to the line output transistor. So it seemed obvious that the power supply was producing an excessive h.t. output. We suspected the feedback path from the chopper transformer to power supply chip and checks here showed that the $68\,\Omega$ fusible resistor R606 was open-circuit. When this and the other two items had been replaced the set breathed life again.

No sound adjustment on screen: The on-screen graphics indicated that volume up/down was working. A check showed that the volume adjustment voltage at pin 2 of IC001 was going up and down when asked but the voltage at pin 7 of the TDA1013A audio chip IC201 remained permanently at 6 V. The cause of this was a leaky diode (D201, type ISS133) in the a.f. mute line.

Wouldn't tune: The on-screen display indicated that tuning took place. As a first step the 33 V output from regulator IC004 was checked. It was OK, and was also present at the collector of Q004. At the other end of resistor R013, which is connected to Q004's collector, the voltage was very low. It turned out that D156 (1SS133) was leaky.

Sony KVX2121U (AE1 chassis)

Loss of sync and/or picture: We eventually traced the cause to dry-joints on coil T101. It's on the right-hand side of signals panel A.

Stuck on channel 11: This set also wouldn't respond to either the front controls or the remote control unit. Everything in the remote control receiver section was OK and we felt that the chip simply needed a reset, which occurred after removing the front panel then replacing it because nothing amiss had been found. One of the buttons had perhaps jammed?

Intermittent crackle on sound: Something was clearly arcing, but when we examined the set we found that someone had already had a go at resoldering everything in the audio output stage. The dry-joint was actually at a bridge rectifier. Always resolder the chassis connections that have lightning symbols by them, as dry-joints here cause the same problem. Dry-joints on the audio opamp chip IC251 are also common, causing loss of one or both channels.

Sony KVX2521U (AE1 chassis)

No picture: We soon found that there was no first anode voltage at the tube. As its heaters were alight and the e.h.t. rustled up the line output stage was clearly working. So we checked back to the source of the first anode voltage. This brought me to D803 and R807 (1 kΩ) both of which were faulty. Replacing them restored a good picture.

Set failed with flash and fizz: This set failed with a violent flash on the screen and an ominous fizzing noise. When we powered it via a variac and an isolating transformer it tried to start then cut out. We were relieved to hear the rustle of collapsing e.h.t. at least the line timebase seemed to be OK. While manipulating the variac we noticed a small, twinkling spark beneath the power transformer T601. Its removal didn't reveal anything immediately obvious, but when we examined it with an eyeglass we discovered that there were dry-joints where the windings are connected to pins 17 and 18, with traces of charring. Pin 17 isn't connected to anything external: it's a tie point for separate sections of the transformer's primary winding. When we'd tidied up this little mess, including sleeved joints, we refitted the transformer. A soak test proved that everything was now OK

Intermittent sound in one or both channels: This is quite common with these sets. The usual cause is dry-joints around the audio output chip IC251.

Intermittent of picture and sound: The 12 V supply was lost when the chassis was flexed. Q608 at the front, centre of the chassis was dry-jointed.

Stretched field scan: There was a stretched field scan from the centre to the top of the screen and no field scan from the centre to the bottom. We found that C531 (680 μF) had fallen in value to about 300 μF.

Sony KVX25TU (AE1 chassis)

Intermittent failure of line output transistor: A short-circuited L806 was the cause.

Raster cramped at bottom and stretched at the top: When we removed the back cover we longed to see a PCL805! Instead, however, we saw a 680 μF, 25 V capacitor, C531, with its case split. Could it have been connected to the cathode of something?!

No sound: When we removed the LA4280 audio output chip IC251 it fell in half! Fortunately a replacement produced good-quality sound without further ado. The customer confessed to having connected two external speakers to the set without using proper plugs. Instead he'd pushed the bare leads into the speaker sockets, using drawing pins to hold them in place. Say no more!

Intermittent loss of sound and picture: Normal operation would be restored if the metal section that runs along the centre of the main board was tapped. All three legs of the 7812 12 V regulator Q608 were found to be dry-jointed. For reliability with these receivers we re-solder the chopper transformer's pins, the metal plate's through-board tags and C612 in the power supply.

Intermittent loss of sound: This is normally caused by dry-joints around the audio chip, but the dealer had resoldered these. He hadn't resoldered the connection to the chopper transformer's 40 V pin, however. Doing this cured the fault.

Tandberg

Tandberg CTV3 Chassis

'Fingerprint' patterning on colour: Try resoldering the earth connections to the sound module!

Pumping: This can be caused by failure of any one of a dozen components on the line output panel. High on the list of items to check for this condition is diode CR738 on the power supply panel. An RS BYW56 is a satisfactory replacement for the BY127 used in this position.

Dead set: Fuse F725 had blown apart because the BU126 chopper transistor Q735 was short-circuit. Further checks showed that the BU208A line output transistor had also died. After renewing the above items we switched the set on. The sound came up but there was no raster and the e.h.t. tripler was very hot to touch. A new tripler finally got the set working.

Tatung

TATUNG 140 CHASSIS
TATUNG 145 CHASSIS
TATUNG 160 CHASSIS
TATUNG 165 CHASSIS
TATUNG 170 CHASSIS
TATUNG 180 CHASSIS
TATUNG 190 CHASSIS
TATUNG C SERIES CHASSIS

Tatung 140 chassis

Field roll, hold control at end: Check whether R423 (2.2 MΩ) is open-circuit.

Power supply tripping: Cold checks were carried out on the components in the line output stage but everything here seemed to be in order. So we tried the power supply with a dummy load. It still tripped. Cold checks on the resistors and semiconductor devices cleared them all of suspicion. We then wondered whether to try replacing the TDA4600 chopper control chip or check the various electrolytics in the circuit. Having got this far with cold checks, we decided to persevere and struck lucky first time – our Hameg component tester showed that C808 (100 μF, 25 V), which is connected to pin 9 of the chip, had a slight leak. Replacing it restored normal operation. We've since found that this component causes similar problems in Grundig sets.

Set wouldn't start up: We found that there was a start-up voltage (9 V) at pin 9 of the TDA4600 chopper control chip I801 but no voltage at pin 5 as R805 (390 kΩ) had gone open-circuit.

Field bouce on teletext: This was an interlace fault that was soon traced to a dry-joint at pin 8 of connector M501.

Picture slow to appear, crackling: The owner also said that the problem had been present for some time, and that he thought the switch was faulty! The power supply in this chassis is based on a TDA4600 chip. When you get a fault like this, look no further than the electrolytics associated with this device. In this set they are C807, C808 and C810, all 100 μF, 25 V. Replacing all three restored normal operation at switch-on.

No picture, motorboating sound: This tired looking set came in with no picture and uncontrollable motorboating and screeching from its loudspeaker. The power supply seemed a good place to start, and replacing the electrolytics around the TDA4600 chip is always a good idea. We found that the value of C808 (100 μF) had dropped dramatically. A replacement restored the picture and sound.

Tatung 145 chassis

No picture, humming noise on sound: Check whether C808 (100 μF) in the power supply is open-circuit.

Tuning drift: This was cured by replacing the M293AB1 tuning voltage synthesizer chip IR03.

Reluctant to start: This set was actually a Finlux 1014M. Once it was up and running it never failed. We replaced all the diodes and electrolytics in the power supply before we found that R816 (100 Ω, 1 W) was intermittent.

Tatung 160 chassis

Wouldn't start: The power supply was OK but there was no line drive. We found that there was no 12 V supply as R508 and R509 (both 12 kΩ) had gone open-circuit, removing the base bias from the 12 V regulator transistor.

Dead set: The set of which this chassis formed a part had been well and truly blown up! The BU508 line output transistor and TDA3651 field output chip had both died and R411 was burnt. Using a mains light bulb as a dummy load we found that the power supply was churning out 285 V instead of the correct 117 V. The culprit was D808 in the set-h.t. sampling circuit: it read 380 Ω both ways.

Dead set, 'demolition syndrome': A type of fault that's becoming more common is the dreaded 'demolition syndrome', when one small failure starts a chain of destructive events. Our first discovery with this dead set was that the line output transistor was short-circuit. We removed it and temporarily connected a 60 W bulb across its emitter-collector connections. The power supply then worked but there was no line drive from the TDA4503 chip because its 11.5 V supply was missing. Regulator transistor Q501 was non-conductive because one of its two 12 kΩ base bias resistors was open-circuit. We replaced both resistors (R507/8) and fitted a new R4050 line output transistor. When we switched the set on it sprang to life – with a bright raster. The 10 Ω filter resistor R201 in the feed to the RGB output transistors was open-circuit. Did a falling 11.5 V supply distort the line drive, turn the RGB output transistors hard on, blow R201 then destroy the line output transistor?

Power supply was tripping: Instead of the h.t. feed to the line output stage pulsing because of an excessive current demand the h.t. voltage was constant and correct. Scope checks revealed that there was no drive at the base of the line output transistor. The cause of this was traced back to the TDA4503 jungle chip I101 whose 12 V supply was missing. R508 (12 kΩ) in the feed to the base of the 12 V regulator transistor Q503 had gone open-circuit.

Failure of SAW filter in i.f. strip: As these receivers age, internal failure of the SAW filter in the i.f. strip has become an increasing problem. The symptom, either permanent or intermittent, is a ghosting effect, the spurious image being displaced about 1 cm (with a 20 inch tube) to the right of the main one.

Intermittent loss of signals: While the tuner can be responsible, on several occasions we've found that the resistors which supply the 33 V regulator are the cause. In a recent case one of these 3.9 kΩ resistors (R007) had a burn mark around it. The symptoms were normal pictures following a slight drift/sudden signal loss etc.

Slight ghosting: It looked as if there was an aerial fault. After checking the aerial system, which was OK, we replaced the tuner. This didn't make any difference. A replacement SAW filter (Z101) cured the problem.

Dead set: The line output transistor was short-circuit but when a replacement was fitted the set remained dead. The h.t. supply was present but there was no line drive. This is generated in the TDA4503 chip 1801, whose supply at pin 22 was missing. Tracing back from this point we found that the 12 V regulator transistor Q501 had a supply at its collector but no voltage at its base. This is obtained from the h.t. line via R507 and R508 (both 12 kΩ). One of these resistors was open-circuit, but they should be replaced as a pair.

Line collapse with smoke: The line scan coupling capacitor C406 (0.39 μF, 250 V) had split in half. This was a 20 inch set – its value is 0.33 μF with 16 inch sets.

Tatung 165 chassis

Set stuck in standby: Because QR06 (BF391) had gone short-circuit, the 25 V memory supply voltage had increased and the microcontroller chip had died. R007 and R011 (both 3.9 kΩ) were also open-circuit. When the 25 V supply had been restored and a new microcontroller chip had been fitted everything seemed to be fine – except that programme position 1 kept losing its memory. A phone call to Tatung Technical was needed. The nice man suggested adding a 10 μF, 35 V capacitor across RR34 (5.6 kΩ). This solved the problem.

Power supply appeared to trip at switch-on: But a check showed that the 117 V supply was steady. The set could be switched in and out of standby with the remote control unit, but in the 'on' condition the display pulsed bright then dim. We soon found that there was no line drive because the 11.5 V supply was missing at pins 7 and 22 of IC101. It comes from transistor Q501, whose base is biased by R507, R508 and the 12 V zener diode D503. R507 (12 kΩ) was open-circuit.

Low gain: The cause was R102 (4.7 kΩ) which was open-circuit. This resistor biases the base of the SAWF driver transistor Q101.

Peak white raster: The cause was the TDA3565 colour decoder chip I501.

Tatung 170 chassis

No sound, no LED display and a bright raster: At switch-on there was no sound, no LED display and a bright raster which quickly went dark with a half-inch bright band across the top. Absence of the display suggested a missing l.t. rail and we found that the 6 V supply had dropped to only 2.2 V. The 110 V and 18 V supplies were OK, and no excessive load on the 6 V line was apparent. R820, the 0.22 Ω resistor in the 6 V supply, was the first suspect but the rectifier diode D811 (BA157) was in fact the cause of the problem. It had developed a high forward resistance, although a cold check revealed only a small increase to $18\,\Omega$ instead of the usual $15\,\Omega$ or so.

Various intermittent faults: The sync, brightness level, sound, tuning and various other things were affected. We've known the teletext panel to be responsible for intermittent faults in this chassis, so we resoldered the plug and socket on the text PCB. This made no difference at all – in fact the trouble was even worse. The faults were now much more intermittent and could be rectified by only slight pressure on any part of the main panel. After much panel flexing, prodding, soldering and tapping the cause of the fault was traced to the soldered rivets that connect the outer chassis to the main board. After desoldering, tightening and resoldering these rivets the intermittent faults had been cleared.

Line output stage wouldn't start up: The channel indicator on the front worked, and the set could be switched into standby. We decided to replace the two transistors (Q401/2) in the line driver stage. This made no difference. There was a line drive waveform at the base of Q401, but it was a little cramped and low in amplitude. This seemed to point to capacitor trouble, but the few capacitors involved were all OK. What we eventually found was that R426 ($1\,k\Omega$), which is connected between the base of Q401 and chassis, had gone high in value. Q402's base is also connected to chassis via a $1\,k\Omega$ resistor (R427). After replacing both of these resistors the line drive had been restored and the set worked normally.

Damage in electrical storm: Every time we have an electrical storm we get these sets in for repair. My tip is to replace IR02 (PCD8572) and I001 (SAB3035) as a pair before getting too deeply involved in investigating the causes of obscure faults. In 90 per cent of cases replacement of these two chips will be all that's required.

Tatung 180 chassis

Lightning damage: Three of these sets came in after a storm with thunder and lightning. They all had the same fault symptoms – no sound and vision, with the LED lit – and in each case the cause was the SAB3035 chip IC1001 on the front panel. Roll on the next storm!

Dead set: This set was dead because the line output transistor was short-circuit. We fitted a replacement then looked for some reasons for the failure of the original one. After attending to a few possible dry-joints, we decided to start up the power supply with a 60 W bulb as the load instead of the line output stage. As the h.t. was correct, we reconnected the line output stage and confidently switched on. Bad move! The line output transistor once again turned into a wire link. Further checks in the line output stage revealed that C433 (6.8 nF, 2 kV), one of the capacitors in the EW diode modulator circuit, was leaky.

Inability to tune in stations: This set was a Proline NV2400/NV2700. The complaint was the inability to tune in stations coupled with a delay in responding to channel change commands. We found that the 33 V supply to the citac chip had disappeared because one of the two 10 kΩ resistors (R005 and R006) that supply the 33 V stabilizer had gone open-circuit. It's a good idea to replace them both – they are located next to the tuner.

Picture brighter on right: This set's picture was brighter at the right-hand side. The usual cause of such problems is the reservoir capacitor for the h.t. supply to the RGB output stages (C426 in the 180 chassis), but this turned out to be OK. By carrying out scope checks we traced the cause of the fault to C508 (0.1 μF), which couples the luminance signal to pin 8 of the TDA3562A colour decoder chip. It was virtually open-circuit.

Tatung 190 chassis

Dead set: With a dead set you may find R802 open circuit and R803 discoloured. Replace them both (both are 15 kΩ, 0.5 W types).

Sound but no picture: We found that the line output stage wasn't working because there was no line drive at pin 26 of the TDA4505 chip IC101. Further scope checks around this chip showed that the line oscillator frequency at pin 23 was nearly 1 MHz instead of 15 625 Hz. The 2.7 nF timing capacitor C111 had gone open-circuit, a replacement restoring normal operation.

Dead, mains fuse OK: Check the two 15 kΩ start-up resistors R802 and R803. They tend to go open-circuit.

No results: When the back had been removed a tripping noise could be heard coming from the power supply area, but the standby light was not illuminated. We tried disconnecting the supply lines from the chopper transformer, but the tripping persisted. While checking in this area we noticed that the h.t. reservoir capacitor C814 (47 μF, 250 V) had leaked on to the main panel. A replacement failed to cure the fault, however. Eventually, while resoldering in the area of the mains bridge rectifier's reservoir capacitor C808 (100 μF, 385 V), we heard it bubble under the heat of the iron. On removing it much 'gunge' could be seen. We cleaned up the mess and fitted a replacement, after which the set worked fine.

Tatung C series chassis

No remote control operation: Although the 5 V supply was present, there was no output from the remote control receiver chip IC750. The TFMS4360N originally used in this position has been replaced by a TFMS5360 with a new mounting bracket. Fitting this restored normal operation.

Folded picture or vertical bar: If the line scan produces a folded picture for the first 5 minutes, or there's a vertical white bar down the centre of the screen, check R413 (18 Ω, 0.5 W) which tends to go high in value. It provides the 18 V feed to the line driver stage. Replace it then ensure that the h.t. voltage is correct – 115 V with 14 and 20 inch models, 109.5 V with 15 and 21 inch models. We've had this fault with several models, including the Goodmans 2050R.

Reverts to standby in high brightness scenes: One of these sets would cut out and revert to standby on high brightness scenes. When we monitored the h.t. voltage we found that it dropped as the brightness increased. The cause of the fault was traced to CE803 (220 μF, 16 V) which had dried out. It's the reservoir capacitor for the supply to the TDA4605 chopper control chip IC801. Since it is mounted next to a hot resistor, it's advisable to fit a high-temperature component in this position.

Toshiba

TOSHIBA 140E4B
TOSHIBA 145R7B
TOSHIBA 175T9B
TOSHIBA 210T6B
TOSHIBA 2112DB
TOSHIBA 217D9B
TOSHIBA 2500TB
TOSHIBA 261T4B
TOSHIBA 285T8BZ
TOSHIBA C2226

Toshiba 140E4B

Snowy picture: It looked as though the tuner was faulty, but a replacement made no difference and there was also a line of vertical 'rope' interference on the left-hand side of the screen. The BF324 i.f. preamplifier transistor Q801 was the cause of the low gain, but the rope effect was still noticeable with a weak signal. Mr Tosh told us to juggle the green jumper lead under the PCB beneath the tuner to minimize the effect. He was right. Such are the wonders of high technology . . .

Dead set or intermittent dead set: Check the pins of the line output transformer. The cause of the trouble is dry-joints here.

Arcing noise and loud speaker buzz: At the same tune as the audio accompaniment the picture would reduce in height and width, what was left of the display being best described as a combination of line tearing and field fold-over. We found that every pin of the line output transformer was dry-jointed and that there were several suspect joints in the power supply cucuitry. A blanket resoldering job in these areas put matters right. The line output transformer pin that had actually caused the symptoms was pin 3: when we desoldered and cleaned it we found that there were signs of burning around the hole in the PCB. We had to scrape this clean prior to resoldering.

Picture disappeared after seconds: Good sound and a good picture were present when this set was switched on. After a few seconds, however, the picture disappeared. We had a quick tap around (the blunt end of a screwdriver, wisely wielded, produces amazing results) and the picture reappeared. A check on the print side of the panel then revealed a nice dry-joint at pin 2 of the line output transformer. After this and a few more suspect connections had been resoldered the set worked perfectly.

No line sync: One of these sets displayed a symptom that's rare these days, no line sync. A quick check showed that there was no feedback waveform from the line output transformer at pin 35 of IC501. Resistor R402 (27 kΩ), which is connected to pin 10 of the transformer, was open-circuit.

Toshiba 145R7B

Unusual sound fault: There was pitch, not volume, distortion. Some voices were OK but others would distort. The cause of the trouble was that the sound mute transistor Q860 would sometimes conduct due to a bias of some 0.6 V at its base – this should be present only during channel change or when tuning the set. A leaky 1N4148 diode (D203) was producing this bias.

No results: There were no results with no output from the LED and the 5 V supply was missing. Ra25 was found to be open-circuit, a replacement putting matters right. Don't try to make the chassis tracks correspond to the circuit diagram, however! There's an error on the diagram: Ra25 is not connected to the h.t. line as shown, in fact it obtains its supply from the bridge rectifier end of R801. We suggest amending the diagram in case this causes trouble.

Would not go into standby correctly: This set worked all right until it was put into standby. The sound was then muted but the picture remained. We found that the standby switching transistors Q802 and Q803 weren't operating because R833 (120 kΩ) was open-circuit.

Toshiba 175T9B

Only 2 inches of field scan lines: The rest of the screen was blank. We removed the back and had a tap around the field timebase area. Up came the full picture. On closer inspection we found a perfect dry-joint on C306, which is connected to pin 3 of the TDA2579-N6 sync/timebase generator chip Q340. After resoldering this joint we were rewarded with full scanning.

Field collapse after a few hours: There was a good picture when this set was switched on, so we left it running. After a few hours there was field collapse: not the normal straight line but a nice wavy one, a clue that even Inspector Clouseau couldn't miss. We went straight to the field scan section and found a perfect dry-joint on the scan coil plug connector. After some surgery the set was given a lengthy test and proved to be OK.

Remote control didn't work: We found that the set was also stuck on channel 2, and that none of the front controls worked. Operation via the remote control unit was established by desoldering the front control keys (SA01). Before checking on the price of a control key assembly we melted the plastic rivets open to gain access to the thick-film assembly. There were clear signs of liquid contamination here. We removed the assembly, cleaned off the contamination and reassembled it. The results were perfect. We've done several of these repairs since.

Power supply produces odd whine: The cause was traced to C749 (680 nF, 2 kV) which was short-circuit. It's connected across the line output transistor.

Toshiba 210T6B

Dead, appeared to trip at switch-on: We disconnected the 112 V supply to the line output stage but this made no difference – in fact the 112 V supply was missing, but the standby and remote control supplies were present. Transistor Q803 is used to kill the oscillator section of the chopper chip IC801 for standby or for the electronic trip action. In standby it's driven by QR01 which is in turn driven by the optocoupler DR10. QR01 was leaky collector-to-emitter but replacing it made no difference. The basic cause of the trouble was that the optotransistor in DR10 was faulty. Replacing DR10 restored normal results.

Dead set, standby light permanently on: First check that the 5 V (always) supply is present at the collector of QA15 (BC557A). If this supply is present but there's no voltage at the base and emitter of this transistor the M50747–205SP microcontroller chip ICA01 is faulty.

Pale, indistinct raster: Picture brightness didn't alter when the brightness control was adjusted. We suspected that a sort of reflected brightness within the tube was the cause of the trouble, as a result of the raster being deflected off the viewing area. This proved to be the case, the culprit being the field scan coupling capacitor C316 (4700 μF, 25 V). This fault can also cook up R327, which is in the supply to the field output chip.

Toshiba 2112DB

Failure of line output transformer: The component in question is T461, part number 23236427. It seems to be quite a common occurrence in this model. After replacing it always check the standby operation as the standby switching transistors Q843 and Q845, which are in the power supply, are often damaged.

Would go to standby after 30 minutes: We couldn't instigate the fault in the workshop, but found that pin 64 of the microcontroller chip appeared to be dry-jointed. After resoldering it and giving the set a long soak test we decided that all was now well.

Blank screen: We soon discovered that there was no on-screen display and no remote control action. According to that wonderful Toshiba help manual, the thing to do in this event is to check the 5 V supply to the text panel. If it's missing, replace the 5.6 V zener diode DF80. The advice was spot on. Thanks Toshiba.

Toshiba 217D9B

Lightning damage to audio section: Electrical storms have produced some real problems. This Nicam set had taken a blast and suffered damage to its audio department. There was no sound at all, but the fact that the Nicam indicator was lit suggested that maybe the damage wasn't so bad after all. While carrying out various d.c. voltage checks we found that the mute line at pin 17 of the TA8720AN audio switching chip QV01 was high at 6 V. When the mute line was earthed the sound came up loud and clear. Following the line back to source we came to Q609 (BC557A) which was short-circuit base-to-emitter, thus applying some 5 V to the mute line and turning QV01 off. Replacing this transistor restored normal sound.

Intermittently excessive brightness: The cause was traced to a cracked solder joint at the line output transformer's auxiliary earth pin. This connection earths the low voltage end of the voltage divider network that includes the first anode potentiometer.

Field scan non-linear when cold: The top of the picture was stretched and the bottom cramped. It slowly improved as the set warmed up. We suspected the field scan coupling capacitor or a supply reservoir capacitor, but no. The surprising culprit turned out to be the 2.2 μF ramp integration capacitor C303, which is connected to pin 31 of IC501. It would have made an excellent thermometer.

Toshiba 2500TB

Chroma patterning from cold: This was a nasty one: the deeper the saturation the worse the patterning, which also varied tremendously with the setting of the colour level. It was most prevalent in red and blue, and was present with video as well as r.f. inputs. As the set warmed up the fault cleared. There were black lines in the chroma, and diagonal swathes of white bars ran through it all. As the set was only a couple of months old we contacted Toshiba to check on whether there were any known problems. Indeed there were – the fault can be caused by pick-up from the teletext oscillator, and there's a modification involving replacement of ICF01 and fitting two diodes on the text PCB. But when we looked this had already been done! Scope checks showed that there was noise on the d.c. colour control line, i.e. at pin 7 of the do-everything chip IC501. The decoupling capacitor C515 (22 μF) was found to be very low in value when cold, a replacement curing the fault.

Chopper transistor failure: Should the ON4408 chopper transistor Q802 fail, check whether R810 (330 kΩ, 1 W) is open-circuit. Its part number is 24377334. Note that the text in the manual states that the +B voltage is 120 V. This is incorrect for the 2500TB, in which the +B voltage should be set at 145 V, measured across C451. R851 is the set +B control.

Toshiba 261T4B

Curious bending of the verticals: The degree of bending varied with picture content and contrast changes. We were about to start checking the power supply rails when we noticed that there was very slight lack of width. Adjusting the width control increased the width, but pincushion distortion began to appear. After checks on the semi-conductor devices on the EW board failed to reveal anything amiss we tested the d.c. resistance of the EW coils L361 and L362. L361 is quoted as having a resistance of 3.3 Ω and the reading was slightly low. To be on the safe side we ordered them both. Replacing the coils one at a time proved that L361 was at fault, with short-circuit turns.

Set went off to standby: We soon found that the line output transistor was short-circuit. A replacement was fitted and checks were made for shorts in the line output stage. As everything seemed to be OK we crossed our fingers and switched on. The sound came on but there was no picture and a nasty rattling noise came from the line output transformer. A check on the h.t. showed that it was low at 70 V. We suspected a short in the transformer, but everything ran very cool. The next step was to check the various supply voltages on the secondary side of the transformer, to see if any were being loaded down. The 12 V and 8 V lines were low and the 140 V line was at zero. Rectifier diode D143 was OK but the safety resistor R481 was open-circuit. Replacing this restored the picture and after setting up the focus etc. we were just about to refit the back when we noticed that there was slight line tearing. This could be made to come and go by flexing the PCB. The cause turned out to be dry-joints around the line driver transformer.

Loud noise from the line output transformer at switch-on/off: This set could be switched on and off but all one got was a rather loud grating noise from the line output transformer. A check showed that the h.t. at the transformer was low at only 50 V. We suspected the transformer but found that the 2.4 Ω resistor in series with its supply (R481) read 200 Ω, although its colours were perfect. A replacement restored normal operation.

Line output transistor failed, but no picture after replacement: If on fitting a replacement for Q404 there is still no picture, low e.h.t. and the 140 V supply low, check whether R481 (2.4 Ω safety type, part number 24983249) is open-circuit. Because of the way in which the combined line output/chopper transformer works, the 140 V supply will read approximately 60 V when R481 is open-circuit.

Toshiba 285T8BZ

Failure to start from standby when hot: We had been warned that this set intermittently failed to start from standby when hot. On the bench it wouldn't start at all. Checks revealed that there were about 1000 back securing screws, also that the standby/on switching worked to the extent of the opto-isolator DR01 being turned on, but the power supply wasn't having any of it. The voltage at the collector of the photo-transistor in DR01 was negative instead of 7.3 V. A d.c. check at the other end of R814 brought the set to life. It wasn't the meter either, just the probe was enough to do it. So where's the open-circuit high-value resistor? R810 (270 kΩ) checked high, a replacement putting matters right.

Screeching noise from speakers at switch-on: This Nicam set produced a loud screeching noise from the loudspeakers at switch on. It lasted for about a second then disappeared. We scratched our heads a bit and decided to replace the TDA4601 chopper control chip. This seemed to improve matters but the fault persisted. We isolated various supply lines and used the workshop power supply but were unable to trace the cause of the fault to any particular stage. A check with the circuit then showed that there's a 1000 μF capacitor connected to the audio B+ rail. It's mounted close to the audio output chip. Fitting a replacement provided a complete cure. The circuit reference is C638 (1000 μF, 25 V).

Field overscan: Here's a warning about jumping to conclusions. We had two of these sets in with very similar fault symptoms but different causes. The first one came in under warranty with acute field overscan – about 200 per cent. Even with the vertical amplitude control turned to minimum the field scan was excessive to the extent of about 50 per cent. The linearity wasn't much to write home about either, with cramping at the bottom of the screen. A look at the circuit diagram showed that the TA8659N chip IC501 on the main board is responsible for most of the signal processing. It produces field drive pulses at pin 29 for feeding to pin 1 of the TDA8170 field output chip IC303 on the power/deflection panel. Height and linearity are controlled by feedback to pins 31 and 32 respectively of IC501 from RC networks in the scan current path. There's a height control (R351) connected to pin 31 but no linearity control. Our first suspects were the zener diodes D814 and D815 connected to pins 29 and 31 of IC501. We've had them cause scan problems in other Toshiba chassis but this time they were OK. The voltages at pins 31 and 32 were low, pin 32 being at 2.7 V instead of 6.6 V. As the components in the feedback network all seemed to be OK, when checked we began to suspect IC501 itself. As this is a 64-pin device, however, we decided to check IC303 and its few peripheral components first. All were OK. After a brief word with Toshiba to confirm our suspicions we ordered a replacement TA8659N chip. Fitting this restored correct field scan amplitude and linearity. The second set came in a few days later. We were about to order another TA8659N chip when we noticed that this time the overscan wasn't quite so bad – only about 150 per cent. So we started to check the discrete components. Everything was all right until we came to transistor Q363 (BC547A) on the power panel. Its emitter voltage was high at 6.6 V instead of 0.3 V. When it was removed for test we found that there was emitter-collector leakage. A new transistor restored normal operation. One odd thing: the circuit shows that there's a direct connection between the collector of this transistor and pin 31 of IC501, but under the fault condition there was a difference of nearly 0.5 V between the two points although a cold resistance check produced a reading of zero ohms.

Reduced width and line fold-over: The symptoms were accompanied by an arcing sound. The fault could be rectified by tapping the cabinet. A visual check on the power supply/deflection panel showed that there was noticeable discolouration beneath the line scan correction capacitor C363. So C363 was refitted in conveniently placed adjacent holes. This seemed to provide a complete cure – or did it? The set worked until text was selected, revealing severe raster distortion. We then found that the controls associated with the TEA2031A EW scan correction control chip IC361 had no effect. Scope checks showed that a field-rate sawtooth waveform entered this chip, but there was no EW drive output at pin 5. This output drives the EW modulator diodes via the loading coil L362. Full of confidence we ordered and fitted a new TEA2031A chip. But the symptoms remained as before. There was one slight change, however, a most peculiar line-rate waveform, with no field-rate component, could be seen at pin 5 of the chip. Remembering problems with the EW modulator in the ITT CVC30 chassis we decided to remove L362 and recheck the waveform. It was now correct. So L362 could well be responsible for the problem. As we didn't have a replacement we hooked in an ITT loading coil (L24 from the CVC30 chassis) to prove the point. Everything now worked correctly. Close examination of L362's winding showed that there was slight darkening, but it was not very obvious. Fitting the correct replacement completed the job. One can only surmise that if the two faults were linked then the arcing associated with C363 had probably caused the failure of L362.

Field collapse: We found that the TDA8170 field output chip IC303 had a diagonal crack across its body. As the 27 V supply was present at pin 2, a new chip was fitted. About 2 hours later, while the set was being soak tested, it returned to the field collapse condition. This time the 2.7 Ω feed resistor R327 had burnt out. We decided to refer to my trusty but tattered 'rotten faults' notebook. Sure enough there was a note about a modification. We removed D305 and fitted a 47 Ω, 1 W resistor in its place, then refitted the diode on the copper side of the board. We also replaced the 6.8 V zener diode D303, and of course R327. There were no more problems after this.

Toshiba C2226

No line sync: Check whether R402 (15 kΩ) is open-circuit.

Incorrect field linearity: The cause was C309 (2.2 μF, 50 V).

No field scan at switch-on: The field then took about half an hour to reach full amplitude. The cause was C317 (2.2 μF, 50 V).

Comes on from standby after 1 hour: The cause was traced to the 1000 μF, 25 V electrolytic capacitor C932 on the subpanel next to the teletext PCB. Its value had fallen to 400 μF. This is the first power supply fault we've had with these sets – nearly all previous faults have been field cramping of some sort.

Triumph

Triumph CTV8209

No results: After replacing an open-circuit $2.2\,\Omega$ surge limiter and a faulty TDA4600 chopper control chip the set worked for two days after which the picture went. E.h.t. was present but the c.r.t. heaters were out due to a faulty line output transformer. We phoned Mastercare for a price and were told that they don't do spares for some Triumph models, of which this was one – we were directed to another company (Jackson Products) who were able to supply the transformer at a very reasonable price. When fitted the set was restored to normal working order – for 2 hours, after which we had field collapse. Replacing the TDA3651 field output chip made no difference and there were no open-circuit resistors. We then noticed a TDA4503, and recalled that this series of i.c.s incorporates the field generator. Sure enough a replacement produced a full raster. Not bad going as we'd no manual or circuit diagram!

Intermittent power supply failure: This popular 14 inch portable took quite a while to sort out – every now and then it failed to start. If any part of the chopper chip was touched electrically the set would immediately start up. After many days we found that R119, which is connected to pin 4 of the chip, had increased in value from $150\,\mathrm{k}\Omega$ to about $270\,\mathrm{k}\Omega$. Fitting a replacement cured the problem. Incidentally this set is very difficult to tune with the existing tuning head and fitting a replacement makes very little difference.

Failure to come on sometimes, or reverting to standby: There were periods when the set would work for days at a time without trouble, however. The cause of the problem was dry-joints on the pins of the line driver transformer.

No sync: Although this set appears to be identical to the Fidelity F14 the component reference numbers don't match and a lot of the circuitry is different. One set that came in had no sync at all. A scope check at pin 5 of the TDA4503 chip showed that there was no signal here. Line pulses from the collector of the BU508A line output transistor are fed to this pin via the two 22 kΩ, 1 W resistors R138/9 and an RC integrating circuit. R139 was open-circuit. If you get the same fault with the F14 check R109 which consists of ten 5.6 kΩ resistors in series. The signal at pin 5 of the TDA4503 chip should consist of a line-frequency sawtooth waveform at approximately 6 V peak-to-peak.

Loss of memory when cold: The fault was difficult to trace without the correct manual. R160 (470 kΩ) was open-circuit.

Bright raster with raster lines: This Triumph set is actually the Fidelity CTV14 (later ZX3000 chassis version) in disguise. The fault with this one was a bright raster with flyback lines. We quickly traced the cause to the fact that the RGB output transistors were biased hard on. The zener diode that returns the emitters of the output transistors to chassis was short-circuit. Once this was replaced we were rewarded with a picture that was normal apart from the fact that there was no red content. The red output transistor TR11 (BF869) was open-circuit always.

Wouldn't tune to lower group A: Group B and higher channels were not affected. The cause of this unusual fault was eventually traced to R157 (2.2 MΩ) which was open-circuit.

Television index/directory and faults discs plus hard copy indexes and reprints service

Index disc

Version 8 of the computerised Index to *Televison* magazine covers Volumes 38 to 49 (1988–1999). It has thousands of references to TV, VCR, CD, satellite and monitor fault reports and articles, with synopses. It also contains a TV/VCR spares guide, an advertiser's list, a compendium of useful web resources and a directory of trade and professional organisations. The software is quick and easy to use, and runs on any PC with Microsoft Windows. Price is £36 (supplied on a 3.5" HD disc). Those with previous versions can obtain an upgraded version for £16. Please quote the serial number of the original disc. See the CD-ROM package and upgrade offer below.

Fault Report discs

Each disc contains the full text for television, VCR, monitor, camcorder, satellite TV and CD fault reports published in individual volumes of *Television*, giving you easy access to this vital information. Note that the discs cannot be used on their own, only in conjunction with the Index disc: you load the contents of the Fault Report disc on to your computer's hard disc, then access it via the Index disc. Fault Report discs are now available for:

Vol 38 (Nov 1987–Oct 1988);   Vol 39 (Nov 1988–Oct 1989);
Vol 40 (Nov 1989–Oct 1990);   Vol 41 (Nov 1990–Oct 1991);
Vol 42 (Nov 1991–Oct 1992);   Vol 43 (Nov 1992–Oct 1993);
Vol 44 (Nov 1993–Oct 1994);   Vol 45 (Nov 1994–Oct 1995);
Vol 46 (Nov 1995–Oct 1996);   Vol 47 (Nov 1996–Oct 1997);
Vol 48 (Nov 1997–Oct 1998);   Vol 49 (Nov 1998–Oct 1999).

Price £15 each (supplied on 3.5" HD discs).

Fault-finding guide discs

These discs are packed with the text of vital fault-finding information from *Television* – fault-finding articles on particular TV chassis, VCRs and camcorders, Test Cases, What a Life! and Service Briefs. There are now three volumes 1, 2 and 3. They are accessed via the Index disc. Price £15 each (supplied on 3.5" HD discs).

Complete package on CD-ROM

The Index and all the Fault Report and Fault Finding Guide discs are available on one CD-ROM at a price of £196 (this represents a huge saving). Customers who have the previous CD-ROM can upgrade on CD-ROM for £46 (other customers call for a quotation). Please quote the serial number of your disc when you order.

Reprints and hard copy indexes

Reprints of articles from *Television* back to 1986 are also available: ordering information is provided with the Index, or can be obtained from the address below. Hard copy indexes of *Television* are available for Volumes 38 to 49 at £3.50 each.

All the above prices include UK postage and VAT where applicable. Add an extra £1 postage for non-UK EC orders, or £5 for non-EC overseas orders. Cheques should be made payable to SoftCopy Ltd. Access, Visa or MasterCard Credit Cards are accepted. Allow up to 28 days for delivery (UK).

SoftCopy Limited
1 Vineries Close, Cheltenham, GL53 0NU, UK
Telephone: 01242 241 455
Fax: 01242 241 468
e-mail: sales@softcopy.co.uk
Web site: http://www.softcopy.co.uk

Printed in the United Kingdom
by Lightning Source UK Ltd.
134318UK00001B/47/A

9 780750 646333